Cognitive Systems Engineering

Cognitive Systems Engineering

An Integrative Living Laboratory Framework

Edited by

Michael D. McNeese
The Pennsylvania State University
University Park, PA, USA

Peter Kent Forster
The Pennsylvania State University
University Park, PA, USA

CRC Press
Taylor & Francis Group
Boca Raton London New York

CRC Press is an imprint of the
Taylor & Francis Group, an **informa** business

CRC Press
Taylor & Francis Group
6000 Broken Sound Parkway NW, Suite 300
Boca Raton, FL 33487-2742

Library of Congress Cataloging-in-Publication Data

Names: Forster, Peter Kent, editor. | McNeese, Michael, 1954- editor.
Title: Cognitive systems engineering : an integrative living laboratory framework / compiled by Michael D. McNeese, Peter K. Forster.
Description: Boca Raton : Taylor & Francis, CRC Press, 2017. | Includes bibliographical references.
Identifiers: LCCN 2016055807 | ISBN 9781498782296 (hardback : alk. paper) | ISBN 9781498782319 (ebook)
Subjects: LCSH: Social psychology. | Cognitive science. | Work environment. | Teams in the workplace. | Cooperation. | Communication in organizations.
Classification: LCC HM1011 .C63 2017 | DDC 302--dc23
LC record available at https://lccn.loc.gov/2016055807

Visit the Taylor & Francis Web site at
http://www.taylorandfrancis.com

and the CRC Press Web site at
http://www.crcpress.com

Special Memoriam

Dr. David L. Hall (a former dean and professor, College of Information Sciences and Technology, The Pennsylvania State University, University Park, Pennsylvania) initially began work with me as the coeditor on this book before his sudden passing in December 2015. We were all deeply saddened and shocked by his untimely death. Dave was a man of keen intelligence, deep wisdom, and had the gentle and kind soul that many of us strove to be around. He was both a friend and mentor to us and is greatly missed. Most of the chapters and authors of this book were influenced by Dave in many different ways and were dedicated to his vision of research. Because we are all elements of the mosaic he has left behind, it is our firm conviction to dedicate this book in his memory and honor.

Dedication

For the past two years, my wife Judy has stayed strong and persevered while dealing with cancer and chemotherapy while I worked through various elements of this book. Today she is cancer free and healed. Therein, I give thanks to this blessing through this dedication. I dedicate this book to her amazing faith, hope, and love that show through in her thoughtful actions. After 36 years of marriage, the inner light of Christ that meld us together into one still magnifies brightly and leads our paths.

—Michael D. McNeese

Book dedications are always difficult because there are so many people who play an important role. From my perspective, there are three people, in addition to Dave Hall memorialized earlier. First, my coeditor, without whose perseverance, guidance, and friendship, I would not be writing this. Thank you, Mike. Second, Dr. Richard D. Brecht, who recently retired from the University of Maryland, College Park, Maryland. Dick has been a consultant to me on many subjects, and my work on this book is a reflection of his commitment to broadening one's intellectual knowledge. Last, but certainly not the least, I also need to recognize the support of my wife, Sue. Her love, companionship, and understanding are foundational to all of our successes together including this book.

—Peter Kent Forster

Contents

Preface.. xiii
About the Editors xvii
Contributors xix

SECTION I   Introduction and Historical Precedence

Chapter 1   Introduction 3

Michael D. McNeese and Peter Kent Forster

SECTION II   The Living Laboratory: Learning in Action

Chapter 2   The Living Laboratory in the Context of Educational Research 31

Xun Ge

Chapter 3   Distributed Teams in the Living Laboratory: Applications of
Transactive Memory... 47

Vincent Mancuso

Chapter 4   Digital Architectures for Distributed Learning and Collaboration.......67

Jeff Rimland

Chapter 5   A Cognitive-Systems Approach to Understanding Natural Gas
Exploration Supply Chains: A Purview and Research Agenda 81

Michael D. McNeese and James A. Reep

SECTION III   Theoretical Perspectives in Cognitive–Collaborative Systems

Chapter 6   Methodological Techniques and Approaches to Developing
Empirical Insights of Cognition during Collaborative
Information Seeking... 105

Nathan J. McNeese, Mustafa Demir, and Madhu C. Reddy

Chapter 7   A Case Study of Crisis Management Readiness in
                 Cross-National Contexts: Applying Team Cognition Theories
                 to Crisis Response Teams... 131

                 Tristan C. Endsley, Peter Kent Forster, and James A. Reep

Chapter 8   Emotion, Stress, and Collaborative Systems................................... 161

                 Mark S. Pfaff

SECTION IV   Models and Measures of Cognitive Work

Chapter 9   Measuring Team Cognition in Collaborative Systems:
                 Integrative and Interdisciplinary Perspectives 181

                 *Susan Mohammed, Katherine Hamilton, Vincent Mancuso,
                 Rachel Tesler, and Michael D. McNeese*

Chapter 10   Modeling Cognition and Collaborative Work 199

                 Rashaad Jones

Chapter 11   Cognitive Architectures and Psycho-Physiological Measures 221

                 Christopher Dancy

SECTION V   Scaled-World Simulations

Chapter 12   Evaluating Team Cognition: How the NeoCITIES Simulation
                 Can Address Key Research Needs 241

                 *Katherine Hamilton, Susan Mohammed, Rachel Tesler,
                 Vincent Mancuso, and Michael D. McNeese*

Chapter 13   Scaled-World Simulations in Attention Allocation Research:
                 Methodological Issues and Directions for Future Work 261

                 *Dev Minotra, Murat Dikmen, Anson Ho, Michael D. McNeese,
                 and Catherine Burns*

Chapter 14   How Multiplayer Video Games Can Help Prepare Individuals
                 for Some of the World's Most Stressful Jobs.................................... 277

                 Barton K. Pursel and Chris Stubbs

SECTION VI Knowledge Capture, Design, Integration, and Practice

Chapter 15 Police Cognition and Participatory Design 299

          Edward J. Glantz

Chapter 16 Intermodal Interfaces 313

          Arthur Jones

Chapter 17 Bridging Theory and Practice through the Living Laboratory:
The Intersection of User-Centered Design and Agile Software
Development 327

          D. Benjamin Hellar

Chapter 18 Incorporating Human Systems Engineering in Advanced
Military Technology Development 341

          *Patrick L. Craven, Patrice D. Tremoulet,
and Susan Harkness Regli*

SECTION VII The Future of the Living Laboratory

Chapter 19 The Future of the Integrated Living Laboratory Framework:
Innovating Systems for People's Use.. 363

          Michael D. McNeese and Peter Kent Forster

Author Index.. 371

Subject Index 381

Preface

Early in my career (approximately 42 years ago), I was working as a young designer, and one of my projects was to come up with an informed design of a monitoring system of lights and alerts for bus drivers that would help them to be aware of the vehicular system as it was in operation picking up children along a route. This was in fact my first quest with an actual system design that involved (1) human factors engineering, (2) situation awareness, and (3) a human-centered approach that involved a system with some risk. As I quickly realized, the constraints imposed by the vehicle, the driver's considerations, and especially the context of operations are all very interdependent and important. The bus driver's consideration of awareness and how to utilize visual (and auditory) alerts to ensure the safety of the children on the bus was very much the primary problem identified in this design situation. Human factors at the time (1974) were pretty much about displays, controls, and perhaps the design of meaningful alert messages. This was certainly viable, but my sense was that both the cognitive and contextual bases for design were absent. During this time, guidelines rather than theory often defined human factors design. As I continued in my career later on in the U.S. Air Force—standards and guidelines continued to be the zeitgeist of the practitioner—usually wrapped within the Human Factors (HF) procedure policy (a standard) that was coincidental with a system's acquisition process. One could specify other requirements (such as a traditional task analysis or function allocation) that a contractor typically performed (only if required). Although many human factors programs have shown some profit and advancement with this type of approach, it struck me more as microfactors that left plenty of uncertainty and unfinished design on the table. As a practitioner, I was not pleased at all, and it left a sense of not be fulfilled or satisfied with this state of affairs. That was the conceptual beginning of this book.

As I moved from the applied human factors position in the Air Force into a research position as an engineering research psychologist—I literally moved across the spectrum of human factors with some enlightened hope. I was very interested in theory and experiments involving perception, cognition, and the cerebral processing involved in human activity in complex systems. Certainly this was a move away from the applied human factors and the guidelines approach wherein a scientist would conduct a study (almost always quantitatively and empirical) and obtain statistical probability less than .05, then discuss the claims and considerations of the study. However, there were still troubling issues present. One of the main issues is the gap between the theory and the results of experiments and the actual practice of human factors in an applied setting. Another problem was a bias for treating quantitative data analysis as the only approach available. I have referred to this as a data-centric viewpoint. So many of the studies performed, though sound experimentally, were performed within the laboratory of the white tower often with the added caution "generalizability is not determined yet." Hence the gaps that I had experienced earlier as a practitioner continued to be present. The output of the new state-of-the-art research was still not informing practice. Yet, there were other problems, which are still extant today. As mentioned, many of the studies trying to understand pilot behavior or command and control teams in the 1970s, 1980s, and 1990s strictly adhered to a cognitivist-only view wherein cognition was framed as only in the head and not distributed across context (what the head is inside of). This entrenchment created a sort of biased effect for understanding human–machine interaction/ human–system integration that somewhat reflected theoretical perspectives of the times. But as theories changed, there was great expression and emphasis placed on contextualistic perspectives that added insights into context, distributed teamwork, expertise, and the idea that quality data analysis could also have worth.

As times passed, I had a primary need to integrate my views of the world even more so when it came to research and practice. One of the drivers that I thought was missing in human factors came directly from my cognitive science background while working with my advisor, Dr. John Bransford, at Vanderbilt: learning and problem solving. I felt that much of the work across the spectrum in human factors was devoid of learning (i.e., many tasks were approached from static, rigid perspectives) and prior results in research failed to inform guidelines, analysis, and practice in any meaningful way. In addition, many studies were not problem centric (i.e., they did not derive from the problems that novices or experts actually experienced in the field). As I continued in the human factors areas at the Air Force Research Laboratory, many of these struggles persisted and I felt that human factors were very isomorphic both in theoretical research and in practical applications. In addition, our own work within the Air Force Pilot's Associate program in the late 1980s had found that learning and problems were really predicated on working with knowledge elicitation of differing kinds of experts who worked across systems as part of teams that completely influenced how an expert system was designed. This juncture provided the basis then for fomenting my own view—the living ecosystem perspective on human factors, which would take a multimethodological, integrative approach to spanning the gaps I had seen firsthand during these years. As a result, I first did a paper in 1996 for a macroergonomics conference that described the living laboratory in terms of what it was and why it would be useful for complex systems. Since then (20 years ago), we have utilized the living laboratory in a variety of studies, cases that have been provided to me. The laboratory and group that I direct at The Pennsylvania State University (The MINDS [Multidisciplinary Initiatives in Naturalistic Decision Systems] Group), University Park, Pennsylvania, utilizes this approach in every project and every thesis and dissertation. Hence, we have a lot of data points on the validity of use, feedback from customers and funders, and user-experts themselves on the viability of the approach.

Given the wealth of experience we have obtained and given we have not seen any other particular book in the human factors community that specifically tackled the living laboratory and Living EcoSystem Worldview of human factors, cognitive systems, or teamwork, we felt now is the time to initiate this new endeavor with the hope that students, practitioners, and scientists might learn what we have—that multiple perspectives and triangulated approaches anchored against specific problems can produce a balanced, well-informed, and flexible understanding that accounts for joint influences of information, technology, people, and context. In turn, we have worked with specific authors to engage the best set of materials possible to communicate an enduring knowledge of the living lab framework that develops insights and discovery, and paves the way for a broader enlightenment of the human actors that underlie human factors.

Michael D. McNeese
University Park, Pennsylvania

About the Editors

Dr. Michael D. McNeese is a professor and the director of the MINDS Group (Multidisciplinary Initiatives in Naturalistic Decision Systems) at the College of Information Sciences and Technology (IST), The Pennsylvania State University (Penn State), University Park, Pennsylvania. Dr. McNeese is also a professor of psychology (affiliated) in the Department of Psychology, and a professor of education (affiliated) in the Department of Learning Systems and Performance, at Penn State. Recently he stepped down as the senior associate dean for research, graduate studies, and academic affairs at the College of IST. Dr. McNeese also served as the department head and associate dean of research and graduate programs at the College, and was part of the original 10 founding professors at the College of IST. He has been the principal investigator and has managed numerous research projects involving cognitive systems engineering, human factors, human–autonomous interaction, social–cognitive informatics, cognitive psychology, team cognition, user experience, situation awareness, and interactive modeling and simulations for more than 35 years. His research has been funded by diverse sources (National Science Foundation (NSF), Office of Naval Research (ONR), Army Research Laboratory (ARL), Army Research Organization (ARO), Air Force Research Laboratory (AFRL), National Geospatial Intelligence Agency (NGIA), and Lockheed Martin) through a wide variety of program offices and initiatives. Before moving to Penn State in 2000, he was a senior scientist and the director of collaborative design technology at the U.S. Air Force Research Laboratory (Wright–Patterson Air Force Base, Dayton, Ohio). He was one of the principal scientists in the U.S. Air Force responsible for cognitive systems engineering and team cognition as related to command and control and emergency operations. Dr. McNeese earned his PhD in cognitive science at Vanderbilt University, Nashville, Tennessee, and an MA in experimental-cognitive psychology at the University of Dayton, Dayton, Ohio, and he was a visiting professor in the Department of Integrated Systems Engineering, The Ohio State University, Columbus, Ohio, and a research associate at the Vanderbilt University Center for Learning Technology. He has written more than 250 publications in research/application domains including emergency crisis management; fighter pilot performance; pilot–vehicle interaction; battle management command, control, and communication operations; cyber and information security; intelligence and image analyst work; geographical intelligence gathering, information fusion, police cognition, natural gas exploitation, emergency medicine; and aviation. His most recent work focuses on the cognitive science perspectives within cybersecurity, utilizing the interdisciplinary living laboratory framework as articulated in this book.

Dr. Peter Kent Forster is the associate dean for online and professional education and a member of the graduate faculty in the College of Information Sciences and Technology (IST) at The Pennsylvania State University (Penn State), University Park, Pennsylvania. He is a member of IST's Multidisciplinary Initiatives in Naturalistic Decision Systems (MINDS) Center and the Center for Enterprise Architecture, working on risk and crisis management including using simulations and tabletop exercises to improve command and control. He earned a PhD in political science (international relations) at The Pennsylvania State University, has an affiliate status with the School of International Affairs, and is the cochair of the NATO (North Atlantic Treaty Organization)/OSCE (Organization for Security and Co-operation in Europe) Partnership for Peace Consortium Combating Terrorism Working Group (CTWG), which brings academics and practitioners together to develop policy recommendations on counterterrorism strategy and tactics.

Dr. Forster's primary areas of research and teaching interests include terrorism/ counterterrorism, risk and crisis management, international relations and national security, and homeland security. Most recently, Forster developed and facilitated a tabletop exercise involving representatives from 40 countries under the auspices of the CTWG. These events are being repeated in a number of Partnership for Peace Consortium (PfPC) countries. Since 2010, he has been the principal investigator on a grant, exploring integrating processes and technologies to improve law enforcement's situational awareness of major issues such as human trafficking. He also oversees a research project on improving the understanding of how extremist organizations recruit, vet, and integrate Americans.

Dr. Forster has been involved in security sector reform initiatives including defense institution building in the Caucasus and South central Europe as well as consulting on national distance education initiatives in central Asia and the Caucasus. He is a coauthor and contributing author to books on NATO's military burden sharing and intervention and cognitive systems including *Multinational Military Intervention* (with Stephen J. Cimbala and Peter Kent Forster), *Policy, Strategy & War: The George W. Bush Defense Program* (edited by Stephen J. Cimbala), and *Cognitive Systems Engineering* (edited by Michael D. McNeese and Peter Kent Forster, forthcoming). He has published articles on technology and terrorism, understanding distributed team cognition, homeland security, and American foreign policy and interests in central Asia and the Caucasus.

Dr. Forster has extensive experience in online education, including program design, development, and implementation. He consulted on developing national online programs in central Asia and corporate program nationally and internationally, and continues to teach online for Penn State. Dr. Forster teaches courses on crisis and risk management, cybercrime, terrorism, and war; counterterrorism, the impact of information on twenty-first century society, war and conflict, and the international relations of the Middle East.

Contributors

Catherine Burns
Department of Systems Design
 Engineering
University of Waterloo
Waterloo, Ontario, Canada

Patrick L. Craven
Human-Cyber Solutions, LLC
Moorestown, New Jersey

Christopher Dancy
Department of Computer Science
Bucknell University
Lewisburg, Pennsylvania

Mustafa Demir
Department of Human Systems
 Engineering
Ira A. Fulton Schools of Engineering
Arizona State University
Mesa, Arizona

Murat Dikmen
Department of Systems Design
 Engineering
University of Waterloo
Waterloo, Ontario, Canada

Tristan C. Endsley
Charles Stark Draper Laboratory
Cambridge, Massachusetts

Peter Kent Forster
College of Information Sciences and
 Technology
The Pennsylvania State University
University Park, Pennsylvania

Xun Ge
Department of Educational
 Psychology
The University of Oklahoma
Norman, Oklahoma

Edward J. Glantz
College of Information Sciences and
 Technology
The Pennsylvania State University
University Park, Pennsylvania

Katherine Hamilton
College of Information Sciences and
 Technology
The Pennsylvania State University
University Park, Pennsylvania

D. Benjamin Hellar
Next Century Corporation
Annapolis Junction, Maryland

Anson Ho
Department of Systems Design
 Engineering
University of Waterloo
Waterloo, Ontario, Canada

Arthur Jones
Department of Supply Chain and
 Information Systems
The Pennsylvania State University
University Park, Pennsylvania

Rashaad Jones
Department of Information Technology
Georgia Gwinnett College
Lawrenceville, Georgia

Vincent Mancuso
Lincoln Laboratory
Massachusetts Institute of Technology
Cambridge, Massachusetts

Michael D. McNeese
College of Information Sciences and
 Technology
The Pennsylvania State University
University Park, Pennsylvania

Nathan J. McNeese
Department of Human Systems
  Engineering
Ira A. Fulton Schools of Engineering
Arizona State University
Mesa, Arizona

Dev Minotra
Department of Systems Design
  Engineering
University of Waterloo
Waterloo, Ontario, Canada

Susan Mohammed
Department of Psychology
The Pennsylvania State University
University Park, Pennsylvania

Mark S. Pfaff
Collaboration and Social Computing
  Division
The MITRE Corporation
Bedford, Massachusetts

Barton K. Pursel
Teaching and Learning with
  Technology
The Pennsylvania State University
University Park, Pennsylvania

Madhu C. Reddy
Department of Communication Studies
Northwestern University
Evanston, Illinois

James A. Reep
College of Information Sciences and
  Technology
The Pennsylvania State University
University Park, Pennsylvania

Susan Harkness Regli
University of Pennsylvania
  Health Care
Philadelphia, Pennsylvania

Jeff Rimland
College of Information Sciences and
  Technology
The Pennsylvania State University
University Park, Pennsylvania

Chris Stubbs
Teaching and Learning with
  Technology
The Pennsylvania State University
University Park, Pennsylvania

Rachel Tesler
Westat, Inc.
Rockville, Maryland

Patrice D. Tremoulet
Tremoulet Consulting, LLC
Moorestown, New Jersey

Section I

Introduction and
Historical Precedence

1 Introduction

Michael D. McNeese and Peter Kent Forster

CONTENTS

Orientation 4
Motivation for the Book 5
  Historical Imperatives and Flow 5
    Prelude 5
    Early Work in Ecological Settings and Cognitive Design Practice 7
    Collaborative and Human–Autonomous Systems 8
    The AKADAM Initiative 8
    Developing Team Cognition and Collaborative Systems 9
    Moving Forward with Sociocognitive Simulations and Technology 11
    The MINDS Group as It Has Evolved at The Pennsylvania
    State University 11
Audience for the Book 13
  Why the Book Is Necessary 14
  Topics and Coverage of the Book 14
  A Progression of Human Factors/Cognitive System Terms 15
The Relevance of the Living Laboratory Framework 17
  Basic Philosophy 17
    Ecological in Nature 17
    Experience Matters in Fields of Practice 18
  Elements of the Living Laboratory Framework 20
    Knowledge as Design as Living Ecosystem 23
Structure of the Book 24
Concluding Remarks 26
References 27

ADVANCED ORGANIZER

- Agent–environment transactions represent mutual activities in cognitive systems.
- Design of systems that fails to take into consideration the cognitive–contextual aspects of the user frequently results in errors or mistakes in performance.
- The living laboratory highlights the principle that intentional design must produce effective adaptation to the environment to produce sustainability, resilience, and reliability.

ORIENTATION

I first heard the term *living laboratory* from one of my conversations with an Air Force Research Laboratory (AFRL) division chief in the early 1990s as we discussed ideas about how to study decision making and design decision aids within the command and control domain. Indeed the term has a separate history as a generic term that colors the way it has been considered in our research motivations (see Wikipedia, Living Laboratory, https://en.wikipedia.org/wiki/Living_lab). Having said that, the way this book uses the term *living laboratory framework* (LLF), as applicable within the realm of cognitive systems engineering CSE, has specific meaning and gravitas. The division chief's original quest suggested that research and design must be looked at not just from an ivory tower of theory and lab studies, but from the perspective of a challenging and dynamically changing real world where humans and machines work things out in the midst of ongoing work requirements. The implication being that laboratory research needed to be grounded in the context of work, applicable to designs that enabled humans (embedded within multiple systems) to do *cognitive* work, and reflect the mutual interplay of *agent–environment transactions* wherein power, insights, and effectiveness could be produced. Too frequently, design did not consider the context, the human, the dynamics of change, and what interaction actually meant. In turn what were supposed to be systems that supported *cognition* actually turned out to be outright failures that produced the inability to control and monitor work. Such a state of affairs contributed heavily to what we know as *human error* or the conquests of work that ended up with slips, mistakes, accidents, and even catastrophic failures (Norman, 1988). Although these initial tenets are still true, what we mean and we connote with concepts such as *context*, *agent*, and *cognitive* has changed substantially over the past 25 years, often through the introduction of new technological sophistication.

The initial interpretation of a living laboratory was that poor design often results from limited uses of static guidelines, ancient experimental studies, or traditional wisdom. In contrast, intentional design must dynamically *come alive* based on the mutual transactions of the agent within the environment. In this sense, *come alive* means being able to adapt to circumstances and situations, to sustain operations in balance to maintain veridicality, reliability, and validity, and to be resilient when change happens. Adaptation, sustainability, resilience, and effectiveness are qualities that need to be addressed when practitioners engage in the practice of CSE. As one reads through the various sections and chapters within this book, it is our hope that the LLF unfolds a pathway toward such qualities and opens up new awareness of *what can be.*

MOTIVATION FOR THE BOOK

HISTORICAL IMPERATIVES AND FLOW

The influences spanning across CSE that have led to this book have emerged from many corridors and perspectives, stretched across passages of time (approximately 30 years). This trajectory might be categorized in the following four phases:

1. Early work in *ecological psychology, social science, and design practice.*
2. Formulations in *cognitive science and human–autonomous systems.*
3. Underpinnings in *team cognition and collaborative systems.*
4. Refinements through *sociocognitive simulation and technology development.*

Although I have been involved with many talented and wonderful people and different work groups throughout the years, the evolution of what has come to be our current practice (and indeed the representation of chapters within this volume) has not been straightforward but rather nonlinear, even chaotic at times. I have chosen to label the distinct niche of CSE that we have engaged in—albeit changes and refinements over the years—as the *integrated* LLF. We will explain the concepts and ideas underlying this approach shortly, but first it is insightful to talk about the history and developmental insights of CSE as we have known it and come to practice it in our laboratory, collectively known as the MINDS Group.*

Prelude

Much of the work that has led to this book has emerged from the broad research area of human factors. It has necessarily been coupled with understanding cognition and in turn how cognition could impact the design of machines, interfaces, tools, systems, and/or environments that invariably intersect individual work and team work. My own degrees include titles such as *experimental-cognitive psychology* and *cognitive science.* The HFES† subtechnical group that I am in is titled *cognitive engineering and decision making* (which is the largest technical group of the society). In addition, one of the books I have edited in the past includes the term *cognitive systems engineering* (McNeese & Vidulich, 2002). Indeed, cognition seems to be pervasive and

---

* **The MINDS** (Multidisciplinary Initiatives in Naturalistic Decision Making) **Group** at The Pennsylvania State University (https://minds.ist.psu.edu/): This represents the collective research group I have directed in toto for the past 17 years. This group has utilized the living laboratory approach for cognitive systems engineering in various instantiations.
† HFES – Human Factors and Ergonomics Society.

holds an esteemed place as a foundational concept in the human factors community of practice. As cognition is a state that is experienced, felt, or noted by the mind, we can say that in fact it is one of the phenomenal bases of human factors (see McNeese, 2016, for a further review on this topic). At the HFES annual meeting each year, we increasingly see evidence of the historical imperative of cognition impacting design through different concepts, constructs, frameworks, models, processes, approaches, and outcomes. For example, terms such as *mental models, situation awareness, cognitive ergonomics, attention processes, knowledge/skill acquisition, cognitive task analysis, transactive memory*, and *macrocognition* are elaborated year-in and year-out within various papers.

As cognition can make note of itself (metacognition), is sensed by *us* and others, and can be thought about as we recall it or change it in midstream (dynamically reconstructed), it is often historically portrayed as an internal, intrinsic, introspective, microscopic, and as inside-the-head phenomenon. However, that is not the full story, as we would like to turn cognition on its head, so to speak. A retrospective look at cognition over the past 50 years reveals a changing landscape of connotative formalisms—Weltanschauungs that mirror positivist, interpretivist, or alternative approaches to frame the place that cognition has in life and work.

As cognition is elaborated by multiple factions with differing goals and cultural lens, flip-over (Whitaker, Selvaraj, Brown, & McNeese, 1995) inevitably occurs creating connotative and interpretive layers of definition, reason, and power. The way *cognitive* is used and employed actively defines what it is and what it means to the user. One of the ways cognition has been used and flipped for use to understand and change people's behavior is through (what has been termed by many) *cognitive engineering* (see Hollnagel & Woods, 1983; Rasmussen, 1986). In this case, one way to conceptualize this book (as a volume on cognitive engineering) is to ask the following questions: (1) Where can cognition make an impact on design? (2) How can a design conform cognition for a particular use or situation—good or bad? (3) How can a complex system be engineered to be integrative with its environment and the people who work with the system in that environment? Cognition flip-over occurs in CSE in that cognition is interpreted to be distributed more externally in the built and social environment—emerging in a dynamical way through the transactions of individuals. Aptly stated by Mace (1977) "Ask not what's inside your head, but what your head is inside of." During the early-to-mid 1970s, this kind of flip-over influenced my own understanding of cognition and what it meant for engineering design was being disrupted and simultaneously transformed through learning. This is exemplified by the following quotes:

> People think in conjunction and partnership with others and with the help of culturally provided tools and implements.—Gavriel Salomon, 1997 (p. xiii).
>
> Cognition observed in everyday practice is—stretched over, not divided—among mind, body, activity, and culturally organized settings.—Jean Lave, 1988 (p. 5).
>
> The emphasis on finding and describing *knowledge structures* that are somewhere *inside* the individual encourages us to overlook the fact that human cognition is always situated in a complex sociocultural world and cannot be unaffected by it.—Ed Hutchins, 1995 (p. xiii).

As an example of alternative landscapes, CSE has typically taken a different view of inside-the-head cognition, and some of the primary schools of thought are heavily influenced by ecological psychology and affordances (Norman, 1988; Rasmussen, 1986; Vicente, 1999). Like many others, I have been influenced by ecological theories as well, but more specifically in terms of situating and representing cognition within specific environmental, contextual, and cultural surrounds, often referred to in the research literature as *situated cognition* (Brown, Collins, & Duguid, 1989) and/or *distributed cognition* (Hutchins, 1995). Therein, a lot of the LLF is predicated on ideas and seed concepts that propagated within these areas.

Early Work in Ecological Settings and Cognitive Design Practice

My first encounter with designed environments,* people's behavior, and how they think or do things came through work on a senior honors synthesis at The University of Dayton, Ohio (McNeese, 1977). This thesis consisted of an ecological study of how groups use space (personal space and orientations at different architectural or situated contexts within the university setting—library, student union, etc.). This way of considering how people act and think about their actions within the context of space was very much coupled with individual/social awareness and how designed environments afford different use and experiences. The study involved qualitative methods and represented research ascribed to environmental psychology (Altman, 1975), ecological settings (Barker, 1968), and human–social cognition (Fiske & Taylor, 1991). Although not specifically representative of the kinds of quantitative cognitive studies I would later be engaged with, this study was important as it (1) represented an initial foray into ecological dynamics within built contexts, (2) explored how social psychological variables interact with the ecology, and (3) studied how systems can be designed to enhance the intended function of an environment (i.e., *use*). Although this was very early work, it laid foundations for the melded, integrative, and interdisciplinary nature of social psychology, collaboration, human-centered design, and ecological settings—components still facile for CSE in our lab. In particular, the emphasis on studying collaboration in context, and how design influences behavior, set the initial seeds for the living laboratory conceptualization.

Much of the formative years in cognitive systems and the ecological foundations came from the philosophical direction and training of my advisor, Dr. John Bransford (while he was at Vanderbilt University, Nashville, Tennessee), in the areas of problem solving—problem-based learning—and information seeking. Much of my formal training was in cognitive psychology and design, but my intent was to do experimental work to better understand phenomenon in terms of ecological connections of a human embodied in a context. The theories associated with distributed cognition (Hutchins, 1995), ecological psychology (Gibson, 1979), and problem

---

* Note this was my first encounter with a research study involving ecological settings and design work. As mentioned within the preface—my first encounter with human factors and design per se came through work as a designer on a school bus monitoring display system circa 1973 with the Automatic Control Device Corporation, Dayton, Ohio.

solving/learning (Bransford & Stein, 1984) started to permeate how I framed cognitive systems within real-world environments (fields of practice). As such, they were very much in play as forces that helped meld a specific methodological framework for the pursuit of cognitive-engineered systems.

Collaborative and Human–Autonomous Systems

Continuing with this line of thought, as one considers where the current framework came from, one important influence was placement in an early research role (as a USAF [U.S. Air Force] scientist) responsible for the government cognitive science oversight of the USAF Pilot's Associate program in the mid 1980s. This position enabled yet another distinct change in thinking about cognition—basically that it could be mediated among humans, intelligent agents, and interfaces—within the context they were required to perform in. This came about in 1986 while conceptualizing how humans and artificial intelligent systems might unfold within a complex fighter pilot environment (McNeese, 1986). This notion was important for what I termed *mindware inferencing*, and how collaborative-based cognitive systems could develop sensemaking about how their action was relevant in the unfolding context, and in turn how the ecology of interagency drove cognition to be distributed outward. At the time I referred to this as *macrocognition* and *macroawareness* to signal the increasing role of context, how interdependent layers of information are distributed across an environment (the information surround), how human and computational agents represent a new type of teamwork (human–autonomous interaction) within the aviation domain. This provided a lens on which CSE could be shed in a new light—what might be referred to as a *collaborative ecosystem* where domain knowledge became very important for designing intelligent systems that transact, transform, and adapt in context in order to survive and be sustainable. This set up a seed to begin thinking about cognition using a living ecological metaphor, and to enable methods that could bring this to pass. This marked a variation in my own zeitgeist with respect to cognition (cognitivist to ecological–contextualist/social constructivism) as it unveiled the future path I would take in my own research agenda in the next 30 years. Engineering agents to adapt with humans to formulate unique types of teams that can evolve with the changes in the context they live in. In a crude way, this was the primordial beginning of collaborative ecosystems as a *living laboratory*.

The AKADAM Initiative

During the late 1980s into the mid 1990s my group at the Paul M. Fitts Human Engineering Division, AFRL developed a specific initiative and approach for the Pilot's Associate program (which we called *advanced knowledge and design acquisition methods—AKADAM* (see McNeese et al., 1990; Zaff, McNeese, & Snyder, 1993), which modeled some of the first principles and notions of what would eventually become a partial basis for the LLF (McNeese, 1996). The program was predicated on ideas involving *knowledge as design* as articulated by David Perkins (1986). The basis of this approach focused on understanding a *user's experience* in terms of both conceptual and functional knowledge as a foundation to develop designs for specific contexts of use. AKADAM utilized techniques in concept mapping (conceptual knowledge representation) and IDEF (functional knowledge representation) to understand information systems and to inform human-centered design (through the use of picto-literal representations such as design storyboarding).

Theoretically, AKADAM integrated cognitive, computer, and design science as a nexus that would be applied in various domains. Therein, AKADAM really substantiated and tested methods for knowledge elicitation and design storyboarding to reveal cognitive aspects of systems within specified contexts. Continuing with the experience gained with earlier efforts in ecological settings and design, AKADAM maintained a sustained coupling with contextualistic-social constructivist perspectives that framed cognition as a phenomenon that integrated inside-the-head knowledge with distributed, external situations/events. Around the time of our 1995 publication summarizing the AKADAM work—Ed Hutchins came out with his monumental book on *distributed cognition* (Hutchins, 1995) that continued to reinforce how we thought about cognition, but also highlighted the point that CSE would need to aim not only at individual work but collaborative, distributed work that increasingly was supported by information technologies (e.g., computer-supported cooperative work, Schmidt & Bannon, 1992).

In essence, AKADAM was an early CSE approach that produced human-centered designs specifically adapted for human–autonomous interaction (i.e., for the Pilot's Associate) that pointed toward coordination/cooperation. While this approach was applied in human–autonomous interaction initially, it was also adapted and applied in other complex domains as well (see McNeese, Zaff, Citera, Brown, & Whitaker, 1995).

Developing Team Cognition and Collaborative Systems

Simultaneous with creating and applying AKADAM, I was very much absorbed into three related theoretical areas that bridged cognitive science, social psychology, and team cognition to support two distinct areas within the AFRL community: (1) command, control, communications, and intelligence (C^3I) and (2) collaborative design technologies (CDT). As one can see, this is where theory intersected practice and design work, continuing themes of research from the other phases of work I was involved with. When I transferred into the Paul M. Fitts Human Engineering Division[*], I was immersed into the C^3I community and part of a group called COPE (C^3 Operator Performance Engineering). Working with the COPE program provided another forerunner of experiences that led to the LLF perspective of CSE. Actually one of the roles I took up in this group was looking at team cognition through the lens of cognitive science, specifically in the form of quantitative research

---

[*] My initial work in the U.S. Air Force was as an engineering psychologist (civil servant) working for the Human Factors Branch of the Aeronautical Systems Division, Wright–Patterson AFB, Ohio (1981–1984). This was an excellent opportunity, as part of the work consisted of being a chief human factors engineer for real-world system programs offices (e.g., the KC-10 was one plane wherein I was responsible for all human factors considerations), whereas another aspect was to be a research scientist within their crew station design facility where I was heavily involved in the human–computer interaction designs and assessment for the F-16 aircraft cockpit.

experiments utilizing team-based simulations (TRAP—team resource allocation problem, Brown & Leupp, 1983; Kimble & McNeese, 1987; Wilson, McNeese, & Brown, 1987; CITIES, Wellens & Ergener, 1988; adapted DDD, McNeese, Perusich, & Rentsch, 1999) that investigated sociocognitive science constructs such as team-situation awareness, team mental models, information sharing, and team performance. The CITIES simulation was the initial work that launched our newer simulation—NeoCITIES—a client-server architecture-based simulation in which multiple team cognition research studies have occurred (see McNeese & McNeese, 2016). The emphasis on simulation set forth another key seed for the LLF. In fact, it provided the medium to connect many of the elements of the LLF together that allowed innovative design ideas to be tested before introduced as interventions into fields of practice.

As time marched on, I was also able to forge paths between computer-supported cooperative work (CSCW) and design teams after being appointed as the director of the AFRL Collaborative Design Technology Laboratory. This provided the stimulus to integrate and merge the research from AKADAM and human centered design with team cognition and the development of collaborative systems in unique ways. This initiated the early beginnings of the LLF as an integrated set of methods to explore and build cognitive systems for complex domains that could in turn be tested in scaled world testbeds (simulations) prior to employment.

At Wright–Patterson and/or within the greater Dayton and Columbus Ohio Human Factors community, I have been heavily invested and owe a lot. Although there are both similarities and differences between naturalistic decision making (Zsambok & Klein, 1997) and cognitive engineering research and practice, at times they have had a major cross-pollinating effect. As I was working as an engineering research psychologist at AFRL at WPAFB, I cannot over estimate the contributions and influence made to my own thinking by such luminaries as Dr. Gary Klein (then of Klein Associates, Inc.), Professor John Flach (Wright State University), Dr. A. Rodney Wellens (University of Miami), and Dr. David Woods and Dr. Emily Patterson (The Ohio State University). I did a sabbatical as a visiting professor at Ohio State working with David, who shaped a lot of beliefs and principles about CSE that still hold true today. Likewise my collaborations and friendship with the late Rod Wellens will always be valued and remembered.

A lot of the research I am currently supervising with my PhD students is directly linked to the theoretical ideas and operational measurements of group situation awareness that Rod articulated in the early days (Wellens, 1993). Likewise my government and support contractor colleagues provided great insight and challenged my perspectives. In particular, Dr. Randall Whitaker and Dr. Brian Zaff (then of Logicon Technical Support Services—LTSI) and Professor Clifford Brown (Wittenberg University, Ohio) composed my research group while I was the director of the Collaborative Design Technology Laboratory, and really helped to infuse foundational knowledge within the areas of CSCW and team cognition. My colleagues at the Fitts Human Engineering Division also provided intellectual stimulation and support within the cognitive systems area. In particular, Donald Monk, Dr. Robert Eggleston, Dr. Michael Vidulich, and Gilbert Kuperman were highly instrumental in my development as a professional. They sharpened my knowledge of human factors and cognition within the aviation domain. Special thanks to Dr. Michael Vidulich who was my coeditor for the first book I did on cognitive engineering for the aviation community (McNeese & Vidulich, 2002). Special thanks also to Dr. Kenneth Boff who supported the visions of cognitive systems as our Chief Scientist and as our division chief at AFRL.

Moving Forward with Sociocognitive Simulations and Technology

During the formative years of cognitive engineering (while at Wright–Patterson AFB), the seeds were thrust into motion for establishing an integrative framework to apply to various fields of practice. Although the original framework was first published over 20 years ago, the actual utilization of the LLF has been ever-present while as a Professor at The Pennsylvania State University. I moved from AFRL to Penn State in 2000 (17 years ago) and established what was originally called the user science and engineering (USE) Laboratory. That eventually morphed into the MINDS Group at Penn State. MINDS stands for Multidisciplinary Initiatives in Naturalistic Decision Systems. As director of The MINDS Group, I have overseen a number of LLF efforts and implementations, being strategic for a number of PhD dissertations produced in our group as well as actively applied to different research grants.

Hence, this final phase of development and history—sociocognitive simulations and technology development—put into place a strong emphasis on scaled worlds and innovating new technological designs that could be embedded within a given scaled world for test, evaluation, and refinement. While originally doing fieldwork studies for C³I, the original CITIES simulation (Wellens) provided an analogical work setting focus—*emergency crisis management.* As years went by we redesigned and structured a newer simulation termed NeoCITIES, which has been an anchor for the research conducted while at The Pennsylvania State University. NeoCITIES has also spawned other spinoff simulations in the area of cybersecurity (see Tyworth, Giacobe, Mancuso, McNeese, & Hall, 2013) to create new opportunities and insights in research.

The MINDS Group as It Has Evolved at The Pennsylvania State University

Although the preceding historical movements laid foundations for the LLF and defined the subsequent methodological perspective, our articulation of a living laboratory necessarily means it (1) *evolves* with changing forces to be relevant, (2) provides *useful outcomes*, and (3) highlights *learning* to move data and information toward solutions. As articulated problems compose the fabric of societal advancement and exist as the fulcrum in the LLF. Identifying, solving, and exploring problem spaces generate information seeking that colors in the tradeoffs that represent viable solutions. The general mantra of the LLF is actually the same as a top value communicated at The Pennsylvania State University—that of *Making Life Better.* The following statement exemplifies this:

> As the knowledge needs of society rapidly change and expand, higher learning has a more important role to play than ever before in advancing the quality of life. Penn State's commitment to students, to outreach, and to progress touches the lives of most Pennsylvanians and improves the quality of life for all (PSU Budget Office Communication).

The MINDS Group as mentioned in the previous section represents an emerging living laboratory within itself, extensible and adapting as a function of the demands put before it. A living laboratory is made of individuals who form a knowledge-based collective that expands interdisciplinary knowledge, promotes mutual learning, and targets problems to make life better for contexts that are addressed. In some ways a living laboratory is akin to a living document in that it is continually updated, changed, adapted, and adjusted to better serve the original purpose for its existence—but also evolves to accomplish new purposes that arise along the way through the collective efforts of those who work together. This highlights the living ecosystem characteristics mentioned earlier. One of the primary reasons for composing this volume was to show the collective output of the MINDS Group as it has grown as an intact living laboratory over the past 17 years—by presenting some of the learning, activities, products, and innovations that have transgressed albeit in many different ways. Many of the chapters that compose this volume represent the distillation of the collective mindset of the MINDS Group. Although many of the contributors have moved forward to new venues and careers, a lot of their worldviews, perspective, awareness, heart, and training have been developed as active participants and change agents in the LLF.

Note that the following students in the MINDS Group have contributed to the collective knowledge of CSE through the LLF, and have produced dissertations under the advisement of Dr. Michael McNeese—Dr. Rashaad Jones, Dr. Edward Glantz, Dr. Mark Pfaff, Dr. Arthur Jones, Dr. Vincent Mancuso, Dr. Chris Dancy (Dr. Frank Ritter and Dr. McNeese—coadvisors), Dr. Tristan Endsley, (Dr. McNeese and Dr. Forster—coadvisors), James Reep, (PhD candidate and current student, Dr. McNeese, advisor). Dr. David Hall provided dissertation advisement to Dr. Ben Heller and Dr. Jeff Rimland. Dr. Hall and/or Dr. McNeese served on the dissertation committee of most of these students either as a member or coadvisor. Dr. Madhu Reddy (formerly of The Pennsylvania State University, College of IST and now at Northwestern University) provided dissertation advisement to Dr. Nathan McNeese, a member of the MINDS Group who has also contributed to this volume. It is our great honor to have all of these MINDS Group members (and former students) produce chapters for this book and therein collectively represent The Pennsylvania State University MINDS Group perspectives on CSE. Furthermore, the work continues with Samantha Weirman (Dr. McNeese and Dr. Forster as coadvisors) who is pursuing her PhD and will be contributing to the body of knowledge.

We also acknowledge the following dissertations that have contributed to both the MINDS Group (as members advised by Dr. McNeese and/or Dr. Hall) and to the collective knowledge of CSE and the LLF: Dr. Ivanna Terrell, Dr. Erik Connors, Dr. Andrew Reiffers, and Dr. Nicholas Giacobe.

In addition, we also point out the contributions made to the LLF through postdoctoral scholars who have been part of the MINDS Group over the years: Dr. Isaac Brewer, Dr. Katherine Hamilton, and Dr. Mike Tyworth. I also want to pay special tribute to MINDS Group members who have been my Research Assistants and/or Teaching Assistants along the way—(Dr. Xun Ge, my first TA and RA at The Pennsylvania State University; Dr. Patrick Craven; Dr. Bimal Balakrishnan; Dr. Tyrone Jefferson; Dr. Lori Ferzandi; Dr. Elena Theodorou; Dr. Priya Bains) who have made indirect contributions within this volume.

Finally The Pennsylvania State University faculty who have served as advisors and mentors to the above researchers have been extremely important to facilitate the kind of active learning environment we have put in place. Their knowledge, devotion, and encouragement have made this book what it is. The late Dr. David Hall, Dr. Susan Mohammed, and Dr. Peter Kent Forster have been valuable components of the MINDS Group and have provided the vision necessary for students to achieve high levels of success. Also special thanks must go to Professor Clifford Brown (one of my longtime colleagues and mentors) who spent his sabbatical from Wittenberg University, Ohio working with the MINDS Group developing scenarios and research studies for the NeoCITIES simulation.

AUDIENCE FOR THE BOOK

This book has been written for cognitive systems engineers, information scientists, and human factors professionals who are typically involved with complex systems that include the necessary integration of information, technology, and people within a given work context. It provides useful theoretical imprints, methodological approaches, and practical findings for professionals involved with cognitive systems research and development. In addition, the book provides a viable product for use in the classroom for faculty and students enrolled in courses that are pertinent to cognitive and information sciences, industrial/cognitive systems engineering, human factors—human-systems integration (HSI), human–computer interaction (HCI), user experience (UX), and usability. The volume also provides relevance for iSchools (Information Schools) wherein the information society is connected to socio-technical systems, context, information seeking, human-centered design, and collaboration as studied through a panoply of philosophies relevant to science and the building of systems.

The type of work presented is typically interdisciplinary, can be observed, interpreted, and analyzed from multiple perspectives, consists of layered-interdependent systems and system of systems, involves ill-definition and uncertainty, yet has routine elements as well, is emergent, temporal, may contain elements of information overloading from multiple sources, requires information fusion across various sources (distributed cognition), is typically predicated on successful individual and team work, is reliant on interdependencies with various technologies, is understood through the application and use of multiple methodologies, and is usually comprehended through data-information-knowledge-based analytics. The philosophy that pervades this book's understanding of complex systems and environments is one that is

- Human centered
- Problem focused
- Team enabled
- Design supported

In summary, the book is designed for professors, graduate students, specialists, and practitioners who are currently engaged in human-centered design or other applied aspects of modeling, simulation, and design that require joint understanding of theory and practice. As the book synthesizes aspects of information science and human factors through the LLF it fills a niche that is rather sparse at the moment. Furthermore, practitioners in given domains/fields of practices utilized within specific chapters may have a focused interest in work relevant to their concerns (e.g., emergency crisis management, cyber security). Faculty teaching specific research methodologies in cognitive systems and/or human factors would utilize the book to show a given genre of research methods of relevance and how they are used in an integrative manner. The book may be used as both reference and as a textbook for college coursework (fourth year undergraduate or graduate student work).

WHY THE BOOK IS NECESSARY

Many books in human factors specialties and professions downplay information and cognitive science, computational foundations, and the information technology aspects of distributed work. In turn the truly interdisciplinary components that are required to competently address complex systems in today's contemporary society are not really dealt with or acknowledged very well. Likewise, many in the information and cognitive science community have given human factors, cognitive engineering, and HSI short shrift especially downplaying teamwork and the emerging context that underlies human-centered design. The consequences—therein—is that research and design requires a broad transformative leading edge approach that supplies integrative understanding for *learning and discovery*. Unfortunately each of these respective areas tends to be polarized from the others and provides isomorphic limited perspectives. Limited perspectives arise mainly in theories, methods, and design practices.

This overall issue is seen in many areas today where computation and cognition and context must exist as an ecology of layered and functional relationships that facilitate dynamic change and adaptivity in order to assuage the emerging complexity that ensues when information, technology, and people have to function and work together in an efficient and effective way. To the extent that the information society that lies before us is human centered—then an effective framework to generate ecological precedence is very much necessary. Therein, we are writing this book to introduce, explain, and provide studies, examples, and cases that demonstrate the use of an integrated LLF in order to establish a cohesive, feasible, and mixed methodological approach for complex interdisciplinary systems.

TOPICS AND COVERAGE OF THE BOOK

As indicated in the previous paragraph, complex systems are omnipresent in everyday life with the introduction and presence of information technology within areas such as social networks, the Internet of Things, online banking, big data, traffic control and transportation, smart cities, logistics, military aviation, military command and control, robotics, big data informatics, distributed learning, augmented reality, emergency crisis management, intelligence, and cyber security, to name just a few prominent industries. Because distributed work almost always involves people in planning, decision making, operations, safety, and financial interests, there is a prime need for (1) human-centered understanding and (2) implementing cognitive-systems level considerations within information science/information technology designs. As these needs require timely integration of interdisciplinary (and often nested) topics the LLF provides a holistic and sustainable approach to leverage researcher and designer activities toward distributed work concepts that are sound, testable, as well as implementable for revision, resilience, and adaptation.

The book chapters cover what an integrated LLF and approach consist of through relevant and current research topics, application of apropos methods, and development of specific cases as studies; and specifically will answer the following questions: Why is the living laboratory important for implementing cognitive systems within a given context? How can I utilize the living laboratory for various fields of practice and work domains? How does human system integration/team system integration emerge within a dynamic, living environment? How is cognition understood in terms of information science and living systems metaphors? How can information, technology, and people be synthesized in a meaningful way using multiple methodological approaches? What is the basis for interdisciplinary research and design where human-centered knowledge and technology development, theory and practice, models and use, scenarios and simulations and are all considered holistically and kept in proper balance to insure that problems are both learned and solved within the constraints of many impinging variables and factors.

A Progression of Human Factors/Cognitive System Terms

As one begins to read the book chapters it will become obvious that the use of terminology and lingo can connote similarities in meaning yet also represent subtle differences in the application of the author's terms to a given research or design topic. Not only is this true for this book, but it is becoming a problem for disciplines in general where authors or researchers even use the same term but dispatch different intentions in the meaning. This book is absolutely focused on the topical content related to cognitive and collaborative systems extant in complex domains, and methods that result in understanding, developing and building, and testing systems *in situ*. But let us calibrate some thresholds to help situate meanings (at least the way we see it). The LLF framework represents a given utilization of CSE methods. As a reader you will discover that similar terms often are incorporated in the human factors communities of practice in government, business, industry, education, and academia.

As a foundation this book utilizes the term *human factors* to represent a wide purview of designing systems for human use incorporating as many perspectives as necessary or allowable. Human factors, hence, is a comprehensive broad term used to cover many areas. Since I have been involved in human factors (now spanning over 44 years) my first exposure to this area was through two sources: (a) the *Industrial Psychology* book (Tiffin & McCormick, 1965) which was used for an initial course in industrial psychology taught by Dr. John Reising in 1973 and (b) the first edition the book *Human Factors in Engineering and Design* (McCormick, 1976) which was used in an independent study course in human factors with Dr. David Biers (University of Dayton). Note both of these books were associated with

Ernest J. McCormick, a monumental superstar of human factors engineering. The first book focused more on psychology while the second book emphasized components of engineering and design. These books set up human factors as a discipline that actually subsumed human factors engineering in breadth although not necessarily in time. I subsequently came across and own the historical original text *Human Engineering* (McCormick, 1957), which I discovered is short for "Human Factors in Engineering" (p. 1).

As cognition started to become the main phenomena for understanding human behavior (what cognitive psychologists study) it also became a basis for designing for human use. The term *cognitive systems* engineering includes this notion as demonstrated by the work of Jens Rasmussen (1986), and Erik Hollnagel and David Woods (1983) but also connotes the idea that cognition is not only inside the head but is in fact distributed across complex systems (what Norman refers to *user-centered systems*, See Norman & Draper, 1986, and referenced by Hutchins, 1995 as distributed cognition*). Our book locks on to this strong notion of cognitive systems rather than just seeing cognition as a smaller subset of systems engineering as connoted CSE.

In 1982 I was exposed to the necessity of human factors being applied to computer systems as this was necessary in my research and work at the Aeronautical Systems Division at Wright–Patterson AFB, OH. The work as communicated earlier in the history section was heavily coupled to cognitive processes, which often were jointly carried out or assisted by computer systems. The work I did entailing specification of the HCI in the F-16 cockpit with Richard Geiselhart, ASD/ENECH and Dr. Richard Shiffler, ASD/ENECH exemplifies a reformulation of how pilots use information within computer systems to enhance their flying abilities. The work we were involved with at the Air Force Research Laboratory in redesigning military command posts (with Donald Monk and Maris Vikmanis) to incorporate the latest digital technologies and to build human-computer interfaces thus extending human capabilities is also representative of HCI. Our work continued in HCI and even expanded into teams supported with collaborative computing technologies (*computer-supported cooperative work*). Although one can look up specific definitions as to how these areas are defined and what they do, I have decided to classify them under the general term *human factors*. I am sure many would argue against this while others might argue for it but for the purpose of this book it is a useful construction.

As one looks at the current zeitgeist within the human factors umbrella, areas of emphasis may fall out according to whether one gravitates toward a certain societal association Human Factors and Ergonomic Society (HFES), Association of Computing Machinery (ACM), or Institute of Electrical and Electronic Engineers (IEEE). Recently, we see the prominence of terms like *human systems engineering* (HSE) or HSI that signify certain emphasis points within the overall gestalt. Even more prominent is the focus on UX and how this is really driving many human

---

* We also make reference here to the broad notion of activity theory which has been relevant also in social constructions of what we mean by an integrated living laboratory framework, and hence acknowledge the importance of activities and their cultural/historical foundation (see Nardi, 1996).

factors and *human-centered design* practices today especially within industry and software-related companies. All of this can be framed as a positive response as human production, safety, and well being is strengthened.

While each book chapter reifies meaning within the context of the topic, research, and perspective undertaken, these areas together represent a substantial investment on the future of human beings as they interact and transform life in the contemporary digital world. What one does as a human factors/cognitive systems practitioner may tend to focus on one of the above italicized emphasis areas but many of these areas are intrinsically relatable, complementary, and not mutually exclusive. The chapters presented in this volume innately cover CSE through the lens of the integrate LLF.

THE RELEVANCE OF THE LIVING LABORATORY FRAMEWORK

Many of the chapters that compose this volume are in fact emanating from members of the MINDS Group at The Pennsylvania State University. The unique compilation of chapters presented in this volume represents the state of the art as we have uncovered it (1) within our understanding of CSE, (2) produced through the transformative vision of the integrated LLF, and (3) using multiple integrated methodological approaches to complex systems and environments. While other phases and facets of cognitive system engineering have been encountered over the years (as elaborated in the first section of the chapter), this book represents a collective volume on what the LLF is, and how it has been used for specific challenge problems. It has been possible through the inspiration, hard work, and *mindset* of the authors who have brought their ideas, research, and aspirations together by utilizing the LLF. We will now look at some of the specifics underlying the LLF.

BASIC PHILOSOPHY

Through the passage of time we have outlined the historical foundations that have been responsible for the emergence of CSE and the initiation of the LLF as a means to accomplish methods of CSE. At this point, we will describe the philosophy and basic elements of what the LLF is. The explanation below uses an overlay model to describe planes of development that can be accomplished in building cognitive systems that work.

Ecological in Nature

This book is focused on articulating all aspects of the LLF as applicable to human factors, cognitive systems, and collaborative work. It provides an interdisciplinary formation of human factors from a living ecosystem perspective (hence the name—living laboratory), wherein the environment is very important to understand human intention and activity. The approach affords a flexible and leveraged use of mixed research methodologies to comprehensively address complex systems in a thoughtful—yet useful way. Living systems require dynamic and timely integration of multiple sources of data, information, and knowledge in order to evolve and adapt to the context they exist in while they pursue intentions and desires. Resilience and sustainable operations are high-level values in approaching human factors from a living ecosystem perspective. In turn the living laboratory reflects the values of (1) multiple perspectives and (2) system of systems transformation with the intent to translate knowledge into designs that afford effective use. This requires theoretical knowledge to be integrated with real-world fields of practice and human expertise through the use of various tools, techniques, and processes. HSI is accomplished by utilizing a problem-based learning approach within the LLF. Therein, mixed methods are structured/integrated to leverage understanding and development of problems, issues, and constraints that are revealed in a context.

Experience Matters in Fields of Practice

The foundation of the living laboratory is based on the idea that *experience matters*, but it is not all that matters when one takes an interdisciplinary/transformative view of information, technology, and people. Experience builds on itself through *human–agent interaction** set within a specified context of use. Take, for example, an area within our research labs at Penn State—emergency crisis management. Experience emerges cojointly with complex situations that involve human actors (often a distributed team) using technological artifacts (e.g., analytical social networks) to track changes within the context. In this case, an agent may be another human (human-to-human interaction results in typical formulations of teamwork), or a system, tool, aid, computational agent (a newer form of teamwork), or a combination therein. In addition, the context may be multilayered in that it could include movements in and out of the physical, social, cultural, and cognitive environments that specify changes in the situation as the agent becomes aware of it. Experience requires sensemaking, awareness, and a trajectory of what is coming next (expectations/beliefs/hypothesis formulation). Experience can be informed by the context of action, by cognitive learning within a context, through the culture present in the context, by providing opportunities of exploration via a scaled world of agents where testing what you know is put to use and adapted through many cycles, and through sound designs that create ecological balance in human–agent interaction (cognitive systems engineering/human-centered design).

The living-laboratory approach begins by focusing on experience that is acquired in a context or field of practice when a human addresses a set of problems, situations, or cases—often through the use of technologies that amplify performance, enhance learning/cognition, or overcome limitations/expand capabilities. The LLF researcher's major quest is to identify and define problems in a way that user's experience becomes readily available and is understood to the maximum extent possible. Note this is how we nuance UX within the living world.† Of course because real-world

---

* Note here that this directly comes from the historical precedence of *collaborative-and-human autonomous systems* (as described from Section 2.1.2 of this chapter) and has continued as an emphasis in current work. We refer to human and computational agents as broadly defining human–agent interaction (human-to-human agent, human to computational agents, or computer-to-computer agents) within collaborative systems.

† As mentioned previously, many of the semantics of like-minded areas (e.g., user experience) can be connoted differently yet often traverse very similar phenomenological paths.

contexts do not often present every detail or nuance of problem complexity, methods must also be available beyond just observation (more on that to come). Users come in many shapes and sizes, with differing levels of cognitive and physical proclivities. Experience is the knowledge obtained (either theoretical or practical) as a result of addressing various problems, issues, or constraints that is spent into the future (i.e., used or accessed again when similar situations or problems emerge—often without being told to do so). Experience is contingent on perception, memory, situation awareness, attention, language, and other cognitive variables that a user can utilize to address problems within a given system in a context of use. Therefore, *knowledge as design* must intersect UX through examination of problems in the context of use.

As inferred above, the researcher's quest is to thoroughly understand use and UX (for individuals and/or collaborative activity) as problem spaces are encountered. Hence, the lynch pins in the LLF entail how intentions *fold out* through the joint influences of experience, problems, interaction, and context. Each of these lynchpins is derived from empirical work in research within cognitive science, cognitive systems, and computer-supported cooperative work. Therein, they are not just plucked from thin air but represent empirical evidence that supports how people learn and solve problems (Bransford, Brown, & Cocking, 2004) in living emergent worlds that are often messy, complex, dynamic, multifaceted, and ill-defined. See Xun Ge, Chapter 2 (this volume), as it unpacks problems within the context of educational settings.

Practice provides patterns as to the way we go about acting on intentions and desires, and the way we go about doing work that includes complex relationships among information, technology, and people. Practice is predicated on learning new paths to explore the environment and can be the way that transforms the ill-defined, unstructured, and uncertain problems into ones that provide the user with a more routinized, safe, and expected rate of return. As many problems are new, they require learning, creativity, and exploration for discovery to occur. Therefore, nonroutine problems can exhibit major challenges for actors in order to be solved or negotiated. Many problems can be wicked, nonlinear, and complex while at the same time require knowledge of multiple contexts and constraints that are operative.

Often we contrast theoretical with practical viewpoints artificially separating them into isolated understandings that provide only limited angles. The living laboratory aspires to integrate perspectives on how experience can be informed through four major elements: (1) observational ethnography, (2) knowledge elicitation of subject matter experts, (3) development of scaled world simulations–scenarios, and (4) generation of new innovation technologies (e.g., interfaces, decision aids, work tools, intelligent agents) through workable prototypes. In the course of exercising these specific elements—*models* and *use* start to be built and reified. Mutual learning is acquired through opportunistic use of the elements in the living laboratory, wherein feedback and feed-forward help to substantiate knowledge as design (Perkins, 1986). Through these methodological pursuits, mature knowledge concepts can begin to formulate and be tested and refined. Designs can be put into practice as interventions that further shape cycles of the overall process. As such, human–agent interaction can begin to solidify on a firm foundation and provide balance/resiliency within a system, or system of systems. Therein, the living laboratory approach results in using dynamics to comprehensively encourage interdisciplinary science resulting in movement that adapts to change, see patterns that emerge, and encourage transformative forces to play out for a given context. To quote Woods (1998), "the experimenter becomes designer while the designer becomes experimenter."

ELEMENTS OF THE LIVING LABORATORY FRAMEWORK

Living laboratories emphasize the necessity of studying the context where phenomena exists, but our take on this is that information seeking helps to develop theoretical knowledge and practical knowledge jointly and envelops experience and depth on understanding a given phenomena. Approaching phenomena from multiple perspectives/methods can yield a broader comprehension and integrates sources of understanding to facilitate a stronger base for *knowledge as design* (Perkins, 1986).

Figure 1.1 shows the basic-level coupling at the heart of the framework—what we refer to as the *vertical axis* (shown by the vertical arrow). Figure 1.2 shows the first overlay of the framework: the interrelationship among theory-problems-practice from the perspective of the researcher beginning investigations into living-emergent phenomena.

The living laboratory hence starts with problems that are extant in the phenomena under study in order to expand the researcher's experience with the phenomena. Problems are the basis for discovery in real-world contexts and specific fields of practice, but they can also be informed through the use of theoretical knowledge. As problems present situations that invoke information seeking, exploration, and data discovery in order to resolve them—grand challenges can be identified, set up, and defined, and explored in differing ways—prior to—committing to a solution space. Note in Figure 1.2 that in addition to the vertical axis there is also

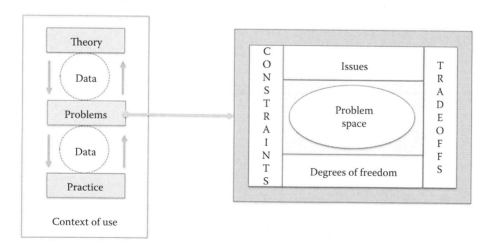

FIGURE 1.1   Basic elements of the living laboratory/Layer I.

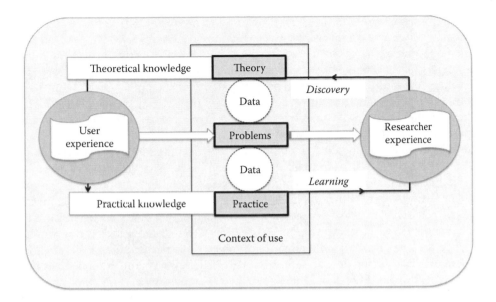

FIGURE 1.2   The living laboratory framework/Layer II.

a *horizontal axis* (shown by a horizontal arrow) within the LLF representing how researchers obtain knowledge about UX in addressing problems to gain researcher experience (RX). Continuing with the layered overlay model of the LLF, Figure 1.3 shows the final development of the integrated living laboratory elements that compose the framework. Although the figure shows an optimal situation wherein all

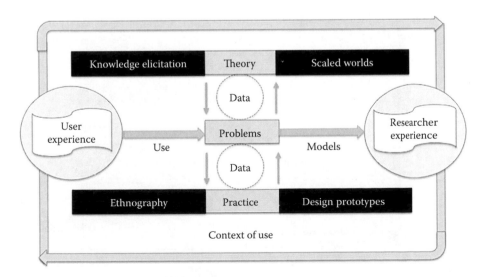

FIGURE 1.3   The living laboratory framework–Layer III.

the elements are utilized in equal proportion to produce UX and RX, the reality of every cognitive engineering project may only utilize a subportion of all the elements that are possible. This is owing to the practical nature of work and the living aspect, wherein demands and constraints make only certain paths possible, as more knowledge is discovered and integrated. For example, a project may only present opportunities for knowledge elicitation, theory, and scaled-world simulation. Figure 1.4 shows the disproportionality of the elements when the LLF is only partially used for this situation. If more sources of knowledge are possible then reliability, validity, and completeness can be elevated, but often real world cognitive system engineering has to be considered within constraints such as time, cost, schedule, availability of experts, and access to environments, as perfect situations do not exist. Also note that in Figure 1.4, the proportionate size of the hexagon varies with the level of effort expended. In this case, the smaller knowledge elicitation element indicates that the availability of experts was limited to 2–3 day period; hence the level of effort is therein proportionate. This particular case shows more of an emphasis on top down research with the major effort being expended through the scaled world, even though the scaled world scenarios and task makeup was informed through theory and knowledge elicitation.

From a top-level view, problems can be approached from different directions: theoretical/quantitative knowledge and practical/qualitative knowledge. The living laboratory utilizes both of these approaches but aims toward integrated solutions. The left side of the Figure 1.3 tends to look at both *what the head is inside of—the context* and *what is inside the head—expertise* (Mace, 1977) from the qualitative methods of ethnography and knowledge elicitation. Alternatively, the right side of the figure tends to represent quantitative/design-based methods, although each of these elements can have quantitative and qualitative analyses present. For example, the development of stories related to UX can lead directly to qualitative design sketches that build up into interactive storyboards, which can then be built as reconfigurable design prototypes that go on to be tested in a scaled world quantitatively. The framework is based on feedback and

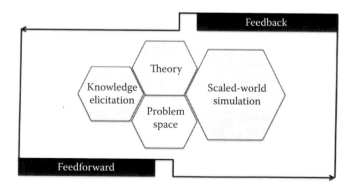

FIGURE 1.4   Example of proportionality of the cognitive systems engineering work.

feed-forward flow to afford integration, holistic understanding, and multiple perspectives of how knowledge influences design within a given field of practice. Therein, prototypes also flow back into practice connecting theory to problems to practice for reiteration and reification.

Knowledge as Design as Living Ecosystem

Living ecosystem's philosophy suggests that cognitive systems are approached in a certain way; wherein they are distributed across cognition and context, embodied in real-world activities, and emergent and situated through successive human–agent interaction necessary for dynamic growth, change that produces adaptation, and therein sustain themselves in order to survive and stay strong in threatening environments.* In the original paper (McNeese, 1996) we refer to this notion as follows:

> The Merriam-Webster dictionary defines ecological as the interrelationship of organisms and their environments; in turn it defines perspective as the capacity to view things with the proper relationships to their value, importance, and basic qualities. Taken together they emphasize the multidisciplinary nature of living systems, their environments, and the reciprocity that has developed between them.

The term *living* connotes an ecological system (ecosystem) wherein human–agent interaction is an emergent set of activities (the give and take) between an agent and an environment, wherein the affordances of the environment are always adjusted in relationship to the capabilities of the agent at a point of time. Interaction emerges in time and space, and is directed through the intentions of the actor–agent, but constrained by what may be available (social, physical, and economic resources), as applied to a given *problem space*. An ecological system is grounded when it is in balance, as the components are greater than the sum of the individual parts (collective induction). When the system is unbalanced, isolated, perturbed, and in conflict, it may not achieve the intentions the agent strives for (i.e., failure or errors are likely imminent). One of the tenets of the living laboratory is to design and secure systems that are adaptive, sustainable, livable, dynamic, and thriving rather than systems that are ad hoc, failing, outdated, and not understood by the human agent. The core principles of the living laboratory are (a) to understand, define, and explore problems from interdisciplinary, multifaceted perspectives; (b) to apply and integrate a multiple methodological lens (different techniques/tools/measurements) to obtain holistic and cohesive knowledge about a problem; and (c) to use this knowledge to design new technologies that enable innovative human–agent interaction that leads to improvements in performance; thereby establishing balance within the ecosystem, enabling transformative effectiveness and efficiency, while increasing resilience in assuaging complexity.

---

* This perspective is highly consistent with the theoretical work on distributed cognition (Hollan, Hutchins, & Kirsh, 2000); situated cognition (Brown, Collins, & Duguid, 1989), naturalistic decision making (Klein, 2008); embodied interaction (Dourish, 2001).

STRUCTURE OF THE BOOK

This book consists of 19 chapters and has been divided into the following 7 sections to make assimilation of the topic content easy to comprehend:

Section I   Introduction and historical precedence (Chapter 1)
Section II   The living laboratory: Learning in action (Chapters 2–5)
Section III   Theoretical perspectives in cognitive–collaborative systems (Chapters 6–8)
Section IV   Models and measures of cognitive work (Chapters 9–11)
Section V   Scaled-world simulations (Chapters 12–14)
Section VI   Knowledge capture, design, integration, and practice (Chapters 15–18)
Section VII   The future of the living laboratory (Chapter 19)

These sections have been structured to represent the various themes and elements that are present within the LLF. In turn, they provide a comprehensive set of topics that span across the LLF and provide the methodological approach that represents this instantiation of CSE.

Within a given section the papers will present research and topics relevant to the general topical content. Although it is the case that the chapters were stratified by topic, many of the chapters could have been viable for other sections as well, given the scope of information presented.

Academics have been accused of conducting research that is devoid of real-world application. The LLF directly contradicts that perception. The strength of this book is the concrete examples of how the LLF can cross the Rubicon that sometimes separates research from real-world problems. These chapters allow the readers to examine the application of methods and tools to improve their understanding of cognitive engineering systems in expanding foundational knowledge while contributing and maintaining relevance through the provision of solutions to real-world problems. They give substance to the principles of human centered, problem focused, team enabled, and designed-supported research.

The initial chapters examine learning in action, which focuses on the role of the LLF in improving group and individual learning. Dr. Rimland describes the application of a technical solution to support distributed problem-based learning. He specifically focuses on federated opportunities (i.e., note taking and exams), and exemplifies the LLF through its application aimed at enhancing student engagement in introductory computer programming and technology courses in Penn State's College of Information Sciences and Technology. Through the application, physical objection as cognitive artifacts to enhance lasting memories, Dr. Mancuso explains how team cognition and trans-active memory may be used to access outside knowledge and thus improve team performance. In his research, he notes the limitation constrained by the individual's internal knowledge in spite of technological sophistication that has improved the set of cognitive artifacts available. Rather than examining student tasks, Dr. Ge explores the challenges of technology, which simultaneously acts as scaffolding and data collection tools when conducting education research. She concludes that tools may be purposed differently to manage the external influence of stakeholders and other contextual constraints. Dr. McNeese and Reep conclude this section with an examination on the integration of technology with the supply chain to improve efficiencies and sustainability in the natural gas industry. Although deviating slightly from the more pure CSE approaches, this chapter incorporates ethnographic and knowledge-elicitation methods to understand the realities of the industry. It then provides a functional analysis of the presence or absence of collaborative systems and offers a framework for additional research on how technologies might be used to improve efficiencies.

The second section expounds on theoretical perspectives of cognitive–collaborative systems and links the theoretical and applied through the LLF. Drs. Nathan McNeese and Reddy and Demir proposed that collaboration among teams is becoming increasingly important, but collaborative information seeking remains the purview of social and technical research rather than focusing on cognition. Using collaborative information seeking as a framework, They offers perspectives on multiple cognitively-based methods and approaches to better understanding the environment. Drs. Endsley and Forster and Reep used the LLF to assess cognition and teamwork in a crisis situation. Crises create a complex environment that reduces the time in which individuals and teams must respond. Through case studies, the researchers provide a framework for not only understanding crises and cognition, but also examine team-situational awareness and performance in a cross-cultural environment. Stress, particularly in crisis situations, often has been identified as a key factor in operator error; however, research on stress, according to Dr. Pfaff, often portrays it as a *black box*. His chapter addresses the importance of using multiple methodological approaches to understand the nexus of stress, emotion, and cognition. Using the LLF, he substantiates the validity of the multimethodological approach and provides a basis for more reliable interdisciplinary research into the impact of stress.

Measuring and modeling cognition is the concentration of third section. Here, concepts such as information sharing, situation awareness, and team mental models are used to recognize relationships between inputs and outputs in behavior and performance. Drs. Mohammed, Hamilton, Mancuso, Tesler, and McNeese reviewed four studies that explore how various interventions (e.g., storytelling or temporal models) might influence team cognition. Through these methods, the authors expand the literature relevant to the LLF by empirically investigating information, situation awareness, and team mental models in a set of studies; *infusing* time into assessments of the characteristics; exploring the benefits of ad hoc stories and team and individual reflection on performance to improve shared understanding and thus improved performance. In *Modeling Cognition and Collaborative Work*, Dr. Rashaad Jones uses the LLF to create decision-making models using Fuzzy Cognitive Mapping (FCM) and again models the relationships between inputs and outputs to produce concept maps or decision-aids to improve the decision making. In this case of knowledge elicitation, a LLF foundation identifies the decision space and viable decision-aids for a variety of settings. Dr. Dancy's chapter concludes this section by inspecting unified theories of cognition from computational systems (i.e., cognitive architectures) and psycho-physiological measures to study the effects of system design on both predicted performance and changes in cognitive processes.

Often discussed in the introductory chapter, NeoCITIES, a computer-simulated task environment used as the basis for much of the research by MINDS, is the focus of the scaled-world simulation section. In Dr. Hamilton et al.'s chapter, the authors identify both the importance of team cognition across sectors (e.g., defense, medicine, and sports) and the challenges of time and labor that hinder research in this area. They offer that the simulation-based study is a solution to this problem, stressing that NeoCITIES enables the researcher to study multiple forms of team cognition within simulations that are central to the LLF. NeoCITIES is a crisis-management simulation and Dr. Minotra et al. adapt the simulation to study cyber security using a mixed methodological approach, examining the impact of workload and information predictability during crisis events on performance. While demonstrating the value of the LLF, this study resulted in the development of a dual-task attention research testbed, but also recommended adapting scaled worlds to more deeply explore microcognitive tasks. The section concludes with a look at the ubiquitous video game, which, besides providing entertainment, offers a dynamic testbed for research. Dr. Pursel and Stubbs explore how the LLF may be used to study the video phenomenon within the context of Massively Multiplayer Online Games (MMOGs). The authors focus on MMOGs as learning environments assessing behaviors and skills learned in specific games, and the extent to which these skills may be applied to operating in a crisis management environment.

The experimental sections conclude with the *Knowledge Capture, Design, Integration, and Practice* section that further confirms how the LLF is used to enhance performance in addressing real-world problems. In his chapter, Dr. Glantz uses the LLF to investigate cognitive work in high-risk socially distributed problem spaces. Through the application of ethnography and knowledge elicitation with domain experts, the author developed an understanding of the police working environment and then used participatory design to develop a suitable artifact to enhance cognition. In many respects, this chapter represents the living laboratory at its most fundamental level. Dr. Art Jones investigates the intermodal Human–Computer Interaction (HCI) in crises environments in which decision-critical information is only available in analog and subjective formats. The LLF is used to explore the challenge of developing and evaluating interfaces that are able to quickly and accurately collect and display data that have already been interpreted by human users. Using applied experience, Dr. Hellar furthers the theme of applying theory to solve real-world problems. Confronted with the intelligence community's *big data* challenge (e.g., large quantities of data but very strict parameters of the mission), the LLF is applied to define the problem and break it down into its constituent parts allowing the development of agile and innovative solutions. Drs. Craven, Tremoulet and Harkness Regli conclude this section by asserting that improved application is achieved through experimentation in a real-world setting such as an industrial space.

CONCLUDING REMARKS

You are about to begin a journey in an innovative way to think about CSE that elaborates the basis for and examples of the integrated LLF methodological approach to CSE. Each chapter provides topical content in a way that engages current thinking, engages multiple methodologies to uncover knowledge as design, and provides unique insights into contemporary cognition as a living ecosystem that evolves sustained operations according to the need, survival, and environment. The chapters utilize specific learning helps to assist the reader in comprehension of the topic being considered.

REFERENCES

Altman, I. (1975). *The environment and social behavior: Privacy, personal space, territory, and crowding.* Monterey, CA: Brooks/Cole Publishing Company.

Barker, R. G. (1968). *Ecological psychology: Concepts and methods for studying the environment of human behavior.* Stanford, CA: Stanford University Press.

Bransford, J. D., Brown, A. L., & Cocking, R. R. (2004). *How people learn: Brain, mind, experience, and school.* Committee on Learning Research and Educational Practice. Washington, DC: National Research Council.

Bransford, J. D., & Stein, B. S. (1984). *The ideal problem solver. A guide for improving thinking, learning, and creativity.* New York, NY: Freeman.

Brown, J. S., Collins, A., & Duguid, P. (1989). Situated cognition and the culture of learning. *Educational Researcher, 18*(1), 32–42.

Brown, C. E., & Leupp, D. G. (1983). *Team performance with large and small screen displays* (Report No. AFAMRI. TR-85.0:13). Wright-Patterson AFB, OH: Air Force Aerospace Medical Research Laboratory (DTIC No. 158761).

Fiske, S. T., & Taylor, S. E. (1991). *Social cognition* (2nd ed.). New York, NY: McGraw-Hill.

Gibson, J. J. (1979). *The ecological approach to visual perception.* Boston, MA: Houghton Mifflin Company.

Hollnagel, E., & Woods, D. D. (1983). Cognitive systems engineering: New wine in new bottles. *International Journal of Man-Machine Studies, 18*, 583–600.

Hutchins, E. (1995). *Cognition in the wild.* Cambridge, MA: MIT Press.

Kimble, C., & McNeese, M. D. (1987). *Emergent leadership and team effectiveness on a team resource allocation task* (Report No. AAMRL-87-TR-064). Wright-Patterson Air Force Base, OH: Armstrong Aerospace Medical Research Laboratory.

Lave, J. (1988). *Cognition in practice: Mind, mathematics, and culture in everyday life.* Cambridge, MA: Cambridge University Press.

Mace, W. M. (1977). James J. Gibson's strategy for perceiving: Ask not what's inside your head, but what your head's inside of. In R. Shaw & J. Bransford (Eds.), *Perceiving, acting, and knowing: Towards an ecological psychology* (pp. 43–65). Hillsdale, NJ: Lawrence Erlbaum Associates.

McCormick, E. J. (1957). *Human engineering.* New York, NY: McGraw-Hill.

McCormick, E. J. (1976). *Human factors in engineering and design.* New York, NY: McGraw-Hill Publishing.

McNeese, M. D. (1977). *How the JFK union spaces people out: An ecological study of personal space.* Senior Synthesis. Dayton, OH: University of Dayton, Department of Psychology.

McNeese, M. D. (1986). Humane intelligence: A human factors perspective for developing intelligent cockpits. *IEEE Aerospace and Electronic Systems, 1*(9), 6–12.

McNeese, M. D. (1996). An ecological perspective applied to multi-operator systems. In O. Brown & H. L. Hendrick (Eds.), *Human factors in organizational design and management - VI* (pp. 365–370). Amsterdam, The Netherlands: Elsevier.

McNeese, M. D. (2016). The phenomenal basis of human factors: Situating and distributing cognition within real world activities. Special section: Future directions of CEDM. *Journal of Cognitive Engineering and Decision Making, 10*, 116–119.

McNeese, M. D., & McNeese, N. J. (September, 2016). *Intelligent teamwork: A history, framework, and lessons learned.* Proceedings of the 60th Annual Meeting of the Human Factors and Ergonomics Society, Washington, DC, 19–23.

McNeese, M. D., Perusich, K., & Rentsch, J. (1999). What is command and control coming to? Examining socio-cognitive mediators that expand the common ground of teamwork. In *Proceedings of the 43rd annual meeting of the human factors and ergonomic society* (pp. 209–212). Santa Monica, CA: Human Factors and Ergonomics Society.

McNeese, M. D., & Vidulich, M. (Eds.). (2002). *Cognitive systems engineering in military aviation environments: Avoiding cogminutia fragmentosa.* Wright-Patterson Air Force Base, OH: Human Systems Information Analysis Center (HSIAC) Press.

McNeese, M. D., Zaff, B. S., Citera, M., Brown, C. E., & Whitaker, R. (1995). AKADAM: Eliciting user knowledge to support participatory ergonomics. *The International Journal of Industrial Ergonomics, 15*(5), 345–363.

McNeese, M. D., Zaff, B. S., Peio, K. J., Snyder, D. E., Duncan, J. C., & McFarren, M. R. (1990). *An advanced knowledge and design acquisition methodology: Application for the pilots associate.* AAMRL-TR-90-060. Wright-Patterson Air Force Base, OH: Armstrong Aerospace Medical Research Laboratory.

Nardi, B. A. (1996). Activity theory and human computer interaction. In B. A. Nardi (Ed.), *Context and consciousness: Activity theory and human-computer interaction* (pp. 1–8). Cambridge, MA: The MIT Press.

Norman, D. A. (1988). *The design of everyday things.* New York, NY: Basic Books.

Norman, D. A., & Draper, S. W. (Eds.). (1986). *User centered system design.* Hillsdale, NJ: Lawrence Erlbaum Associates.

Perkins, D. N. (1986). *Knowledge as design.* Hillsdale, NJ: Lawrence Erlbaum Associates.

Rasmussen, J. (1986). *Information processing and human-machine interaction: An approach to cognitive engineering.* New York, NY: North Holland Publishers.

Salomon, G. (1997). *Distributed cognitions: Psychological and educational considerations.* Cambridge: Cambridge University Press.

Schmidt, K., & Bannon, L. (1992). Taking CSCW seriously: Supporting articulation work. *Computer Supported Cooperative Work, 1*(1–2), 7–40.

Tiffin, J., & McCormick, E. J. (1965). *Industrial psychology* (5th ed.). Englewood Hills, NJ: Prentice-Hall.

Tyworth, M., Giacobe, N., Mancuso, V., McNeese, M., & Hall, D. (2013). A human-in-the-loop approach to understanding situation awareness in cyber defense analysis. *EAI Endorsed Transactions on Security and Safety, 13*(2), 1–10.

Vicente, K. (1999). *Cognitive work analysis.* Mahweh, NJ: Lawrence Erlbaum Associates.

Wellens, A. R. (1993). Group situation awareness and distributed decision making: From military to civilian applications. In J. Castellan (Ed.), *Individual and group decision making: Current issues* (pp. 267–291). Hillsdale, NJ: Lawrence Erlbaum Associates.

Wellens, A. R., & Ergener, D. (1988). The C.I.T.I.E.S Game: A computer-based situation assessment task for studying distributed decision making. *Simulation & Games, 19*(3), 304–327.

Whitaker, R. D., Selvaraj, J. A., Brown, C. E., & McNeese, M. D. (1995). *Collaborative design technology: Tools and techniques for improving collaborative design* (Report No. AL/CF-TR-1995-0086). Wright-Patterson Air Force Base, OH: Armstrong Aerospace Medical Research Laboratory.

Wilson, D., McNeese, M. D., & Brown, C. E. (1987). Team performance of a dynamic resource allocation task: Comparison of shared versus isolated work setting. In *Proceedings of the 31st annual meeting of the human factors society* (Vol. 2, pp. 1345–1349). Santa Monica, CA: Human Factors Society.

Woods, D. D. (1998). Designs are hypotheses about how artifacts shape cognition and collaboration, *Ergonomics, 41*, 168–173.

Zaff, B. S., McNeese, M. D., & Snyder, D. E. (1993). Capturing multiple perspectives: A user-centered approach to knowledge acquisition. *Knowledge Acquisition, 5*(1), 79–116.

Zsambok, C. E., & Klein, G. (Eds.). (1997). *Naturalistic decision making.* New York, NY: Lawrence Erlbaum and Associates.

Section II

The Living Laboratory
Learning in Action

2 The Living Laboratory in the Context of Educational Research

Xun Ge

CONTENTS

Introduction..32
Purpose...33
Theoretical Perspectives 33
The Living Laboratory Example 1: Developing a Web-Based Cognitive
Support System 35
    Description of the System 35
    Iterative Design Process in Context 35
        Conceptualization..36
        Developing, Modifying, and Validating the System........................36
        Refining and Expanding the System 36
    Inquiry Methods and Research Tools...36
    The Researchers' Roles..37
    Learner Contribution to the System 37
The Living Laboratory Example 2: Students-as-Designers Learning
Environment 38
    Description of the Learning Environment: Students-as-Designers....................38
    Iterative Design of the Students-as-Designers Learning Environment...............38
        Conceptualization..38
        Ethnography Study 1: Conceptualization and Initiation 38
        Ethnography Study 2: Modification and Redesign 39
    Inquiry Methods and Research Tools...40
    The Researchers' Roles..40
    Working with Stakeholders 41
Conclusion 41
Review Questions..42
References...42

ADVANCED ORGANIZER

- What are the main challenges for the twenty-first century education faced by educational researchers? What do these challenges imply to educational research?
- Why is the living laboratory framework meaningful or significant to educational research?
- How does the design research approach in education intersect with the living laboratory approach in cognitive system engineering? In what ways are they in common?
- What are researchers' roles in designing and developing innovative educational sociotechnical systems?

INTRODUCTION

One of the most essential skills in the twenty-first century is problem solving, particularly ill-defined and ill-structured problem solving (Jonassen, 2011). We encounter ill-structured problems every day in our life. These are problems with vague goals, unclear contextual information, less specified situations, or missing information needed to solve the problems (Chi & Glaser, 1985; Jonassen, 1997; Sinnott, 1989; Voss & Post, 1988). Examples of ill-structured problems include high school freshmen trying to map out a study plan for their high school curricula and educational researchers attempting to develop a computer-supported collaborative learning environment. The complexity of ill-structured problems poses difficulties for students, such as identifying sources of information, representing problems, organizing information for the problem to be solved, and monitoring their understanding and solution process (Feltovich, Spiro, Coulson, & Feltovich, 1996). Therefore, it is critical to reengineer our educational systems to help students develop critical thinking, self-directed learning (Hmelo-Silver, 2004; Hmelo-Silver & Barrows, 2006), and adaptive problem-solving skills through designing student-centered open learning environments (OLEs) (Bransford, Brown, & Cocking, 2000; Hannafin, Land, & Oliver, 1999). OLEs are learning environments encouraging self-inquiry, divergent thinking, and heuristics-based learning in ill-defined, ill-structured domains (Hannafin et al., 1999), represented by various learning approaches, such as problem-based learning (PBL) (Barrows & Tamblyn, 1980), anchored instruction (Cognition and Technology Group at Vanderbilt [CTGV], 1993), and communities of practice (Barab & Duffy, 2000; Wenger, 1998).

Compared with a traditional learning environment, in which students passively receive and process information from the teacher, students in an OLE assume more autonomy and responsibility for their own learning, whereas a teacher plays the role of a facilitator in their knowledge construction and creation process. In such a new learning and instructional paradigm, multiple stakeholders (e.g., students, the teachers, and educational researchers) contribute to and shape the design, development, and implementing of an OLE driven by problems. Furthermore, educational researchers play multiple roles (e.g., designer, facilitator, and researcher) in an attempt to understand ill-structured problems and investigate complex dynamics *in situ* (i.e., the natural setting of schools) (Brown, Collins, & Duguid, 1989). Such questions arise: What are the meaning and process of educational reengineering, as found in OLEs, through the lens of cognitive systems engineering? How does design interact with research in the process of developing an innovative educational socio-technical system? What tools can be used to help us frame and examine the process of cognitive systems reengineering in education?

PURPOSE

This chapter focuses on the living laboratory framework, a tool that can help us answer the aforementioned questions. Following a brief literature review of the living laboratory perspective and design research in education, this chapter is intended to demonstrate, through two examples of educational sociotechnical systems, how the living laboratory approach is contextualized in educational design research. The main characteristics and processes of the living laboratory are identified and examined. The goal of this chapter is to help readers develop understanding of the significance of the living laboratory framework in educational settings and how it can be adapted in various design-research contexts involving various levels of complexity.

THEORETICAL PERSPECTIVES

The living laboratory is a research framework suggesting that cognition and collaboration come about through situated actions that arise during the course of events occurring in a particular context (McNeese et al., 2005). Based on Suchman's (1987) perspective, McNeese and his colleagues (2005) contended that cognition is constructed by social processes and situational contingencies, and that design and research should be conducted in the qualitative and naturalistic setting of social and cultural context. The living laboratory perspective is in complete agreement with the perspective of situated cognition in education advanced by Brown et al. (1989), who argued that situation is indivisible from cognition, and that contexts play an active role in shaping human cognition. The situated cognition perspective emphasizes the importance of learning through participating in a situated activity and in a practice of social communities and the construction of identities in relation to those communities (Barab & Duffy, 2000; Wenger, 1998). Along the same line, Brown (1992) called on educational researchers to conduct *design research* in the naturalist setting of classrooms and the rich context of school culture, rather than in strictly controlled laboratories dominated by traditional experiments.

As an alternative research approach in education, design research is an emerging paradigm for investigating learning *in situ* through systematic design and study of instructional strategies and tools (Brown, 1992). It is an approach to both design of instructional contexts and construction of data-driven theory, in which researchers play the dual roles of both designer and researcher (Brown, 1992; Brown & Campione, 1996; Cobb, Confrey, diSessa, Lehrer, & Schauble, 2003). In design research, educational researchers seek to understand learning phenomena in a complex socio-cultural system in an ill-defined context of day-to-day schools, where situations of learning are not fixed or immutable, but rather open to redesign by the collaborative efforts of educators and researchers (The Design-Based Research Collective, 2003). The perspective of design research is in line with that of the living laboratory, which embraces the sociotechnical systems, including the social aspect of people and society, and technical aspect of organizational structure and processes. In the context of education, the social aspect involves culture at various levels, various stakeholders (e.g., students, teachers, parents, administrators, and policymakers), curricula, community, and society, whereas the technical aspect in education may be interpreted as organization of school systems, classroom culture, learning environments, and the processes to develop OLEs, learning technologies, and tools for educational reengineering. The design research approach intersects with the living laboratory approach, which integrates design, research, scaled-world simulations, and observational methods to address problems, challenges, and constraints that arise in the context of work. Both approaches emphasize the cyclic process of practice–theory and input–output within a holistic system, during which different components interact and inform each other: technology, data, user, and group-centered elements (McNeese et al., 2005).

Situated in the context of education, the living laboratory involves multiple stakeholders (e.g., researchers, students, teachers, parents, and administrators) and various components (e.g., technology, data, and resources) related to OELs. In the design-research process, educational researchers are charged with multiple responsibilities: (1) designing student-centered learning environments (e.g., problem-based, project-based, inquiry-based learning environments) to support students real-world problem-solving, critical thinking, and self-directed learning skills; (2) conducting design research to address educational needs, challenges, and problems; (3) evaluating the impact of the innovative design through iterative process of collecting and analyzing data, and (4) using the results to inform future research, modify, and improve a new sociotechnical system or learning technology.

Through two examples of design research, this chapter is an attempt to demonstrate how the living laboratory framework is applicable to educational design research. The first example is about the development of a web-based cognitive support system, and

MIDWAY BREATHER

- What are the main characteristics of the living laboratory in the context of education? Or how is the living laboratory framework contextualized in educational research?
- How is the cyclic process of design research actualized in the context of education?
- What inquiry methods and tools are used in the living laboratory of design research in education?
- What are the roles of researchers in the educational living laboratory?
- Who are the stakeholders in developing and shaping the sociotechnical systems in OLEs?

the second example is about an ethnographic study of students-as-designers for an authentic game design project. The two examples are presented and discussed regarding (1) the cyclic process of practice and theory, (2) inquiry methods and research tools employed for design research, (3) learner contribution to the system development, and (4) researchers' roles and their relationship with other stakeholders.

THE LIVING LABORATORY EXAMPLE 1: DEVELOPING A WEB-BASED COGNITIVE SUPPORT SYSTEM

DESCRIPTION OF THE SYSTEM

Although ill-structured problem solving has been recognized as an important learning skill and activity in OLEs, it has been well documented that learners often have difficulty applying or transferring knowledge they have learned from the school-based context to the real-world context without being prompted or hinted explicitly (Bransford et al., 2000; Gick & Holyoak, 1980). Research shows that merely providing students with ill-structured problems does not necessarily help them develop effective problem-solving skills.

To facilitate students with problem-based learning, a database-driven, web-based cognitive support system was developed to scaffold complex, ill-structured problem-solving processes through facilitating students' development of metacognitive awareness and self-regulation (see a full description of the system in Ge & Er, 2005; Ge, Planas, & Er, 2010). This learning system is mainly characterized by the question prompt mechanism involving different types of prompts, such as *procedural, elaborative, and reflective* prompts. The *procedural prompts* serve as a problem-solving blueprint, directing students' attention to important aspects of problem solving and helping them to monitor their problem-solving processes. The *elaborative prompts* (e.g., *What is the example of ...? Why is it important? How does it affect...?*) force learners to elaborate their thinking and formulate explanations. These types of questions may help to enhance learners' metacognitive awareness and foster their self-monitoring and self-regulation processes (Lin & Lehman, 1999). The *reflection prompts* (e.g., *What is our plan? Have our goals changed? To do a good job on this project, we need to...*) are designed to encourage reflection on a meta-level that students do not generally consider (Davis & Linn, 2000). In addition, the system also consists of a digital library containing various cases that are indexed according to topic, domain, and level of difficulty, as well as a database used to save and store learners' responses that can be retrieved and displayed on the screen depending on the needs.

ITERATIVE DESIGN PROCESS IN CONTEXT

The development of this cognitive support system was a result of several design and research iterations with data collection in various contexts. It went through an iterative process of three major stages that were situated in specific domains or integrated with a specific curriculum: (1) conceptualization, (2) developing, modifying and validating the system, and (3) refining and expanding the system.

Conceptualization

This system originated with a study (Ge & Land, 2003) investigating the effects of question prompts and peer interactions on scaffolding students' ill-structured problem solving in a PBL environment, based on a critical literature review and conceptualization. The question prompts were first introduced in an undergraduate course in information science and technology to facilitate students in the processes of representing problems, constructing arguments, monitoring, and evaluating solutions. Data were collected during the student's PBL activities, using quasi-experimental method supplemented with observations and think-aloud. The results of the study indicated that question prompts made positive effects on individual's problem-solving performance in both problem-solving outcomes and processes (see Ge & Land, 2003). The study also suggested that the peer-interaction process itself must be guided and scaffolded, using strategies such as question prompts, to maximize the benefits of peer interactions.

Developing, Modifying, and Validating the System

The findings from Ge and Land's (2003) study prompted the researchers to incorporate question prompts in the design of the web-based cognitive system to scaffold student's ill-structured problem solving and peer interactions. In the process of developing the system, a series of quantitative or qualitative inquiry methods were conducted to specifically evaluate the effects of question prompts and validate the mechanism of question prompts in supporting ill-structured problem solving (i.e., Ge, Du, Chen, & Huang, 2005; Ge et al., 2010; Kauffman, Ge, Xie, & Chen, 2008). These studies were conducted in various disciplines, domains, contexts in the university setting, including students (undergraduate or graduate) from curriculum and instruction, instructional design and technology, and pharmacy. The results consistently supported the previous research (e.g., Ge & Land, 2003) that question prompts had significant effects in scaffolding student's ill-structured problem-solving in the following processes: problem representation, generating solutions, constructing argument, and monitoring and evaluation.

Refining and Expanding the System

With the findings from each of the studies, the system was validated and refined, and new components were added to the system. For example, the mechanisms of peer review and expert view were later added to enhance the existing question-prompt mechanism to address the needs for scaffolding peer interactions in PBL (Ge et al., 2010) and provide cognitive apprenticeship (Collins, Brown, & Newman, 1989). Based on another study (Kauffman et al., 2008), procedural prompts were added in the later version of the system (Ge et al., 2010).

INQUIRY METHODS AND RESEARCH TOOLS

The iterative design research was achieved through both quantitative and qualitative methods. Multiple case studies allowed the researchers carry out in-depth investigations regarding the effects of different scaffolding conditions on learners with different prior knowledge levels (Ge, Chen, & Davis, 2005), whereas the quantitative method helped the researchers to evaluate the effects of the cognitive support system by comparing learner's problem-solving outcomes with and without the system support (Ge et al., 2010; Kauffman et al., 2008). Various tools were used to collect data to achieve different purposes. For instance, the think-aloud technique was used to examine students' cognitive and metacognitive processes during ill-structured problem-solving tasks in order to evaluate the effectiveness of the cognitive support system (Ge et al., 2005). In the other two studies (Ge et al., 2010; Kauffman et al., 2008), written reports were used to collect data on learners' problem-solving performance on a larger scale; online observations were made to analyze students' reflections prompted by the system; and interviews were conducted to investigate students' experience and perceptions regarding the cognitive support system.

The Researchers' Roles

Throughout the process of the system development, the researchers played the dual roles of a designer and a researcher, which was a challenge in itself. As designers, they conducted analysis of needs, learners and tasks, determined the nature of the problems, and identified goals and scope of the problems. In their researchers' roles, they collected observation data, took field notes, conducted think-aloud protocols, and analyzed participants' problem-solving reports and other data. Often the two roles were interchanged during different phases of design research, which presented challenges to the researchers in terms of time management, adjusting timelines, dealing with unexpected issues (e.g., conceptual or technological) that occurred during the development of learning technology.

The researchers also had to work with other stakeholders in the design-research process. For example, they worked with domain experts to elicit their information and problem-solving processes; with instructors to select instructional units, generate question prompts, develop problem scenarios, and develop rubrics for assessing learners' problem solving performance. In addition, they also worked with developers in designing interfaces and databases, from multiple views (e.g., a learner, an instructor, and a researcher) (Ge & Er, 2005).

Learner Contribution to the System

Most often users in educational settings are learners. They contributed to the development of the system through participating in the research activities (e.g., written reports and interviews), which provided accumulated data for evaluation to inform the development of the system. Some specific data called to the attention of the researchers to make specific modifications to the system. Ge et al. (2005) found that question prompts had different effects on learners with different knowledge levels. For example, in the absence of specific-domain knowledge, question prompts were not helpful in activating a learner's prior knowledge or relevant schema (Ge et al., 2005); for students who perceived themselves as more competent or confident, question prompts were perceived not only as redundant but also interfering with their thought flow during their problem-solving processes (Ge et al., 2005). Based on these findings, the researchers added the preassessment to the system, which would automatically direct learners to case studies with compatible difficulty levels (Ge & Er, 2005).

THE LIVING LABORATORY EXAMPLE 2: STUDENTS-AS-DESIGNERS LEARNING ENVIRONMENT

DESCRIPTION OF THE LEARNING ENVIRONMENT: STUDENTS-AS-DESIGNERS

Instead of a sociotechnical system discussed in Example 1, this example illustrates a student-centered, project-based, and technology-rich learning environment. It is intended to demonstrate how the living laboratory is reflected through the design research in a naturalist setting of a day-to-day classroom of a high school computer class. Through two ethnographic studies (Ge, Thomas, & Greene, 2006; Thomas, Ge, & Greene, 2011) of iterative design, this example showcases how an ethnographic research approach can be used as a powerful tool to examine and interpret the rich social and cultural aspects of a learning environment. This ethnographic study features the evolution of a technology-rich learning environment, particularly how the researchers, teachers, and students worked together to overcome challenges during the cultural shift from the traditional learning environment to an OLE. In this environment, high school students played the role of instructional designers in creating learning activities and resources for a virtual digital learning environment, which was designed to be used by younger students of elementary school students.

ITERATIVE DESIGN OF THE STUDENTS-AS-DESIGNERS LEARNING ENVIRONMENT

Conceptualization

What prompted this study was the need to address the concern over the issue of *inert* knowledge in the educational system, that is, knowledge that can be recalled when people are specifically prompted, but cannot be spontaneously used to solve problems (Bransford et al., 2000; Gick, 1986; Gick & Holyoak, 1980). Research shows that project-based learning, one of the OLEs (Blumenfeld et al., 1991; Land & Greene, 2000), in which students play the role of designers in authentic learning environments, can help to address the issue of knowledge decontextualization and foster knowledge transfer. Literature indicated that through designing projects, students' engagement and involvement increased over time; and additionally, their self-confidence and ownership of problem solving also increased over time (Blumenfeld et al., 1991). Further, students deepened understanding of domain knowledge and enhanced their problem solving skills (Erickson & Lehrer, 2000; Lehrer, 1993; Liu & Rutledge, 1997). However, there was little research examining the transition from a decontextualized school environment to an OLE. Therefore, two ethnographic studies were carried out to investigate the impact of project-based learning in a high school computer programming I class (see Ge et al., 2006; Thomas et al., 2011), with the iterative process of theory and practice, and of modification and redesign.

Ethnography Study 1: Conceptualization and Initiation

The first study aimed at exploring (1) how classroom culture shifted from a traditional learning environment to a more authentic and open-ended learning environment and (2) how motivated and engaged the students were in their roles of students-as-designers.

Before the project-based learning environment was introduced, the teacher would first demonstrate an example of the programming codes and skills to be mastered for that day, and the students would need to complete a computer programming task independently by applying the new commands and rules they learned that day. If the students finished their assignment early, they were allowed to play games. Playing games during extra class time served as an incentive for students to focus, engage, and complete their assignments early.

In the new learning environment, the high school students were asked to be designers to create fun educational resources for Quest Atlantis (QA), an educational program with interactive 3D virtual game environments designed to provide meaningful problem-solving experience for elementary and middle school students (Barab, Thomas, Dodge, Carteaux, & Tuzun, 2005). The high school students were told to work with peers or independently. QA involved environmental, societal, and scientific challenges and activities designed to engage students in complex, prosocial *Quests* using the myth of Atlantis as the background story, and the high school students' job was to create the *Quests* for younger student users.

This authentic project-based learning environment was implemented for three weeks. According to the data, the high school students initially benefited from the new open environment, which provided them with an opportunity to design *Quests* for their authentic clients and express their creativity freely. They were motivated and engaged at the beginning and produced some useful products, even though some students did not put in as much effort as they should have. The researchers found that some immutable elements from the previous learning environment presented barriers to peer collaboration, persistence, and motivation for high-level creative work. Some group members were quickly distracted to other things or went back to play games (as was part of the old culture), leaving the other group members members to work on the projects. The researchers also found that the students enjoyed autonomy, but they also took advantage of the autonomy and misinterpreted the flexibility as doing what they liked to do. In other words, rather than being accountable for their group projects, some students used the group time to play games or do homework for other classes.

The first ethnographic study confirmed that the students-as-designers benefited from the OLE in several aspects, including autonomy, creativity, and problem-solving skills. However, some students did not demonstrate persistence or higher level of motivation, and their projects showed minimal level of critical thinking or mental effort. In the follow-up interviews, the researchers found that there was a tension between the old and the new culture. On one hand, the students were motivated about their roles as designers and they also had high degree of metacognitive awareness. On the other hand, their motivation quickly dropped before they had an opportunity to create a high quality project. Part of the reason, as the researchers (Ge et al., 2006) figured, could be the lack of the feedback from the real clients.

Ethnography Study 2: Modification and Redesign

Based on the results of the first ethnographic study, the researchers modified their design and conducted a second ethnographic study in the following year by involving some fourth and fifth grade elementary school students from one local school (see Thomas et al., 2011). The elementary school students served as the clients of the high school students, and their task was to critique and evaluate the prototypes (i.e., the QA *Quests*) created by the high school students. The researchers delivered the prototypes created by high school students to the elementary school, where the fifth grade students were shown the QA Quests created by the high school students. The younger students were provided with an evaluation form with questions, and their feedback was collected on each of the high school student's projects, which was then given to the high school students.

The study revealed that the clients' feedback had a profound impact on the high school students. For instance, several students were quite disappointed with the feedback received, realizing that major changes were needed in order to address the younger students' feedback. One group ended up changing their topic completely, whereas an individual kept his topic, but tried to make his project more game-like where he had created a stand-alone tool (i.e., a calculator) previously. The students seemed to be more engaged and applied more newly learned programming language and skills (e.g., Java) in their projects, in comparison with the students in the previous study. All of the projects had Java applied to create applets as a part of a game, whereas in the first study most of the groups or individuals used PowerPoint to create their projects. In addition, the storylines in the second study seemed more complex, for example, alternative paths were created that required the users to play through different scenarios and make decisions in selecting a path. The researchers observed that students used various strategies to make their projects work. Some students used trials and errors in order to get the new program to work while some others searched Internet sites for Java ideas, although there were still some other students who relied heavily on the teacher for help.

INQUIRY METHODS AND RESEARCH TOOLS

A major concern was the development of appropriate methodological tools for capturing the subtleties of the evolving classroom culture. To be able to fully engage in the ethnographic research, the researchers used a systematic research approach for data collection and analysis and employed a set of technology tools and methods, which the researchers called technology-rich ethnography (TRE) (Ge et al., 2006; Thomas et al., 2011). The data collected using the TRE tool sets included observation notes from the three researchers; interviews and focus groups with students and the teacher; video recordings of daily work; audio-recorded group interactions; chat logs within QA, and the projects created by the students.

Analysis began as the three researchers discussed the daily observations. First, global observations about the classroom were noted, and then specific groups of students were discussed. The observation protocols were completed independently by each of the researchers, and summaries were generated and discussed. The interviews were transcribed and analyzed, along with the other data sources. The three researchers collected, coded, and analyzed all the transcripts to generate themes.

THE RESEARCHERS' ROLES

The researchers played dual roles in the two ethnographic studies. They were instructional designers, teachers, and researchers at the same time. The three researchers went to the class every day for three weeks. They worked with the teacher to facilitate individual students or groups with their project progress, prompting them with questions, providing feedback, and scaffolding them as needed. The researchers also had to address students' motivation and emotional issues throughout the project. In this way, researchers influenced the environment under investigation. The researchers also took field notes and asked them questions as part of the ethnography-research process. An inherent challenge the researchers experienced in the process of ethnography was trying to strike a balance between emic and etic (experience-near and experience-distant) perspectives (Geertz, 1976, 1983). In other words, the researchers must keep a balance of the actions they took in the sociotechnical system while engaging with the meaning they gleaned from their observations (Lincoln & Guba, 2000).

WORKING WITH STAKEHOLDERS

This iteration of studies involved several stakeholders and multiple needs due to the complexity of the projects and the research. First, we coordinated with the high school teacher about using his class as a test bed for our research and discussed with him about how our research could be integrated into his class, and which instructional unit would be appropriate for our research. Then we had to obtain permission from the administrator to conduct a study at the school and receive approvals from the institutional review board (IRB). At the same time, the researchers had to obtain permission from the QA administrators to allow access to QA site and be accountable for monitoring students for their proper online behaviors based on the QA user policy. Further, the researchers also had to identify and contact the administrator of a local elementary school to gain their agreement to participate in this study. In the meantime, the researchers needed to plan and coordinate the timeline for high school students to provide the deliverables and for elementary school students to work on the evaluation and feedback for the high school students.

Through the iteration of the design, modification, and redesign, the researchers were able to observe the evolution of the students-as-designers OLE *in situ* and investigate the development of a highly contextualized learning environment. The flexible, dynamic approach of TRE were employed to observe potential changes in process, and to leverage lessons learned from these observations to inform future design work as well as the development of theory.

CONCLUSION

The two living laboratory examples illustrate the fact that research and design work hand-in-hand, and the purpose of educational research is to engage learners in useful and meaningful learning experiences through carefully designed student-centered learning environments. The ultimate goals of educational research are to improve learning and instruction, and to help students develop critical thinking and problem-solving skills so that they will be able to deal with the challenges presented by the twenty-first century. This chapter indicates that it takes a long process of iterative design-research cycles to improve a system or learning environment, which demands researchers to play dual roles of instructional designer and researcher, working with various stakeholders for different tasks, and accommodating research design for various situations. It also implies that research in educational settings is messy and complex, and that it is an ill-structured problem in an ill-defined field in itself, which calls for researchers to examine the contexts of application and manage complexity of research. In addition, this chapter reveals that various inquiry methods and tools are needed to achieve various research and evaluation purposes.

This chapter demonstrates the practical and theoretical value of the living laboratory framework for educational research through two examples. It also shows special characteristics of the living laboratory contextualized in designing educational sociotechnical systems, including both technological learning system and sociocultural learning system. Although the two examples presented two different inquiry methods for different research purposes, they shared similar key components of the living laboratory, including *group-centered* design (i.e., multiple stakeholders), iterative process of practice and theory and design and research, various inquiry methods and research tools, and multiple roles of the researchers. The living laboratory framework offers a unique tool for educational researchers to think out of the box in the process of reengineering sociotechnical systems in education from the lens of cognitive systems engineering.

REVIEW QUESTIONS

1. What is the interrelationship between design and research in education?
2. In what ways are the two living laboratory examples similar or different?
3. What were the main stages for the iterative design research in the educational living laboratory according to the two examples?
4. What roles did the researchers play, or what responsibilities did the researchers assume, in the educational living laboratory?
5. What were the roles of technology in scaffolding students' ill-structured problem-solving and in collecting data for design research?
6. How were various inquiry methods and research tools used to achieve different research and evaluation purposes?
7. How did various stakeholders contribute to the development of the sociotechnical systems?

REFERENCES

Barab, S. A., & Duffy, T. M. (2000). From practice fields to communities of practice. In D. Jonassen & S. Land (Eds.), *Theoretical foundations of learning environments*. Mahwah, NJ: Lawrence Erlbaum Associates.

Barab, S. A., Thomas, M., Dodge, T., Carteaux, R., & Tuzun, H. (2005). Making learning fun: Quest Atlantis, a game without guns. *Educational Technology Research and Development, 53*(1), 86–107.

Barrows, H. S., & Tamblyn, R. M. (1980). *Problem-based learning: An approach to medical education*. New York, NY: Springer.

Blumenfeld, P. C., Soloway, E., Marx, R. W., Krajcik, J. S., Guzdial, M., & Palincsar, A. (1991). Motivating project-based learning. *Educational Psychologist, 26*, 369–398.

Bransford, J. D., Brown, A. L., & Cocking, R. R. (Eds.). (2000). *How people learn: Brain, mind, experience, and school*. Washington, DC: National Academy Press.

Brown, A. L. (1992). Design experiments: Theoretical and methodological challenges in creating complex interventions in classroom settings. *The Journal of the Learning Sciences, 2*(2), 141–178.

Brown, A. L., & Campione, J. C. (1996). Psychological theory and the design of innovative learning environments: On procedures, principles, and systems. In L. Schauble & R. Glaser (Eds.), *Innovations in learning: New environments for education* (pp. 289–325). Mahwah, NJ: Erlbaum.

Brown, J. S., Collins, A., & Duguid, P. (1989). Situated cognition and the culture of learning. *Educational Researcher, 18*(1), 32–42.

Chi, M. T. H., & Glaser, R. (1985). Problem-solving ability. In R. J. Sternberg (Ed.), *Human abilities: An information processing approach* (pp. 227–250). New York, NY: W. H. Freeman and Company.

Cobb, P., Confrey, J., diSessa, A., Lehrer, R., & Schauble, L. (2003). Design experiments in educational research. *Educational Researcher, 32*(1), 9–13.

Cognition and Technology Group at Vanderbilt. (1993). Anchored instruction and situated cognition revisited. *Educational Technology, 33*(3), 52–70.

Collins, A., Brown, J. S., & Newman, S. E. (1989). Cognitive apprenticeship: Teaching the crafts of reading, writing, and mathematics. In L. B. Resnick (Ed.), *Knowing, learning, and instruction: Essays in honor of Robert Glaser* (pp. 453–494). Hillsdale, NJ: Lawrence Erlbaum Associates.

Davis, E. A., & Linn, M. (2000). Scaffolding students' knowledge integration: Prompts for reflection in KIE. *International Journal of Science Education, 22*(8), 819–837.

The Design-Based Research Collective. (2003). Design-based research: An emerging paradigm for educational inquiry. *Educational Researcher, 32*(1), 5–8.

Erickson, J., & Lehrer, R. (2000). What's in a link? Student conceptions of the rhetoric of association in hypermedia composition. In S. P. Lajoie (Ed.), *Computers as cognitive tools, Vol.2: No more walls* (pp. 197–226). Mahwah, NJ: Lawrence Erlbaum Associates.

Feltovich, P. J., Spiro, R. J., Coulson, R. L., & Feltovich, J. (1996). Collaboration within and among minds: Mastering complexity, individuality and in groups. In T. Koschmann (Ed.), *CSCL: Theory and practice of an emerging paradigm* (pp. 25–44). Mahwah, NJ: Lawrence Erlbaum Associates.

Ge, X., Chen, C. H., & Davis, K. A. (2005). Scaffolding novice instructional designers' problem-solving processes using question prompts in a web-based learning environment. *Journal of Educational Computing Research, 33*(2), 219–248.

Ge, X., Du, J., Chen, C., & Huang, K. (2005). *The effects of question prompts in scaffolding ill-structured problem solving in a Web-based learning environment.* Paper presented at the annual meeting of Association of Educational Communications and Technology, Orlando, FL.

Ge, X., & Er, N. (2005). An online support system to scaffold complex problem solving in real-world contexts. *Interactive Learning Environments, 13*(3), 139–157.

Ge, X., & Land, S. M. (2003). Scaffolding students' problem-solving processes in an ill-structured task using question prompts and peer interactions. *Educational Technology Research and Development, 51*(1), 21–38.

Ge, X., Planas, L. G., & Er, N. (2010). A cognitive support system to scaffold students' problem-based learning in a Web-based learning environment. *Interdisciplinary Journal of Problem-based Learning, 4*(1), 30–56.

Ge, X., Thomas, M. K., & Greene, B. (2006). Technology-Rich ethnography for examining the transition to authentic problem-solving in a high school computer programming class. *Journal of Educational Computing Research, 34*(4), 319–352.

Geertz, C. (1976). From the native's point of view: On the nature of anthropological understanding. In K. Basso & H. A. Selby (Eds.), *Meaning in anthropology* (pp. 221–237). Albuquerque, NM: University of New Mexico Press.

Geertz, C. (1983). Thick description: Toward an interpretive theory of culture. In R. M. Emerson (Ed.), *Contemporary field research: A collection of readings* (pp. 37–59). Prospect Heights, IL: Waveland Press.

Gick, M. L. (1986). Problem solving strategies. *Educational Psychologist, 21*(1&2), 99–120.

Gick, M. L., & Holyoak, K. J. (1980). Analogical problem solving. *Cognitive Psychology, 12*, 306–355.

Hannafin, M., Land, S., & Oliver, K. (1999). Open learning environments: Foundations, methods, and models. In C. M. Reigeluth (Ed.), *Instructional design theories and models: A new paradigm of instructional theory.* Mahwah, NJ: Lawrence Erlbaum Associates.

Hmelo-Silver, C. E. (2004). Problem-based learning: What and how do students learn? *Educational Psychology Review, 16*(3), 235–266. doi:10.1023/B:EDPR.0000034022.16470.f3

Hmelo-Silver, C. E., & Barrows, H. S. (2006). Goals and strategies of a problem-based learning facilitator. *Interdisciplinary Journal of Problem-Based Learning, 1*(1), 21–39. doi:10.7771/1541-5015.1004

Jonassen, D. H. (1997). Instructional design models for well-structured and ill-structured problem-solving learning outcomes. *Educational Technology Research and Development, 45*(1), 65–94. doi:10.1007/BF02299613

Jonassen, D. H. (2011). *Learning to solve problems: A handbook for designing problem-solving learning environments.* New York, NY: Routledge.

Kauffman, D., Ge, X., Xie, K., & Chen, C. (2008). Prompting in web-based environments: Supporting self-monitoring and problem solving skills in college students. *Journal of Educational Computing Research, 38*(2), 115–137.

Land, S. M., & Greene, B. A. (2000). Project-based learning with the World Wide Web: A qualitative study of resource integration. *Educational Technology Research and Development, 48*(1), 45–66.

Lehrer, R. (1993). Author of knowledge: Patterns of hypermedia design. In S. P. Lajoie & S. J. Derry (Eds.), *Computers as cognitive tools* (pp. 197–227). Hillsdale, NJ: Lawrence Erlbaum Associates.

Lin, X., & Lehman, J. D. (1999). Supporting learning of variable control in a computer-based biology environment: Effects of prompting college students to reflect on their own thinking. *Journal of Research in Science Teaching, 3*(7), 837–858.

Lincoln, Y. S., & Guba, E. G. (2000). Paradigmatic controversies, contradictions, and emerging confluences. In N. K. Denzin & Y. S. Lincoln (Eds.), *Handbook of qualitative research: Second edition* (pp. 163–213). Thousand Oaks, CA: Sage.

Liu, M., & Rutledge, K. (1997). The effect of a "learner as multimedia designer" environment on at risk-high school students' motivation and learning of design knowledge. *Journal of Educational Computing Research, 16*(2), 145–177.

McNeese, M. D., Connors, E. S., Jones, R. E. T., Terrell, I. S., Jefferson, T. Jr., Brewer, I., & Bains, P. (2005). *Encountering computer-supported cooperative work via the living lab: Application to emergency crisis management.* Proceedings of the 11th International Conference on Human-Computer Interaction, Las Vegas, NV.

Sinnott, J. D. (1989). A model for solution of ill-structured problems: Implications for everyday and abstract problem solving. In J. D. Sinnott (Ed.), *Everyday problem solving: Theory and application* (pp. 72–99). New York, NY: Praeger.

Suchman, L. A. (1987). *Plans and situated actions: The problem of human-machine communication.* New York, NY: Cambridge University Press.

Thomas, M. K., Ge, X., & Greene, B. A. (2011). Fostering 21st century skill development by engaging students in authentic game design projects in a high school computer programming class. *Journal of Educational Computing Research, 44*(4), 391–408.

Voss, J. F., & Post, T. A. (1988). On the solving of ill-structured problems. In M. T. H. Chi & R. Glaser (Eds.), *The nature of expertise* (pp. 261–285). Hillsdale, NJ: Lawrence Erlbaum Associates.

Wenger, E. (1998). *Communities of practice: Learning, meaning, and identity.* Cambridge, MA: Cambridge University Press.

3 Distributed Teams in the Living Laboratory

Applications of Transactive Memory

Vincent Mancuso

CONTENTS

Introduction 48
   Transactive Memory 49
Applications of the Living Laboratory 52
   Case Study: Transactive Memory in Distributed Cyber Operations 53
Practice to Theory: Transactive Memory in Cyber Operations 53
   Scaled-World Simulation: teamNETS 55
   Reconfigurable Prototype: Distributed Team Cognition 57
      Exploring Transactive Memory in teamNETS 58
   Value of the Living Laboratory 59
Future Considerations 61
Conclusion 61
Review Questions 62
References 62

ADVANCED ORGANIZER

This chapter describes the living laboratory as a holistic and cohesive framework for studying distributed teams and their cognitive activities. Specifically, we will discuss how transactive memory systems can be facilitated by a shared group interface in a simulated cybersecurity task. Transactive memory, put simply, is the knowledge of who knows what in a collaboration, and is a critical cognitive structure in many team decision-making tasks. This chapter begins with an introduction of transactive memory and its role in facilitating team interactions and performance. We then discuss how the living laboratory can be used as a framework to

(Continued)

ADVANCED ORGANIZER (Continued)

study transactive memory. This is demonstrated in exemplar research that applies transactive memory to cyber operations with a practice to theory of living laboratory perspective. This case demonstrates the clear utility of the living laboratory, a cognitive engineering approach, over a traditional usability and human–computer interaction approaches in understanding the usability, utility, and efficacy of a distributed team cognition aid. This chapter concludes with a discussion of the value of the living lab, and future considerations for research and practice.

INTRODUCTION

There are three main cognitive processes of memory, encoding, storage, and retrieval (Tulving and Thomson 1973). Information enters memory during encoding, is archived for future access during storage, and is retrieved during the access stage. At each stage, problems may occur that degrade the accuracy of the memory, such as incorrect encoding. It is also possible that at each stage to have the information fall out of memory, as in the case with forgetfulness. To reduce these errors, humans often rely on cognitive artifacts to create lasting accurate memory (Hutchins 1995). This activity, known as distributed cognition, suggests that physical objects acting as cognitive artifacts, are particularly effective at reducing incorrect coding errors and forgetfulness (Norman 1991, Nemeth et al. 2004). Items such as notepads, meeting calendars, and shopping lists are examples of cognitive artifacts employed in our everyday lives. In the seminal research on distributed cognition, Hutchins (1995) found that these cognitive artifacts were critical in all stages of decision making, coordination, and collaboration, as demonstrated in his time observing large sea vessel operations. Although this research is seminal in the distributed cognition literature, it has become dated as computing and technology have continued to mature. With this maturation, and new technologies, the realm of possible distributed cognitive artifacts has expanded. People are able to leverage technology such as mobile devices, and the cloud, to maintain a larger set of distributed cognitive artifacts that can be easily and readily accessed.

Even with this expanded set of possibilities, cognitive artifacts are only useful for maintaining an individual's internal knowledge, and they are limited by the efficacy of the corresponding linkage in the human cognitive structures. Even a highly detailed cognitive artifact, such as technical notes, or systematic directions, are only as useful as the internal knowledge possessed by the individual to decode and utilize their meaning. For example, a person writing down a complex mathematical formula will not find the notation very useful as a memory aid if they lack the mathematical knowledge to understand and apply its meaning.

A potential mitigation to this is to keep more detailed artifacts, outlining lists, and processes of knowledge; however, these can add to data overload and can be more cumbersome to create and possess. An alternate method, from the field of team cognition, is to leverage the team mind. At its roots, team cognition is concerned with the emergence of a unique entity that forms from "the interplay of the individual cognitions of each team member and team process behaviors" (Cooke et al. 2004). Building off this definition, aspects of team cognition can be broken down into two components, the team processes that emerge as a part of the collaboration, and the knowledge situated within and distributed across the team. Although both components are necessary aspects of team's cognitive behaviors, team knowledge can have similar functionalities as distributed cognition. When individuals lack the requisite internal knowledge to make use of cognitive artifacts, they must access their transactive memory—or knowledge of other's knowledge. As opposed to traditional distributed cognition, distributed team cognition allows individuals access to knowledge, as well as skills and abilities to leverage that knowledge outside of their own expertise. On account of this, numerous researchers have demonstrated a positive correlation between transactive memory and team performance (e.g., Wegner et al. 1985, Wegner 1995, 1987, Hollingshead 2001, Ariff et al. 2012, 2013, Mell et al. 2014).

Transactive Memory

Transactive memory, first introduced by Wegner et al. (1985), is a unique form of distributed cognition with emergent group mind properties not traceable to individuals, but is rather part of a larger and organized social memory system. Rather than relying on external physical cognitive artifacts, transactive memory allows humans to use other's as an external store of information that they themselves do not possess (Moreland and Myaskovsky 2000). When contrasted with a physical cognitive artifact, this type of distributed cognition benefits by not only being a store for discrete information such as facts, but also providing more granular details and skills that can be utilized (Wegner et al. 1985). These two types of transactive knowledge are referred to as higher- and lower-order knowledge. Prior research on transactive memory has shown that typically individuals possess higher-order knowledge on a large number of topics, but rely on others for lower-order knowledge. Performance improves when individuals have knowledge—or transactive memory—of others who retain the lower-order knowledge they lack.

Similar to individual memory, transactive memory formation and utilization is broken down into three stages: directory updating, information allocation, and retrieval coordination (Wegner 1995). The directory-updating stage correlates to the encoding stage, where an individual stores information on others knowledge. This information can then be organized in internal cognitive structures with corresponding metadata that makes it readily available. Finally, retrieval coordination is the process in which the transactive memory system is utilized through connecting an individual to a given specialty or piece of knowledge. In the earliest studies of transactive memory, much of the focus was on the retrieval coordination process. Wegner et al. (1985) found that dyads with a shared history (romantic couples) were able to better coordinate their retrieval and achieve better performance, than dyads with limited or no previous interactions. This paradigm was verified and expanded to show that communication is a key antecedent to the first two steps (Hollingshead 1998a). When teams interact with each other and communicate, they are able to collectively encode different aspects across their team, thus developing trust in each other's expertise (Liang et al. 1995). Specifically, in order for transactive memory to work within the context of distributed cognition, three components of information must be stored: task (T), expertise (E), and people (P), forming what is referred to as a TEP (task, expertise, and people) unit (Hollingshead et al. 2011, Ariff et al. 2012).

To maximize one's ability to utilize transactive memory, teams need to properly store and organize their cognitive structure through the collective encoding of task-relevant hierarchical information (Brandon and Hollingshead 2004). In their studies, Hollingshead (1998b) and Moreland and Myaskovsky (2000) found that team training and communication only helped predict team performance, but not individual performance once performing the task outside the team. Their results show, that the encoding process by which transactive memory occurs exist as a unique cognitive behavior, differentiated from traditional memory structures. Although effective at an individual level, transactive memory cognitive artifacts become more powerful when considered at a larger system level. Transactive memory systems occur when a group takes part in a cooperative division of labor for learning, remembering, and communicating task knowledge (Lewis 2004, Lewis and Herndon 2011). Transactive memory focuses on the organized store of knowledge maintained at the individual level; however, when considering a system, there must also be an interpersonal awareness of the other members involved (Hollingshead 2001). An effective transactive memory system can be broken down into a set of interrelated components that can improve or detract from a team's ability to complete a task. These components, or dimensions, were initially proposed as accuracy, agreement, and complexity (Moreland 1999), and later expanded to four dimensions (Austin 2003). Table 3.1 provides a comparison of the two sets of dimensions and their descriptions.

Using these dimensions, for a team or group to have an effective transactive memory system, there would have to be an adequate amount of knowledge divided

TABLE 3.1

Dimensions of Effective Transactive Memory Systems

Dimension		
Austin (2003)	**Moreland (1999)**	**Description**
Group knowledge stock	Coordination	Combination of the individual knowledge of each person in the group
Consensus about knowledge sources	Coordination	Extent to which the group members agree about who has what knowledge
Specialization of expertise	Specialization	A deeper knowledge base in a narrowly defined area of expertise, and the awareness of who specializes in what
Accuracy of knowledge identification	Credibility	Accuracy to which a member who is identified as having a specialization

among the group, agreement about where the knowledge was stored (in terms of who had it), division of specialty or foci, and sufficient accuracy of knowing who had which specialty. Teams that are able to cognitively distribute and encode transactive knowledge within the above framework are often times more effective than teams that do not (Lewis and Herndon 2011, Ren and Argote 2011).

Although all of these dimensions are important, specialization of expertise has been shown to be the largest contributor to an effective transactive memory system (Austin 2003, Lewis 2004, Littlepage et al. 2008, Michinov and Michinov 2009). Specialization within a team can be further broken down into the concepts of integrated and differentiated structures. In differentiated structures (more specialization), information is distributed across individuals; on the other hand, integrated structure (less specialization) focuses on information that is common to all members. For example, consider a web development team; a team with a differentiated structure would have one pure designer, back-end coder and database engineer, whereas an integrated structure team would have three members that have equal knowledge in all areas. In an early anecdotal assessment of the concept, Wegner (1987) proposed that the potential for knowledge expansion and production is much greater whenever teams possess a differentiated structure rather than an integrated structure. He suggests that since there is a lack of duplication of effort, and each individual brings unique knowledge, the team is able to store information more efficiently and generate more information than one individual could produce. Later empirical research suggests that when teams are more differentiated in their knowledge, they are more likely to improve on their own specializations allowing the team's knowledge coverage to expand by a factor of the number of individuals in the group (Littlepage et al. 2008, Michinov and Michinov 2009).

Early research in transactive memory was powerful in showing the utility of the team mind; however, it was limited by the lack of generalizability. Unlike early research in distributed cognition, we were never able to see transactive memory *in the wild*. More recently researchers have been approaching transactive memory through a qualitative lens to address this gap. Research showing transactive memory at work in open-source software teams (Chen et al. 2013), IS development teams (Hsu et al. 2012), academic project teams (Michinov and Michinov 2009), hospitals (Liao et al. 2015), and management teams (Heavey and Simsek 2014), to name a few, have added to our understanding of how transactive memory operates in real world and complex situations. Although this research has continued to push the bounds of transactive memory research, the organizational and cognitive foci creates a limited environment that may not be holistic or representational of real-world complex work.

As computing and technology become more ubiquitous and integrated into our everyday lives, they have a greater impact on organizational culture. This has allowed teams to transcend across geographic, temporal, and cultural boundaries. These virtual or distributed teams allow organizations to apply multiple perspectives to a problem, offer more flexibility to their employees, and maintain a 24-hour work cycle. With this shift in the organizational paradigm, researchers from numerous domains have become increasingly interested in understanding and supporting these teams.

Although some research has addressed how distributed teams may utilize transactive memory (e.g., Ariff et al. 2012, Chen et al. 2013), they often treat technology and computing as a secondary component of the environment. This may be true in some situations, though, in many organizations, technology has moved to the forefront and become the environment. This problem, the intersection of human behavior, cognition, and technology, is especially suited for being addressed in the living laboratory.

APPLICATIONS OF THE LIVING LABORATORY

As the previous chapters in this book have discussed, the living laboratory is a holistic and ecological approach to design that integrates technology, context, and humans into a cyclical research and development process. This cognitive systems engineering approach differs from previous research in transactive memory, by offering a unique blend that incorporates the technological, organizational, and cognitive aspects of distributed teamwork. A living-laboratory approach utilizes knowledge of humans within a given workplace, to aid in the design and implementation of realistic experiments that are representative of their real work environments, which in turn can be used to test both technology and theory. This approach, allows researchers to leverage the benefits of both experimental and qualitative research, to gain a deeper understanding of the theory, environment, actors within, and their interactions.

Cognitive systems engineering, and especially the living lab are particularly suited for transactive memory research due to their ability to leverage multiple methods in triangulating the entirety of the problem space within a given context or environment. Through the living lab, one can view transactive memory through multiple lenses. We can gain an understanding of the greater environment, and the varying confounds at play through qualitative observations, whereas honing in on the cognitive work through in-lab experimentation and scaled worlds. This perspective not only improves the design of technology, but can also be useful in informing the development of theory. In the subsequent section, we will discuss how the living laboratory framework was applied in the study of transactive memory, and distributed cognitive artifacts, within the context of cyber operations.

MIDWAY BREATHER

- Transactive memory is a critical form of distributed cognition that enables improved team performance through the specialization, coordination, and credibility of knowledge in a team.
- Transactive memory is the knowledge of who knows what, and a transactive memory system is the interplay of multiple individual transactive-memory systems working together in cooperation.
- Major themes in transactive memory literature do not account for technology, and virtually distributed teams in the workspace.
- The living laboratory is well suited to study transactive memory in distributed environment through its blend of qualitative and quantitative research.

CASE STUDY: TRANSACTIVE MEMORY IN DISTRIBUTED CYBER OPERATIONS

Cyber operations have emerged as a critical domain to our national security (Maybury 2015). Traditionally, the majority of work in cyber focuses on issues of computing and technology. More recently, authors have begun to argue for a human-centric approach to the study of cyber operations (Knott et al. 2013, Tyworth et al. 2013, Mancuso et al. 2014). Following this call, research in the fields of human factors, cognitive modeling, algorithms, and visualizations have begun to explore issues of human perception, behavior, and cognition within cyber.

Cyber, at its core is an interdisciplinary domain, in which there is interplay between humans, with technological and organizational constraints. With this, cyber is especially suited to the living laboratory framework. The core premise of this approach is to ground our theoretical insights within the context of the environment of interest, in this case cyber operations. For this research, we have chosen to employ a problem-based approach. Given our understanding of transactive memory, we approached the environment looking for an understanding of the cognitive, team and organizational work, and opportunities for research. Once an understanding of the environment was established, we then took those findings into the laboratory and designed a representational, scaled world, and technological prototypes to study human behavior and cognition within a controlled environment. The following sections discuss our research and findings in more depth.

PRACTICE TO THEORY: TRANSACTIVE MEMORY IN CYBER OPERATIONS

When approaching this problem space, transactive memory in cyber operations, careful consideration must be paid to maintain contextual and ecological validity of the environment. To ensure the generalizability and applicability of the research finding, one must consider and control for the complexities and interdependencies of the cyber environment. Traditionally this is done through qualitative research, using knowledge elicitation and ethnography.

Although extant theory, helped scope our research within the bounds of transactive memory, to better understand how teams generally function in cyber, a more hands-on approach was necessary. In pursuit of this, the research team conducted a set of semistructured interviews with cybersecurity professionals across multiple organizations, as well as an ethnographic observation of a military cyber-defense exercise (Tyworth et al. 2012, 2013). The findings of this research had numerous critical impacts on team collaboration within the environment. For example, in the cyber world, there is an absence of more ecological and contextual anchors that are an integral part of our decision-making process in the physical world (McNeese 2004). As there is no physical world in which events, data, and communications are happening, teams of operators are prone to having divergent mental models and understandings of the environment and situation. This creates a lack of understanding of the *big picture*, or complete cyber-situational awareness, within any individual analyst; rather it is distributed across multiple analysts.

Adding to the divergence is a natural distribution of work across multiple functional domains in cybersecurity. From our observations, we identified at least four distinct functional domains, tactical/intrusion detection, operational/system administration, threat landscape analysis, and management/policy. The data suggests that the amount of overlap in domains is a function of how much their goals overlap. This meant that it was common that operators within one functional domain became so focused on their own goals that they are unaware of the complete picture, and what information analysts operating in another domain required. This lack of common ground caused analysts to file reports that were lacking necessary mission-salient and contextual information, causing wasted and overlapping effort to occur on a regular basis. Table 3.2 describes each of the functional domains in more detail.

Based on this research, these elements, and other complexities from cyber operations (cf. Mancuso et al. 2012), a set of requirements for a representative scaled-world simulation were identified. Interviews with people across functional domains showed that those roles, and their interactions, must be carefully tailed to represent the complexities of the environment. Although each of these roles may, at any given point in time, be working toward a common goal, they also have unique subgoals, methods, and data sets that they work with. On account of these silos, it is often the case that analysts work in isolation with each other, only taking part in loose coordinative activities, rather than more tightly coupled collaborations. Part of the lack of collaboration, is the lack of a common language, which may be the result of the lack of institutional cross training. Although one analyst may generally know what another analyst does, they do not know enough specifics of what they do in order to provide them with the necessary information for decision making. This creates an implicit coordinative boundary, where rather than working together to ensure reports are complete, the analysts simply hand-off reports, with little guidance or accountability. This can result in duplication of efforts, lack of trust, and breakdowns in collaboration.

From this research, and an assessment of previous research on designing simulations, broad simulation requirements can be extracted. We know that roles must be carefully created to represent each of the distinct functional domains in cybersecurity. In many of the functional domains described, one of the biggest issues is

TABLE 3.2
Functional Domains as Described in Tyworth et al. (2013)

Functional Domain	Description
Intrusion/threat analysis	Assessing and mitigating issues where an unauthorized human from an outside computer connects to a machine within the network for inappropriate or malicious purposes
Operational/system administration	Assessing and mitigating issues where software running on a computer is causing damage, stealing information, or other unacceptable behaviors
Threat landscape analysis	Assessing current and emerging trends in the overall cybersecurity landscape
Management/policy	Assessing and mitigating issues where humans inside the network violate policy by exhibiting inappropriate

lack of a common language developed by significantly different jobs. For example, in intrusion/threat analysis, the majority of the data they work with are intrusion detection logs, whereas system administrators may monitor computer process logs, and reports passed to them from humans in the system. Finally, although there is a greater, often invisible, collaboration occurring in cybersecurity teams, the majority of the work is siloed at the individual level. Collaboration within the cyber domain is more a product of transferring reports between the functional domains, with little intertwined collaboration.

From the transactive memory perspective, there are several requirements that must be accounted for. To simulate a depth of knowledge within the task, resources must be developed with the higher-, lower-order knowledge paradigm in mind (Wegner 1987). This will allow players to develop a depth of knowledge in a specific area. Finally, to design a better transactive memory assessment tool, there must be some way of explicitly sharing information that is representative of transactive-memory utilization. This will avoid any problems of invisible work, and allow researchers to better assess a team's transactive memory utilization.

These were then extracted and converted to functional aspects of the simulation and/or experimental structures. The following section will present an overview of the teamNETS simulation, and demonstrate the implementation of these requirements.

SCALED-WORLD SIMULATION: TEAMNETS

Building on the qualitative research, these requirements were incorporated into a scaled-world simulation, teamNETS. Built on the NeoCITIES Experimental Task Simulator (NETS; Mancuso et al. 2012), teamNETS incorporates several of the cognitive and collaborative requirements of team cyber work into a controllable and scalable task. The NETS platform is the newest iteration of the NeoCITIES simulation, which has been an effective testbed for team-cognition research for more than 10 years (e.g., Jones et al. 2004, Hamilton et al. 2010, Hellar and McNeese 2010, Mancuso et al. 2012, Giacobe 2013a,b, McNeese et al. 2014, Endsley et al. 2015). Because of the scalability of NETS, and its ability to rapidly develop and deploy realistic cybersecurity simulations, it offered a perfect platform for studying transactive memory within the context of cyber operations.

In teamNETS, teams consist of three players who are assigned to different specializations within a university cyber response team. These specializations, intrusion and threat analysis, operational and system administration, and management/policy, were designed with considerations to the functional domains of cyber operations found in the qualitative research. Within each of these domains, there was a set of higher-order knowledge that represented general knowledge, and lower-order knowledge that got into more specifics. Within this paradigm, an individual who was a generalist in an area would only possess the higher-order knowledge, whereas in their specialization they would possess the lower-order knowledge as well. This allows the researcher to mimic a real-life cyber operations team. In addition, it opened up the possibility for studying other types of teams, in which the knowledge of different domains was distributed across the team, creating more common ground and collaborative opportunities.

During a simulation scenario, it is the responsibility of the team, to monitor the network for potential threats, and take correct steps to mitigate them as they arise. In order to accomplish these goals, participants utilize a simple user group interface that resembles freely available cybersecurity tools (Figure 3.1).

The main teamNETS interface consists of five components, the location tracker (A), incident report monitor (B), action report monitor (C), team monitor (D), and chat (E). In addition to these components, there are pop-up windows to allow users to transfer information and file action reports. This interface allows users to monitor different locations on the simulated network, investigate or transfer emerging events, file action reports, and monitor individual and team progress in the simulation.

During a scenario, each player receives reports of possible cyber threats that occur on their network of responsibility. When a player recognizes an actual threat, they must use the incident report monitor to read over the event details in the form of a textual description and/or a computer log file. Following this, players have two options, mitigate the threat by filing an action report, or transfer the event to another player who may be better suited to file the report. This functionality was designed to closely mimic the report sharing found in actual cyber operations. After a report is filed, the players receive feedback on the accuracy of the report and whether or not the threat was removed from the network.

The true power of the teamNETS simulation is its ability to measure multiple types of team and individual cognition within the context of a distributed team. At its core, teamNETS is driven by the human performance scoring model (Hellar and McNeese 2010), which allows researchers to assess individual and team performance within the simulation. In addition to performance metrics, teamNETS allows

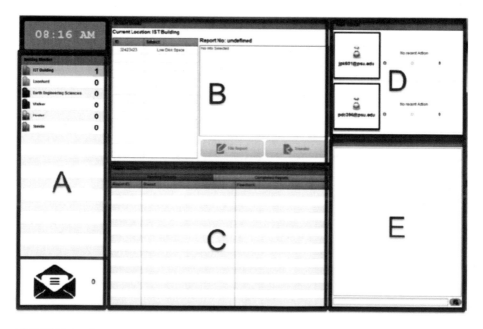

FIGURE 3.1   TeamNETS simulation user interface.

researchers to capture more abstract constructs of team cognition and collaboration. Metrics such as communication patterns and activities can be easily extracted, and used for after-the-fact analyses of the frequencies, processes, and patterns. In addition, teamNETS has built-in metrics for capturing situation awareness, workload, team-mental models, and transactive memory.

RECONFIGURABLE PROTOTYPE: DISTRIBUTED TEAM COGNITION

A major issue identified during the qualitative research was the lack of awareness of other knowledge within the cross-functional cyber teams. Although this was partially a result of the organizational structure, it was also a product of the distributed nature of the work being done. One way of mitigating a lack of interpersonal awareness in distributed work is through shared virtual feedback (Gutwin and Greenberg 2002, Hill and Gutwin 2003). The purpose of shared virtual feedback is to provide others working in the virtual workspace with an awareness of the actions and performance of other team members.

To understand how a mechanism like this can support cyber operators in practice, we implemented the augmented transactive memory interface (ATM-I). Using design guidelines for developing transactive memory aids (Keel 2007), the shared virtual feedback tracks expertise, knowledge, and task information, to highlight interconnected knowledge, and determine the expertise necessary for a given problem. This tool was developed so that it could be used as a supplementary tool, which complemented their cognitive processes rather than replacing them. The final design aimed to recreate TEP units, the building blocks of transactive memory (Brandon and Hollingshead 2004), but in a fashion that allowed quick access to awareness information of others.

In the ATM-I, players receive information that provided them with a real-time evaluation of their team member's performance within each functional domain. Within the information transfer window, users could compare each team member's performance against their own, and the average of the team. Figure 3.2 shows the additions to each interface component as a part of the manipulation.

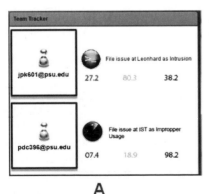
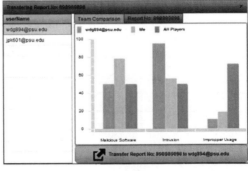

A   **B**

FIGURE 3.2   Team tracker (A) and information transfer (B) with shared virtual feedback enabled.

In the team tracker (A), the shared virtual feedback reports scores that are color coded for each functional domain. As each player files reports, these scores are continuously updated to provide a running average for the scenario. In the transfer report window (B), players are able to select one of their teammates, and get a breakdown of how they compare to their skills across functional areas. The bar chart visualization on the right has three sets of bars, representing each of the functional domains. Within each set, there is a bar that represents the current players skill within that domain (left), the skill of the selected player (middle), and the team average (right). This allows the current player to quickly extract information on each team member's performance and expertise across the three functional domains.

This interface allows the players to distribute their cognition, in this case the awareness of the activities and knowledge of others, onto the shared workspace. The hope is that this allows them to offload the cognitive workload of teamwork (Sellers et al. 2014) onto the system, allowing quick access to the knowledge when needed.

Exploring Transactive Memory in teamNETS

Using the teamNETS platform, a series of controlled laboratory experiments were conducted. Within these studies, two specific independent variables were manipulated, the presence of the distributed team cognition aid, and the type of team knowledge. The presence of the distributed team cognition aid was manipulated by the presence of the ATM-I capabilities in both the team tracker and information transfer windows. This manipulation allowed us to study how the ATM-I could help improve transactive memory and performance for a distributed team. For the team knowledge, we compared teams that were similar to actual cybersecurity teams, to teams who had a more overlapping set of knowledge. This manipulation represented different types of specializations that could occur within a transactive memory system. Realistic cyber teams often possessed more specializations, but had very little overlap. As a comparator, we also tested teams with less specialization and more common knowledge. This allowed us to hone in on the impacts of team composition and knowledge in actual cyber operations.

Unfortunately across all conditions, there was no significant change in performance; however, the findings revealed several factors that may have a major impact on other longitudinal effects not measurable in this study. In conditions where the ATM-I was present, teams had a higher perception of their transactive memory; however, they were unable to utilize it to improve performance outcomes. Results showed that they communicated slightly less than the teams who did not receive the system. This could be a result of illusionary transactive memory (Tschan et al. 2009), where the team thought that they were more aware of the knowledge possessed by others, thus they did not feel the need to communicate as much. Even though there was less communication, which would be a form of explicit collaboration, there was more implicit collaboration in terms of task sharing. In conditions with shared virtual feedback, players more often would defer responsibility to another player. However, an interesting interaction should be noted, in which the shared virtual feedback had a greater impact on transactive memory utilization for teams with lower specialization in their team knowledge. This finding is consistent with seminal research in transactive memory, where a specified memory structure on top of an established one would conflict and lead to performance detriments (Wegner et al. 1985). In the case of this study, the teams with specialized knowledge structures had a very clear division of labor based on their training, and the ATM may have conflicted with their predetermined system. As teams in the lower specialization condition had no predetermined agreement of who was good at what, the ATM interface allowed them to form natural specialties in the different functional domains based purely on their performance. The ATM interface tried to force an explicit transactive memory structure on the team, which acted as interference with the team member's mental models. This resulted in longer time to submit incident reports, and an increase in errors.

A second major finding was the comparison between the two types of knowledge structures. Although there were no performance differences across the two conditions, an interesting outcome was impact of knowledge structures on team communication. In conditions with higher specialized knowledge structures, there was often little to no communication. The teams would simply agree on who was responsible for which type of event at the beginning of a scenario, and stick to that format with no communication. This resulted in a very loose collaboration, in which teams relied more heavily on individual work, and coordinated their actions within a larger collaborative system. Interestingly, these teams, even with little to no communication, an antecedent to transactive memory formation, reported higher perceived transactive memory, than teams with less specialization. Although statistically insignificant, trends in the data and previous research suggest that over a longer period of performance teams with a lower specialization, may have eventually showed greater performance outcomes than teams with greater specialization. This could be because without assigned specializations, the teams were able to naturally form their specialties based on their skill sets and preferences. However, it is also possible that if trained in better communicative practices, the teams with greater specialization would have developed a shared understanding of the problem space, and each other's skills, and improved their performance.

VALUE OF THE LIVING LABORATORY

The greatest value of the living-laboratory approach is its ability to study and understand theoretical and practical issues concurrently, within the context of a given environment. By leveraging qualitative research, we were able to build an environment, and subsequent experiment, to hone in on specific aspects of teamwork, and mitigations, within the context of cybersecurity. This allowed us to identify unique findings that can improve our understanding of technology and teamwork theory.

The fact that the shared virtual feedback did not improve performance, sometimes even hurting the team, was surprising, and has numerous implications from a practical purpose. Today, in modern cyber operations, it is not uncommon for companies to institute collaborative systems, like wikis, shared file repositories, and ticketing systems, in an attempt to disseminate knowledge and awareness across the organization. Although these types of systems can have numerous benefits, it is critical that the organization fully understands their purpose and the people they are trying to support. If haphazardly implemented, there may be negative outcomes, eventually leading to the capability being discarded or ignored. In our study, we found that when teams had a predetermined transactive memory system, they relied too heavily on the shared virtual feedback system, allowing it, rather than human interactions to dictate how they utilized the system. When implementing cognitive aids or intelligent interfaces into software-support systems, the goal of the technology is to enhance the human factors, but more often than not these capabilities are inadequately developed. It is important during implementation to pay special attention to the collaborative operations and dynamic nature of the team to ensure that the tool can augment the collaboration, rather than serve as a means to replace it. This finding serves as an exemplar of the necessity of the cognitive engineering and living-laboratory approach. By taking a holistic approach as studying the presence of a distributed cognitive aid, rather than a traditional human–computer interaction approach that focuses on usability, we were able to extract key findings on not only usability, but also the utility and efficacy of the system. The exploration of the use of the system in context is a critical aspect of this study, and serves as evidence for taking such an approach.

On the other hand, our study also had several findings that can enable a better understanding of transactive memory. From seminal transactive memory literature, we were to believe that teams with established specializations at the beginning of collaborations would be better off than teams with more general knowledge. From a pure performance standpoint, there was no difference, however, after closer examination, teams with more generalized knowledge were able to make up for their lack of depth in each functional domain through communication and information sharing. The main focus of this portion of the study was to investigate the role of transactive memory in distributed teams, a construct that has been ignored or rejected in traditional organizational psychology literature (e.g., Jackson and Moreland 2009). Our findings suggest that teams were able to use the group interface to build a transactive memory system; however, the mechanisms did not support effective utilization. This is consistent with research conducted by Jackson and Moreland (2009) where they found similar outcomes in a longitudinal study of classroom teams. This finding shows the underlying nature and complexity of virtual teams, and the challenges that they continue to bring to distributed interactions.

An examination of the subjective reporting, as well as the behavioral outcomes, provide evidence that transactive memory existed; however, due to the limitations of the distributed environment, teams were not able to utilize the systems effectively enough to improve their performance. In addition to the theoretical outcomes, there is a need for more research on transactive memory usage (possibly in the context of high and low performing teams), this has practical implications, including showing the need for cross training and more research on effective distributed cognition tools for improving transactive memory utilization, with less focus on formation and maintenance.

Finally, through the living laboratory, we were able to develop a new simulation to aid in studying multiple facets of individual and team cognition within the context of cybersecurity. TeamNETS includes complex decision making, and an overall task that is representative of a real-world analyst (cyber or otherwise). Although much of the task was simplistic in this chapter, teamNETS is designed to support more complex data sets, such as IDS Alerts, Firewall data, and Network Access Control Logs.

This would provide a richer, more ecologically valid task, in which researchers can hone in on key performance variables and test more mature technological capabilities.

The living lab supported this research by bringing both transactive memory and cyber operations toward a more interdisciplinary perspective. Previous research in transactive memory took a very traditional, psychological perspective, with little consideration for technology. On the other hand, research in cyber is often solely focused on the technology, with little to no consideration of the human. By incorporating a more holistic model to approach the research, we were able to shine a unique light on both transactive memory as well as cyber operations.

FUTURE CONSIDERATIONS

The research discussed in this chapter represents only one revolution of the living laboratory, and not a continuous cycle. The goal of this research was to understand how transactive memory operates in a *next-gen* environment that is dependent on distributed teams and computing. As seen above, this research began to address these questions; however, it is not complete. The research discussed above is merely a starting point, the lessons and findings should be rolled back into the living lab, so that we can continue to hone in on, and understand these constructs.

Although the living laboratory is visualized as cyclical, it is actually a fluid and malleable process. As shown in this research, while typical studies complete the cycle clockwise, going from ethnography to theory, we took an approach that allowed us to establish a theoretical lens, before approaching practice. The findings here may inform future scaled-world simulation experiments, reconfigurable prototypes, knowledge elicitations, or ethnographies. As we move forward with our research in cyber operations and on transactive memory, we must pay close attention to the questions we wish to answer, and figure out which step in the living laboratory is most apropos to address it. For example, before continuing on with future reconfigurable shared virtual feedback implementations, it might be best to go back into the field, and understand how current systems are used, and conduct specific research on their use so that pain points and positives can be extracted. On the other hand, issues of transactive memory may not be appropriate for the field, due to its invisible nature, so further refinement and maturity may be made in future scaled-world experiments.

CONCLUSION

This research represents an interdisciplinary investigation into the role of transactive memory and distributed cognition in cyber operations. Using the living laboratory, we built a holistic and representational simulation, teamNETS, to control for and hone in on key aspects of the cyber environment that impact transactive memory. This simulation served as the basis for a team-based human-in-the-loop experiment to assess on the role of shared virtual feedback and knowledge structures on team performance and transactive memory. Within this controlled environment, our findings both supported the extant literature, and discovered new evidence that furthers our understanding of both transactive memory, and the utility of shared virtual feedback in distributed collaborations. Even with the simplified context of the current simulation, this research serves as contribution to research and the living laboratory. Our findings reiterate those found in the Tyworth et al. (2013) qualitative research that shows the breakdown of collaboration across functional domains in cybersecurity.

At the heart of the living lab and cognitive systems engineering, is the notion that humans influence technology, technology influences human, and humans influence each other as well. By approaching transactive memory and cybersecurity with this lens, we have been able to find meaning and understanding to address key gaps in the literature. Moving forward, it is critical that we and other researchers take our research and findings, and continue to mature our understanding of the environment and the cognitive and collaborative behaviors within.

REVIEW QUESTIONS

1. Identify three distributed cognitive artifacts that you have used, and describe how you used them to improve your cognition and/or performance.
2. Describe the different components of transactive memory.
3. Describe the two types of knowledge specialization structures that may appear in a transactive-memory system.
4. Discuss some alternate solutions to improve transactive memory that may be more successful as a distributed cognitive aid than the current implementation of the ATM-I.
5. Brainstorm how the findings from the scaled-world simulation study could be used to inform future ethnographic research and knowledge elicitation in cyber operations.

REFERENCES

Ariff, M. I. M., S. K. Milton, R. Bosua, and R. Sharma. 2012. Transactive memory systems: Exploring task, expertise and people (Tep) unit formation in virtual teams: Conceptualization and scale measurement development. *PACIS*, Hochiminh City, Vietnam, July 11–15.

Ariff, M. I. M., R. Sharma, S. Milton, and R. Bosua. 2013. Modeling the effect of task interdependence on the relationship between transactive memory systems (TMS) quality and team performance. *2013 International Conference on Research and Innovation in Information Systems (ICRIIS)*, Kuala Lumpur, Malaysia, November 27–28.

Austin, J. R. 2003. Transactive memory in organizational groups: The effects of content, consensus, specialization, and accuracy on group performance. *Journal of Applied Psychology* 88 (5): 866.

Brandon, D. P., and A. B. Hollingshead. 2004. Transactive memory systems in organizations: Matching tasks, expertise, and people. *Organization Science* 15 (6): 633–644.

Chen, X., X. Li, J. Guynes Clark, and G. B. Dietrich. 2013. Knowledge sharing in open source software project teams: A transactive memory system perspective. *International Journal of Information Management* 33 (3): 553–563.

Cooke, N. J., E. Salas, P. A. Kiekel, and B. Bell. 2004. Advances in measuring team cognition. In *Team cognition: Understanding the factors that drive process and performance*, edited by Eduardo Salas and Stephen Fiore, pp. 83–106. Washington, DC: American Psychological Association.

Endsley, T. C., J. A. Reep, M. D. McNeese, and P. K. Forster. 2015. Conducting cross national research lessons learned for the human factors practitioner. *Proceedings of the Human Factors and Ergonomics Society Annual Meeting*, Los Angeles, CA, October 26–30.

Giacobe, N. A. 2013a. Measuring the effectiveness of visual analytics and data fusion techniques on situation awareness in cyber-security. Doctor of Philosophy, College of Information Sciences and Technology, The Pennsylvania State University.

Giacobe, N. A. 2013b. A picture is worth a thousand alerts. *Proceedings of the Human Factors and Ergonomics Society Annual Meeting*, San Diego, CA, September 30–October 04.

Gutwin, C., and S. Greenberg. 2002. A descriptive framework of workspace awareness for real-time groupware. *Computer Supported Cooperative Work (CSCW)* 11 (3–4): 411–446.

Hamilton, K., V. F. Mancuso, D. Minotra, R. Hoult, S. Mohammed, A. Parr, G. Dubey, E. McMillan, and M. D. McNeese. 2010. Using the NeoCITIES 3.1 simulation to study and measure team cognition. *Proceedings of the Human Factors and Ergonomics Society Annual Meeting*, San Francisco, CA, September 27–October 01.

Heavey, C., and Z. Simsek. 2014. Distributed cognition in top management teams and organizational ambidexterity the influence of transactive memory systems. *Journal of Management* 43 (3): 0149206314545652.

Hellar, D. B., and M. McNeese. 2010. NeoCITIES: A simulated command and control task environment for experimental research. *Proceedings of the Human Factors and Ergonomics Society Annual Meeting*, San Francisco, CA, September 27–October 01.

Hill, J., and C. Gutwin. 2003. Awareness support in a groupware widget toolkit. *International Conference on Supporting Group Work (GROUP)*, Sanibel Island, FL, November 9–12.

Hollingshead, A. B. 1998a. Communication, learning, and retrieval in transactive memory systems. *Journal of experimental social psychology* 34 (5): 423–442.

Hollingshead, A. B. 1998b. Retrieval processes in transactive memory systems. *Journal of personality and social psychology* 74 (3): 659.

Hollingshead, A. B. 2001. Cognitive interdependence and convergent expectations in transactive memory. *Journal of personality and social psychology* 81 (6): 1080.

Hollingshead, A. B., N. Gupta, K. Yoon, and D. P. Brandon. 2011. Transactive memory theory and teams: Past, present, and future. In *Theories of Team Cognition: Cross-Disciplinary Perspectives*, edited by Eduardo Salas, Stephen M. Fiore and Michael P. Letsky, pp. 421–455. New York, NY: Routledge, Taylor & Francis Group.

Hsu, J. S.-C., S.-P. Shih, J. C. Chiang, and J. Y.-C. Liu. 2012. The impact of transactive memory systems on IS development teams' coordination, communication, and performance. *International Journal of Project Management* 30 (3): 329–340.

Hutchins, E. 1995. *Cognition in the Wild*. Cambridge, MA: MIT press.

Jackson, M., and R. L. Moreland. 2009. Transactive memory in the classroom. *Small Group Research* 40 (5): 508–534.

Jones, R. E. T., M. D. McNeese, E. S. Connors, T. Jefferson, and D. L. Hall. 2004. A distributed cognition simulation involving homeland security and defense: The development of NeoCITIES. *Proceedings of the Human Factors and Ergonomics Society Annual Meeting*, New Orleans, LA, September 20–24.

Keel, P. E. 2007. EWall: A visual analytics environment for collaborative sense-making. *Information Visualization* 6 (1): 48–63.

Knott, B. A., V. F. Mancuso, K. Bennett, V. Finomore, M. McNeese, J. A. McKneely, and M. Beecher. 2013. Human factors in cyber warfare alternative perspectives. *Proceedings of the Human Factors and Ergonomics Society Annual Meeting*, San Diego, CA, September 30–October 04.

Lewis, K. 2004. Knowledge and performance in knowledge-worker teams: A longitudinal study of transactive memory systems. *Management science* 50 (11): 1519–1533.

Lewis, K., and B. Herndon. 2011. Transactive memory systems: Current issues and future research directions. *Organization Science* 22 (5): 1254–1265.

Liang, D. W., R. Moreland, and L. Argote. 1995. Group versus individual training and group performance: The mediating role of transactive memory. *Personality and Social Psychology Bulletin* 21 (4): 384–393.

Liao, J., A. T. O'Brien, N. L. Jimmieson, and S. L. D. Restubog. 2015. Predicting transactive memory system in multidisciplinary teams: The interplay between team and professional identities. *Journal of Business Research* 68 (5): 965–977.

Littlepage, G. E., A. B. Hollingshead, L. R. Drake, and A. M. Littlepage. 2008. Transactive memory and performance in work groups: Specificity, communication, ability differences, and work allocation. *Group Dynamics: Theory, Research, and Practice* 12 (3): 223.

Mancuso, V. F., J. C. Christensen, J. Cowley, V. Finomore, C. Gonzalez, and B. Knott. 2014. Human factors in cyber warfare II emerging perspectives. *Proceedings of the Human Factors and Ergonomics Society Annual Meeting*, Chicago, IL, October 27–31.

Mancuso, V. F., D. Minotra, N. Giacobe, M. McNeese, and M. Tyworth. 2012. idsNETS: An experimental platform to study situation awareness for intrusion detection analysts. *2012 IEEE International Multi-Disciplinary Conference on Cognitive Methods in Situation Awareness and Decision Support (CogSIMA)*, Boston, MA, October 22–26.

Maybury, M. 2015. Toward the assured cyberspace advantage: Air force cyber vision 2025. *IEEE Security & Privacy* 13 (1): 49–56.

McNeese, M. 2004. How video informs cognitive systems engineering: Making experience count. *Cognition, Technology & Work* 6 (3): 186–196.

McNeese, M. D., V. F. Mancuso, N. J. McNeese, T. Endsley, and P. Forster. 2014. An integrative simulation to study team cognition in emergency crisis management. *Proceedings of the Human Factors and Ergonomics Society Annual Meeting*, Chicago, IL, October 27–31.

Mell, J. N., D. Van Knippenberg, and W. P. van Ginkel. 2014. The catalyst effect: The impact of transactive memory system structure on team performance. *Academy of Management Journal* 57 (4): 1154–1173.

Michinov, N., and E. Michinov. 2009. Investigating the relationship between transactive memory and performance in collaborative learning. *Learning and Instruction* 19 (1): 43–54.

Moreland, R. L. 1999. Transactive memory: Learning who knows what in work groups and organizations. *Key Readings in Social Psychology* 1: 327.

Moreland, R. L., and L. Myaskovsky. 2000. Exploring the performance benefits of group training: Transactive memory or improved communication? *Organizational Behavior and Human Decision Processes* 82 (1): 117–133.

Nemeth, C. P., R. I. Cook, M. O. Connor, and P. A. Klock. 2004. Using cognitive artifacts to understand distributed cognition. *IEEE Transactions on Systems, Man and Cybernetics, Part A: Systems and Humans* 34 (6): 726–735.

Norman, D. A. 1991. Cognitive artifacts. In *Designing interaction: Psychology at the human-computer interface*, edited by John M. Carroll, pp. 17–38. New York, NY: Cambridge University Press.

Ren, Y., and L. Argote. 2011. Transactive memory systems 1985–2010: An integrative framework of key dimensions, antecedents, and consequences. *The Academy of Management Annals* 5 (1): 189–229.

Sellers, J., W. S. Helton, K. Näswall, G. J. Funke, and B. A. Knott. 2014. Development of the Team Workload Questionnaire (TWLQ). *Proceedings of the Human Factors and Ergonomics Society Annual Meeting*, Chicago, IL, October 27–31.

Tschan, F., N. K. Semmer, A. Gurtner, L. Bizzari, M. Spychiger, M. Breuer, and S. U. Marsch. 2009. Explicit reasoning, confirmation bias, and illusory transactive memory: A simulation study of group medical decision making. *Small Group Research* 40: 271–300.

Tulving, E., and D. M. Thomson. 1973. Encoding specificity and retrieval processes in episodic memory. *Psychological review* 80 (5): 352.

Tyworth, M., N. A. Giacobe, V. F. Mancuso, and C. D. Dancy. 2012. The distributed nature of cyber situation awareness. *2012 IEEE International Multi-Disciplinary Conference on Cognitive Methods in Situation Awareness and Decision Support*, New Orleans, LA, March 5–7.

Tyworth, M., N. A. Giacobe, V. F. Mancuso, M. D. McNeese, and D. L. Hall. 2013. A human-in-the-loop approach to understanding situation awareness in cyber defence analysis. *EAI Endorsed Transactions on Security and Safety* 13 (1–6):e6.

Wegner, D. M. 1987. Transactive memory: A contemporary analysis of the group mind. In *Theories of group behavior*, edited by Brian Mullen and George R. Goethals, pp. 185–208. New York: Springer.

Wegner, D. M. 1995. A computer network model of human transactive memory. *Social cognition* 13 (3): 319.

Wegner, D. M., T. Giuliano, and P. T. Hertel. 1985. Cognitive interdependence in close relationships. In *Compatible and incompatible relationships*, edited by William J. Ickes, pp. 253–276. New York: Springer.

4 Digital Architectures for Distributed Learning and Collaboration

Jeff Rimland

CONTENTS

Introduction 67
Collaborating on Group Programming Assignments 69
Can *Forking* Be Applied to Sharing Ideas? 73
    Idea Mining with Federated Wiki 74
Conclusion 78
Review Questions 78
References 78

ADVANCED ORGANIZER

This chapter explores both the theoretical basis and practical application of network-based architectures that enable collaborative learning. Emphasis is placed on technologies that support distributed problem-based learning (PBL), along with specific examples of federated note taking and evaluations (in which students take quizzes with access to their own notes, the notes of classmates, and collaboratively created artifacts). These examples are presented within the context of both intermediate computer programming courses and an introductory information sciences course, in which federated activities mirror applications at the individual, organizational, and societal levels.

INTRODUCTION

Courses involving the information sciences often require problem finding, definition, and exploration activities before actually solving the assigned problem. Many problems are complex, situated within real-world contexts, and involve in-depth awareness of multiple constraints. Given the landscape of learning, there is evidence that shows many benefits of collaborating in pairs and various sized teams (Cockburn & Williams, 2001), and with members of the public community outside the classroom. In addition, the learning experience evokes a wide gamut of emotional states that must be taken into account when engineering any technological tools that are designed to improve human collaboration (McNeese, 2003). The living laboratory framework (LLF) is predicated on PBL and, in turn, affords an active approach to developing cognitive systems, collaborative tools, and analytical interfaces that can support mutual effectiveness in addressing innovations in the classroom.

For many group activities, conventional collaboration tools provided by learning management system (LMS) applications such as Canvas (http://canvaslms.com)—which include discussion forums, group submission of assignments, and pages that can be edited by multiple students—are sufficient for supporting the necessary group interactions. Other cases require the integration of multiple tools, the repurposing of tools typically used in other areas, or the introduction of new technologies that are currently under development.

GitHub is beginning to gain traction as a leading collaborative tool for computer programming education. Although highly effective and relevant, it still presents many challenges in the classroom setting (Zagalsky, Feliciano, Storey, Zhao, & Wang, 2015). Wikis and other Web 2.0 technologies such as blogs and podcasts have been adopted for use in educating healthcare practitioners (Boulos, Maramba, & Wheeler, 2006), as well as in higher education in general (Parker & Chao, 2007), and have gained general acceptance in many areas of teaching. In addition, there has been research into the role of software systems that support the synchronization of collaborative activities in an educational setting via user notification of significant events in their learning environment (Carroll, Neale, Isenhour, Rosson, & McCrickard, 2003). Although collaboration began in conventional in-person classrooms, collaborative technologies have been studied extensively in the context of online education and Massive Online Open Course (MOOC) settings (Kop & Carroll, 2011) due to the need for replacement of the face-to-face interaction that occurs in a physical classroom, as well as for addressing the feelings of loneliness and isolation that have been reported in online education (Shea, 2006). The challenges and obstacles intrinsic to such collaborative tools themselves have also been examined (Guzdial, Ludovice, Realff, Morley, & Carroll, 2002).

Effective use of these tools (i.e., actual improvement to student learning outcomes) requires cognizance of prior research, understanding of relevant theory, and mindfulness of student interactions—as well as technical understanding of the tools themselves.

This chapter contextualizes prior research, recent technological advances, and anecdotal evidence from the classroom into the living lab framework. An emphasis is placed on the utilization of collaboration tools including GitHub (https://www.github.com), Federated Wiki (https://wardcunningham.github.io/), slack (https://www.slack.com), and Google Docs/Sheets in teaching Information Sciences and Technology (IST) courses at The Pennsylvania State University. They are described as used in both introductory information science seminars and in introductory and intermediate programming courses.

It is important to emphasize that these technologies were only included in classroom activities when they supported a specific teaching or learning objective such as promoting higher-level reasoning and critical thinking, encouraging student learning via teaching their fellow group members, or enriching discussions via collaboratively created artifacts. Any results described herein are based on instructor and learning assistant (LA) observations, student feedback, and informal comparison between multiple course sections.

COLLABORATING ON GROUP PROGRAMMING ASSIGNMENTS

There are many reasons to include group assignments in a programming course (Holland & Reeves, 1994; Williams, Wiebe, Yang, Ferzli, & Miller, 2002; Wulf, 2005). Students learn more deeply through the opportunity to explain concepts to their teammates, are often more comfortable asking their peers for help, and gain experience with team interactions that will help prepare them for working on an actual development team once they enter the professional world. However, as the scenarios below demonstrate, there are significant obstacles that stand in the way of students working together on a programming task in an effective and efficient manner. In-person groups of 3–4 students collaborating on a programming task typically falls into one of the following models of workflow:

1. Each team member works on the entire program in parallel at his or her own workstation. When one student has difficulty continuing with the program, they pause and ask the other students for help. If that student finds the help that they need, they may resume programming in parallel with the rest of the team. If they are unable to understand the required task, they may switch to passively *backseat driving*, or watching their teammates program. This approach is most common in introductory programming courses.
2. The team initially divides the labor among each team member. This is most common in intermediate or advanced courses, and is facilitated by object oriented programming (OOP) languages, such as Java, that encourage the division of the application into multiple files/modules based on functionality or other logical grouping. Each student then works on their assigned areas of code until a useful amount of work has been completed, and then shares their files with the rest of the team via e-mail or a shared drop box.
3. Students break off into pairs, groups of three, or even larger groups sitting around a single workstation while one student types. This approach is most consistent with the Agile programming practice of *pair programming* (Williams, Kessler, Cunningham, & Jeffries, 2000) when performed in groups of two, but quickly becomes unproductive in larger groups.

Each of these approaches has benefits and drawbacks. When using the first approach, each student is essentially writing the entire program on their own, which is beneficial in terms of ensuring exposure to concepts and practice using the techniques necessary for the assignment, but some benefits of team work are negated. There is typically minimal synthesis of ideas because discussion tends to be limited to remediation, and this workflow does little to prepare students for real-world programming environments.

The second approach, while more representative of the professional environment that students are likely to encounter after graduation, has downsides of its own. Using e-mail or Dropbox to share files makes it very easy to accidentally lose work by permanently overwriting previous versions. For example, using the popular NetBeans integrated development environment (IDE) (www.netbeans.org), it is common to export an entire project to a ZIP file and e-mail that file to a collaborator, who can then import that project into their own NetBeans workspace. If two students are working on a project with the same name, it is very easy for one student to accidentally and irretrievably overwrite their own work by importing their teammate's project. In fact, this happens frequently enough in programming classes that instructors can recognize the telltale gesture (hands grasping at hair, mouth agape in disbelief) of instantly overwriting several hours of work from across the classroom. Another problem with this approach is that intermediate programming students may be prone to use software objects (i.e., classes) that lack cohesion (e.g., one class addressing one task) and are strongly coupled (Fowler, Beck, Brant, Opdyke, & Roberts, 1999) to other software objects (i.e., they have a high degree of dependency on other classes), which results in multiple files being edited by each student. This significantly complicates the process of integrating each student's contributions.

The third approach has been shown to work well in pairs (Cockburn & Williams, 2001), but often results in boredom or lack of engagement in larger teams, when students who are not actually sitting at the keyboard can have a difficult time contributing, having their voices heard, or even seeing the monitor.

To remedy many of these challenges and improve on collaborative workflow, many small classroom groups have begun using distributed version control systems such as Git, which are commonly hosted by service providers such as GitHub or GitLab. Distributed tools differ from the centralized version control mechanisms of the past in that each user has their own repository (that may be shared with others or kept private) and the ability to copy or *fork* code between repositories. Service providers such as GitHub provide *freemium* access (Kumar, 2014), which allows students to learn the tools on their own or as part of an in-class formalized structure established by the instructor. As a result, students are able to readily adopt the tools informally as part of their collaboration process. When students adopt GitHub use informally, it has been my observation that they typically use the more simplistic *shared repository* approach, in which one student creates an initial project within their GitHub account, and grants their teammates access to *push* (i.e., upload) changes to their repository, which becomes the de facto repository for the project. This immediately results in the following benefits:

1. It is nearly impossible to overwrite another user's changes in such a manner that they cannot be retrieved.
2. The entire commit history (i.e., who did what, and when) is permanently and automatically recorded along with the code.
3. The code can be readily shared with the instructor, other groups, and other members of the open source community.

Although the first two of these benefits are difficult to contest, the third point introduces some challenges. The default usage of GitHub is to allow the public access to view (but not modify) your code repository. This can create certain challenges that includes the following:

1. Some institutional policies prohibit an instructor from requiring a student to make their work public.
2. Students who are assigned identical programming problems in a class can readily view the work of other students, which can then be modified slightly, and passed off as their own work.

To remedy this, GitHub has made its *Micro* account (that includes five private repositories) free for students and educators (https://education.github.com/). In certain situations, there are still benefits to using public repositories in an academic setting. These may include projects where students are working on similar but different projects, or when the entire class is collaborating on a single project.

Although the shared repository approach is usually sufficient for small groups of 3 or 4 students collaborating on a project, there are times when a greater degree of administrative control is necessary. In an intermediate Java design and development course (IST 311) at Penn State's College of Information Sciences and Technology, GitHub is used to facilitate a final project in which the entire class of 40 undergraduate students collaborates on a single project in order to simulate a more realistic professional software development environment. In this situation, a shared repository would have resulted in an excessively chaotic development environment, so the instructor chose to implement the *Fork and Pull* model of collaboration.

The Fork and Pull model is revolutionary not only to open source software or software development generally, but also to computer-supported collaborative work (CSCW) and computer-supported collaborative learning (CSCL) in general. As will be described subsequently, it can be applied to ideas as well as software code. The Fork and Pull workflow begins with a user discovering open source code on a distributed version control system (such as GitHub) on which they would like to collaborate but do not have permissions to make changes directly to. This code might be related to a massive open source software project such as Mozilla (https://github.com/mozilla), or an in-class assignment in which the instructor wishes to retain some control over the code base in order to maintain a degree of usability and interoperability between modules created by multiple students in the class (e.g., https://github.com/jrimland/SimIST). Although the user who discovers such a code repository does not have permission to modify it in its home directory, they are able to fork (i.e., copy) a version of that code repository (i.e., repo) into their own GitHub account. Once the repo has been copied into their own account, the user (along with any other users to whom they granted permission—such as fellow teammates in a programming class) has permission to make any changes that they see fit. Although the forking mechanism provides a convenient way to view, modify, and improve open source code, it does not, by itself, enable the user to contribute to the primary project. The pull mechanism fills this gap by allowing the user who forked the project to send a request to the owner of the original repository (via a single button click) to evaluate their changes for possible integration with the original project. The Fork and Pull model has been a boon for open source software development, but its classroom use has been limited due to a steeper learning curve and more challenging conflict resolution situations. For example, advanced techniques are required to resolve the conflict that occurs when multiple students modify the same code modules within a single development iteration.

A key aspect of managing this learning curve is to provide lessons specifically tailored to incrementally increasing student comfort with Git and the GitHub environment. In IST 311 (Intermediate Object Oriented Design and Applications), this process began with an introduction to the basics of Git and GitHub, and how NetBeans integration facilitates this process. This also included an introduction to the unique vocabulary of distributed version control systems (e.g., push, pull, repo.) After this introduction, we proceeded through simple in-class examples of GitHub interaction using the shared repository model in small groups. We then completed a group lab in which the deliverable program needed to be submitted in the form of a GitHub repo URL for credit. This assignment provided opportunities for exposure to the process of uploading (pushing) and downloading (pulling) code from a shared repository, as well as the sometimes difficult task of resolving conflicts that arise when multiple users have made competing edits to the same code. Then, the Fork and Pull model is introduced—first on an extremely simple *Hello World* program as an in-class exercise, and then on incrementally more complex problems. In conclusion, the Fork and Pull model enabled the entire class to collaborate on a single final project of sizable complexity, which offered the students a glimpse into the actual workings of a professional software development experience. As importantly, the scale of the collaboration allowed students to experience firsthand the importance of applying solid object oriented design techniques. The importance of using modular code design and cohesive classes, for example, can be easily overlooked when it is simply presented in a lecture. However, experiencing the aggravation and difficulties of resolving merge conflicts that have been brought on by monolithic coding practices allows the students to internalize these practices by linking them with both the functional outcomes as well as the emotional responses of the experience, which has been demonstrated to improve learning and long-term memory (Dolcos & Cabeza, 2002).

MIDWAY BREATHER

- Some teaching and learning challenges require modification of existing educational tools or repurposing tools created for other problem domains.
- Successful group work in programming courses requires scaffolding in terms of team roles, workflows, technologies, expectations, and accountability.
- Distributed version control systems such as GitHub have revolutionized software development, and are slowly being integrated into programming curriculum.

CAN *FORKING* BE APPLIED TO SHARING IDEAS?

The previous section introduced a use of a distributed version control system in the context of an intermediate software design and development class. Although this use is fairly close to the original intent of distributed VCS, it raises the following questions:

- If this mechanism can be used to share and collaboratively develop and improve software code, could it also be used in the same way with all types of ideas?
- Could it be used to bridge gaps of understanding between collaborators in any interdisciplinary challenge?

A key aspect of the LLF is the sharing of ideas, concepts, and knowledge across multiple domains and environments. Reconfigurable prototyping tools (McNeese et al., 2005) are an important means to this goal. This section addresses the above questions within the context of the living-laboratory approach by discussing the classroom use of Federated Wiki in an introductory course in Information Sciences and Technology (IST 110S).

Ward Cunningham invented the wiki as a tool for sharing ideas and collaborating. Years later, the wiki model was applied to an online encyclopedia application, and the extremely popular Wikipedia website was born. Although wiki and Wikipedia both received a great deal of notoriety, Ward noticed that Wikipedia created an environment where there was room for only one version of the truth to exist. Furthermore, gender inequalities and other biases were beginning to make their way into the online encyclopedia (Wagner, Garcia, Jadidi, & Strohmaier, 2015). Each Wikipedia page, although editable by the public, can typically only represent a single side of an argument. For example, if I edit the Wikipedia entry on tennis to state that Pete Sampras was the greatest tennis player in history, and someone else edits that page to say that John McEnroe is the greatest player ever, then we are in competition to have our version of the truth become the canonical *truth* (at least as expressed by Wikipedia). Ward responded to this friction by creating a new type of collaborative platform that blends wiki with several of the key features used in the distributed version control systems previously used for collaborative software development. He essentially created a *GitHub for ideas* (UDell, 2015) and he called it Federated Wiki.

The primary defining feature of Federated Wiki is that it allows each user to host their own pages over which they maintain total editorial control. If they choose so, they may copy or *fork* the contents of another Federated Wiki page to their own server, at which time they are free to edit that page (as it lives on their own server) (see Figure 4.1). The Federated Wiki system automatically provides a link to the original version, and displays the fact that the current work is a derivative of that page.

At first contemplation, it might be difficult to imagine how the silo model shown in Figure 4.1 could improve collaboration over Wikipedia's model. Although each user only has the ability to modify their own wiki, they are freely able to read and fork the wikis of other users (see Figure 4.2).

FIGURE 4.1 Although Wikipedia hosts a single version of each page in a centralized manner, Federated Wiki uses a distributed server model in which each user maintains complete editorial control.

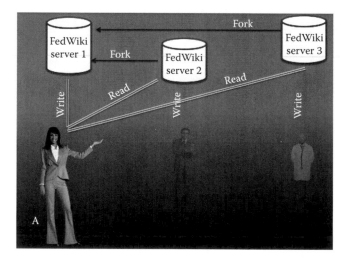

FIGURE 4.2 Federated Wiki users are only able to edit content hosted on their own server, but they are able to read and fork (copy) the contents of other wiki pages.

IDEA MINING WITH FEDERATED WIKI

Rather than encouraging students to take notes in class, I emphasized the concept of *idea mining* (Caulfield, 2014), in which students actively seek out key ideas while listening to a lecture, engaging in a discussion, participating in a group activity, or reading from the online text or a website. Once a key idea is recognized, the student creates a single Federated Wiki page for that idea (see Figure 4.3).

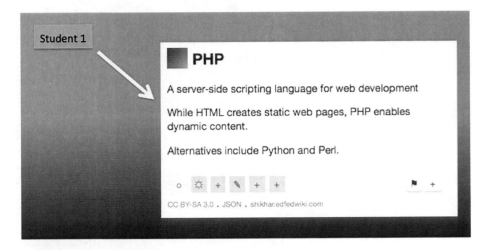

FIGURE 4.3 A student creates a Federated Wiki page describing the PHP scripting language while *idea mining.*

A second student in the class may discover the first student's description of the PHP language, find it useful, and then fork a copy of it to his or her own wiki. Once the page has been forked, the second student may freely edit it and improve on it in any way that they see fit. For example, they might consider it a glaring omission that the first student did not mention that PHP is a component of the LAMP stack (Lawton, 2005). If this were the case, they could update their version of the page as shown in Figure 4.4. It should be emphasized that this change on the wiki of student 2 does not have any effect on the version of this page created by student 1. However, student 1's wiki will now reflect the fact that a newer version of this page is available (via the multicolored boxes shown at the top of Figure 4.4), and provides a link to both newer and older version.

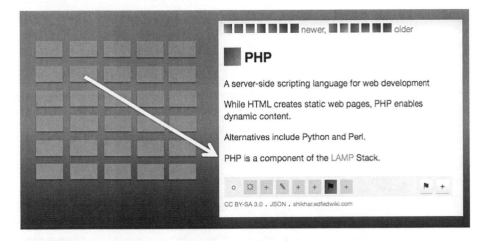

FIGURE 4.4 A forked and expanded version of the initial wiki entry on PHP.

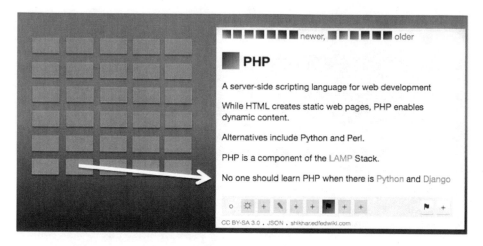

■■■■■■ newer, ■■■■■■ older

PHP

A server-side scripting language for web development

While HTML creates static web pages, PHP enables dynamic content.

Alternatives include Python and Perl.

PHP is a component of the LAMP Stack.

No one should learn PHP when there is Python and Django

CC BY-SA 3.0 . JSON . shikhar.edfedwiki.com

FIGURE 4.5   Disagreement in Federated Wiki versus Wikipedia.

This process of forking and improving leads to an interesting phenomenon. As multiple students make improvements to a page, the edits that are judged to be of the highest quality will be forked by more students than those containing edits that the students feel are less helpful or truthful. This results in a democratized natural selection process in which each modification is effectively up-voted or down-voted by the incidence with which it is selected for inclusion in additional pages (via forking). This is a stark contrast to the Wikipedia model, in which conflicts are resolved by a site administrator (Kittur, Suh, Pendleton, & Chi, 2007). For example, in Federated Wiki, a third student may have different feelings about PHP and fork, and modify the post of student 2 to reflect that (as shown in Figure 4.5).

Student 3 is free to express his or her opinion that no one should bother learning PHP because Django is a superior alternative. If this opinion is up-voted by the community (via forking), then a user will be more likely to stumble on this version of the page than versions that contain conflicting recommendations. The process is completely decentralized, and each user maintains control of pages that are hosted on their wiki.

Federated Wiki has often been described as almost impossible to fully understand and appreciate as a collaborative platform without actually working with it for several days. Notwithstanding, a more relatable explanation of its benefits is attainable by comparing the more conventional collaborative tools used in the classroom with Federated Wiki.

1. *Google Docs*: Google's online document system is, in my experience, the default collaboration method used by most students today. Google Docs work well, are easily shared, and can be submitted to an instructor simply by sharing a URL or inviting them to collaborate. Google Docs fall short in two areas. First, it can be distracting for multiple users to be simultaneously making edits to a single page. Second, all collaborators must either agree on

a single version of the document (similarly to the Wikipedia issues described in earlier sections), or maintain individual documents. Unfortunately, there is no mechanism to automatically link multiple versions of a document on the same topic as appears in Federated Wiki.

2. *Canvas discussion boards and pages*: The Canvas LMS provides various collaborative tools including a discussion forum and *Pages*, which offer wiki-like collaborative editing. The major advantage of utilizing these tools within the context of Canvas (or other LMS) is that students enrolled in a course already have integrated access to these features, so there is no need for an instructor to create separate accounts (as is the case with an externally managed system), or to require them to use an existing personal account (as with Google). Although discussion forums are useful—especially in online classes or MOOC environments, they result in the creation of a single artifact that contains, in an often disjointed manner, the ideas, debates, arguments, and rants of the entire class. Federated Wiki allows each individual user to craft their wiki pages into an artifact that reflects collaboration of their group or community, but is curated and presented in a manner of their choice.

3. *Slack*: Slack (http://slack.com) is a messaging service that has been introduced as a replacement for e-mail. It includes excellent search capabilities, *tagging* of individuals to draw their attention to a specific conversation, ad hoc formation of new channels, and intelligent handling of code (highlighting of keywords specific to a programming language, etc.). It also provides some capability to collaboratively edit documents (although at the time of this writing, this capability seems to exist only in *locking* mode, in which a single user can edit a document at a time after clicking the *Edit* button and locking the resource from edits by other users until they are finished editing). Although I did not find Slack to be a replacement or competition for Federated Wiki or GitHub, it did function as an excellent *back channel* to communicate with teammates or the rest of the class while working remotely.

The key advantages of the Federated Wiki in the classroom are

1. *Learning by creating*: Students learn by cooperatively creating artifacts (rapidly reconfigurable prototypes) that represent the ideas of their groups and communities, but are still ultimately their responsibility to curate and edit.

2. *Deeper sharing of work*: Federated Wiki facilitates both cooperation and collaboration, automatically provides attribution of forked ideas, and versioning information, and allows the development of a social network of users sharing related work.

3. *Discoverable but not obvious attribution*: Although the entire audit trail of forked and modified pages is not a primary concern in the graphical layout of Federated Wiki, it is available and discoverable if needed. This *discoverable but not obvious* attribution system keeps such data unobtrusive when it less

important (while actively sharing and collaborating, for instance), yet makes it available when needed (for the purpose of external publications, etc.).

4. *A public forum*: The public nature of Federated Wiki encouraged a higher quality of work.

CONCLUSION

This chapter used the LLF as a lens with which to examine collaborative aspects of teaching and learning using two very different applications of *forking*—GitHub for collaborating on programming projects, and Federated Wiki for the sharing and building of ideas. Although both GitHub and Federated Wiki provide outstanding capabilities, they are still hindered by steep learning curves for both the students and the instructors. The intended goals of this chapter are to demonstrate both the need for rigorous future research into the efficacy of these tools in the classroom, and to inspire other instructors to brave the obstacles and introduce their students to these technologies that have such tremendous potential.

REVIEW QUESTIONS

1. List the key technologies discussed in this chapter and discuss their benefits and drawbacks.
2. Describe the practical and theoretical differences between *forking* of ideas versus program code.
3. What are the policies regarding the usage of third-party services and publicly-visible course work at your institution?
4. Does Federated Wiki provide any advantages that are unattainable via other technological tools? Please discuss.
5. What is an example of a multidisciplinary challenge that might benefit from idea forking?

REFERENCES

Boulos, M. N. K., Maramba, I., & Wheeler, S. (2006). Wikis, blogs and podcasts: A new generation of Web-based tools for virtual collaborative clinical practice and education. *BMC Medical Education, 6*, 41. doi:10.1186/1472-6920-6-41

Carroll, J. M., Neale, D. C., Isenhour, P. L., Rosson, M. B., & McCrickard, D. S. (2003). Notification and awareness: Synchronizing task-oriented collaborative activity. *International Journal of Human-Computer Studies, 58*(5), 605–632.

Caulfield, M. (2014). Helping the right ideas find one another. Retrieved from https://hapgood. us/2014/12/21/helping-the-right-ideas-find-one-another-fedwiki-happening/

Cockburn, A., & Williams, L. (2001). The costs and benefits of pair programming. *Extreme Programming Examined*, (1), ISBN:0-201-71040-4, 223–243.

Dolcos, F., & Cabeza, R. (2002). Event-related potentials of emotional memory: Encoding pleasant, unpleasant, and neutral pictures. *Cognitive, Affective & Behavioral Neuroscience, 2*(3), 252–263. doi:10.3758/CABN.2.3.252

Fowler, M., Beck, K., Brant, J., Opdyke, W., & Roberts, D. (1999). Refactoring: Improving the Design of Existing Code. *Xtemp 01*, (1), ISBN:0-201-71040-4, 1–337.

Guzdial, M., Ludovice, P., Realff, M., Morley, T., & Carroll, K. (2002, October). When collaboration doesn't work. In *Proceedings of the international conference of the learning sciences*, Seattle, WA, (pp. 125–130).

Holland, D., & Reeves, J. R. (1994). Activity theory and the view from somewhere: Team perspectives on the intellectual work of programming. *Mind, Culture, and Activity*, *1*(1–2), 8–24.

Kittur, A., Suh, B., Pendleton, B. A., & Chi, E. H. (2007, April). He says, she says: Conflict and coordination in Wikipedia. In *Proceedings of the SIGCHI conference on human factors in computing systems*, San Jose, CA, (pp. 453–462). ACM.

Kop, R., & Carroll, F. (2011). Cloud computing and creativity: Learning on a massive open online course. *European Journal of Open, Distance and E-learning*, *14*(2).

Kumar, V. (2014). Making 'freemium' work: Many start-ups fail to recognize the challenges of this popular business model. *Harvard Business Review, 92*(5), 27–29.

Lawton, G. (2015). LAMP lights enterprise development efforts. *Computer, 38*(9), 18–20. doi:10.1109/MC.2005.304

McNeese, M. D. (2003). New visions of human–computer interaction: Making affect compute. *International Journal of Human-Computer Studies*, *59*(1), 33–53.

McNeese, M. D., Connors, E. S., Jones, R. E., Terrell, I. S., Jefferson, Jr., T., Brewer, I., & Bains, P. (2015, July, 22–27). Encountering computer-supported cooperative work via the living lab: Application to emergency crisis management. In *Proceedings of the 11th international conference of human-computer interaction*, Las Vegas, NV.

Parker, K. R., & Chao, J. T. (2007). Wiki as a teaching tool. *Interdisciplinary Journal of Knowledge and Learning Objects*, *3*(1), 57–72.

Shea, P. (2006). A study of students' sense of learning community in online environments. *Journal of Asynchronous Learning Networks*, *10*(1), 35–44.

UDell, J. (2015). A federated Wikipedia. Retrieved April 15, 2016, from https://blog.jonudell.net/2015/01/22/a-federated-wikipedia/

Wagner, C., Garcia, D., Jadidi, M., & Strohmaier, M. (2015). It's a man's Wikipedia? Assessing gender inequality in an online encyclopedia. *Arxiv, 1501.06307*, 1–10. Retrieved from http://arxiv.org/abs/1501.06307

Williams, L., Kessler, R. R., Cunningham, W., & Jeffries, R. (2000). Strengthening the case for pair-programming. *IEEE Software*, July–August, 19–25.

Williams, L., Wiebe, E., Yang, K., Ferzli, M., & Miller, C. (2002). In support of pair programming in the introductory computer science course. *Computer Science Education*, *12*(3), 197–212.

Wulf, T. (2005, October). Constructivist approaches for teaching computer programming. In *Proceedings of the 6th conference on information technology education*. Newark, NJ, (pp. 245–248). ACM.

Zagalsky, A., Feliciano, J., Storey, M., Zhao, Y., & Wang, W. (2015). The emergence of GitHub as a collaborative platform for education. In *Proceedings of the 18th ACM conference on computer supported cooperative work & social computing* (pp. 1906–1917). doi:10.1145/2675133.2675284

5 A Cognitive-Systems Approach to Understanding Natural Gas Exploration Supply Chains
A Purview and Research Agenda

Michael D. McNeese and James A. Reep

CONTENTS

Introduction 82
Background: Natural Gas Field of Practice 83
   The CCRINGSS Initiative 83
Premise 85
Approach 87
   General Progressions 88
Element #1: Ethnographic Fieldwork 90
   Research Overview 90
      Specific Research Agenda Items 91
      Opportunities to Move Research Forward (via CCRINGSS or Otherwise) ... 91
Element #2: Knowledge Elicitation 92
   Research Overview 92
      Specific Challenges/Research Agenda Items 93
      Opportunities to Move Research Forward 94
Element #3: Theoretical Foundations 94
   Research Overview 94
      Specific Challenges/Research Agenda Items 95
      Opportunities to Move Research Forward 96
Element #4: Scaled-World Simulation 96
   Research Overview 96
      Specific Challenges/Research Agenda Items 97
      Opportunities to Move Research Forward 98

Element #5: Design Prototyping and Innovations 98
    Research Overview 98
        Specific Challenges/Research Agenda Items 99
        Opportunities to Move Research Forward... 100
Conclusion 100
Review Questions.. 100
References.. 100

ADVANCE ORGANIZER

This chapter

- Describes the natural gas field of practice and the Collaborative Research on Intelligent Natural Gas Supply Systems (CCRINGSS) initiative.
- Provides a rich description of the living laboratory framework (LLF) and its elements.
- Explains the challenges and potential research directions for each of the living laboratory framework elements.

INTRODUCTION

Over the past 10 years, natural gas and natural gas exploration has expanded beyond expectations with the need to supply the world with additional gas and oil.

"High oil price(s), depleting reserves in the West, and an exponentially growing demand, in particular from Asia" (De Graaff, 2011, p. 262) have fueled growth found within nationalized energy organizations, which have traditionally controlled 80%–90% of global oil and gas reserves (De Graaff, 2011). The natural gas exploitation and exploration community has shown tremendous growth and potential (many new jobs have been generated [Wei, Patadia, & Kammen, 2010], new sources of energy unveiled [Mathiesen, Lund, & Karlsson, 2011], new advancements in technology built [Akhondi, Talevski, Carlsen, & Petersen, 2010], and economic growth stimulated [Weber, 2012]). Yet, there are many issues, concerns, and constraints that loom, especially as related to human-centered perspectives (human–computer interaction, mobile distributed cognition, information, and collaboration technologies). This chapter specifically draws from the focus of end-to-end supply chain management within the natural gas industry as specifically taken up within the new Center for CCRINGSS at The Pennsylvania State University. The CCRINGSS initiative seeks to connect Penn State students and researchers with industry leaders to collaboratively engage in interdisciplinary work. It aims to improve efficiency and sustainability through innovation both at the technological and supply chain management levels. Although not a specifically human-centered design effort per se, this unique initiative offers a lot of possibilities to examine human-centered design principles, which can result in improved individual and team performance in the natural gas environment.

Although there are many broad strokes and mutual outcomes present as objectives within the CCRINGSS program, this chapter specifically reifies elements that are salient for the information and cognitive sciences. This chapter will be of use for theoretical as well as practical concerns, and necessarily takes an interdisciplinary, transformative, and systems-level approach to this particular field of practice.

BACKGROUND: NATURAL GAS FIELD OF PRACTICE

In contemporary society the supply chains that underlie natural gas production are incredibly interdependent on information, people, technology, and the changes that emerge from the context itself. If we first look at natural gas production and exploitation it is very dependent on seeking information about what spot is best for setting up production (*prospecting*), setting up and moving equipment, materials, support systems, and operators to exploit the gas from underground reserves, for example, fracking, as needed (*logistical processes*), distilling a product that is obtainable and viable (*manufacturing*) within a given timeframe with the resources available (*managerial*), distributing information from multiple sources to share awareness and collaborative work, for example, command, control, communications, and intelligence, (*collaborating*), integrating data and information across various sensors, layers, and venues (*information fusion*). As these various windows into the domain are considered, there are requisite levels of understanding, knowledge, and technology that can further inform each one to develop research and design trajectories that expand the state of the art.

THE CCRINGSS INITIATIVE

CCRINGSS is a unique research and educational center that has been collaboratively funded by the General Electric Corporation and The Pennsylvania State University. The original CCRINGSS white paper outlining the focus of the program indicates that, "The Center will develop and foster interdisciplinary research that links advanced methods and models in supply chain management with research in data analytics, visualization, information fusion, and engineering research in water/infrastructure, extraction/stimulation, well pads, and chemical/materials engineering" (CCRINGSS, 2014). The effort espouses a joint four-college effort that looks at the interdisciplinary innovations and key ecosystem drivers underlying the supply chain and operation of natural gas exploitation and exploration to setup research foci that enables state-of-the-art innovation and educational impact. The initiative engages many disciplines and stakeholders that spans across a complex supply and demand network of links, core processes, and markets. The collaborative research that is envisioned brings together (1) key drivers such as people, processes, and organizations, natural gas markets, environment, safety and security, technology enablement with (2) systems level insight and decision support and (3) educational/workforce developments. Within the triangulation of this vision are implications such as shale gas risk, costs, economics, industry, performance, compliance, safety, and sustainability. As this center is a new endeavor,[*] there are many moving pieces that must come into clearer focus, dovetail, and produce an

---

[*] Rolled out in 2015.

infrastructure of people, solutions, and technology. One metaphor for considering the broad expanse of the vision and its implementation is to frame the CCRINGSS initiative as being predicated on *collaborative ecology*.

A collaborative ecology represents a large-scale set of enterprise couplings that incorporate many different elements that have transformative and interactive relationships, many of which are interdependent and changing. An ecological system is heavily integrated with the environment, and what the environment provides for sustainable growth, given a set of intentional purposes. One way to look at this is through the concept of an *affordance*. An affordance in a system represents the possibility for action on an object or environment, whereas an *effectivity* represents whether an agent has the capability to effect the action demanded by the affordance. In turn, affordances and effectivities are always defined in terms of each other and often specified through data-information links. This kind of language is directly related to ecological psychology (Gibson, 1979), and represents a new way to frame the landscape for this particular domain.

Inherently, the *ecological system* comes about to refer to the living nature of things, and how elements depend on each other for survival (survival of the fittest model). In this case, information–people–technologies jointly undergo evolutionary change to produce the intentionality the system is designed for. An ecological system designed for natural gas supply chains is incredibly connected to people–data–information and requisite changes in a number of different environments (transportation, economics, geology, hydrology to name a few). As *collaborative technology* emerges with equilibrations in elements and environments—affordances and effectivities can be established to insure intentions are viable, sustainable, reasonable, and ethical. With new research being put forth to bring about an ecological system with these values and qualities, it is important to establish where research must go, to accomplish new innovation within natural gas supply systems. Part of this chapter underlines that point and generates a research agenda that can be addressed and implemented. However, our focus is squarely within the ideas of human-centered design—as related to collaborative technologies—that result in an ecological natural gas supply system.

An important component of the CCRINGSS work that is tightly coupled with collaborative technologies for natural gas supply chains is an awareness that collaboration is intrinsic to concepts and processes of ecological-based living systems. Collaboration is what creates an emergent synthesis of people, machines, information, technology, and environments—that works. Determining *what works* is bound to whether the collective intentions of the actors in their environments are generated with measures of success (does the supply chain result in advantages in the outputs of the natural gas exploitation/exploration in terms of the values that are relevant for the industry). It is the goal of this chapter to outline research areas, and an agenda to be taken up that if accomplished would certainly result in improvement with these measures of success.

Collaboration is one of the main research topics of the MINDS Group within the College of Information Sciences and Technology (IST) wherein the study of teamwork, technologies, and information is undertaken with an emphasis to *understand distributed—team cognition* (e.g., team-situation awareness, team-mental models collaborative-information sharing, cultural-cognitive impacts), *human-centered design* (e.g., geographic information systems), and *decision augmentation* (e.g., team–agent collaboration, team-decision aids, and collaborative technologies). Many of these same areas are ripe for advancement within the CCRINGSS initiative as well. Therein, much of our collective (the MINDS Group) research efforts in CCRINGSS are centered on various aspects of these research niches, which will be outlined more in the living laboratory approach section of the chapter.

When we consider the collaborative work perspective within natural gas supply chains it is necessarily informational and social in context. Distributed cognition theory (Holland, Hutchins, & Kirsh, 2000) implies that information is distributed across both time (temporality constraints) and place (spatiality constraints), data-information-measures can be aggregated in layers, and be fused together from multiple sources (information fusion), and therein subject to information analytic processes to derive more data-information-measures (therein information is in-formation). Information flows are meaningful to determine bandwidth (targets-sources), propagation strength, distance, and time, information is very much socially distributed across digital worlds (social networks/social media, for example, chatrooms and mobile devices, for example, cell phones and visual devices, for example, cameras at lights, toll booths, businesses, ATMs, and elsewhere) that are interconnected with people everywhere to form new ways of sharing, seeking, retrieving, and saving information. We are now highly reliant on information apps (applications on mobile devices, phones, tablets, and kiosks) to perform many functions, and Wikipedia and YouTube provide new sources of information search. Open architecture computing allows online communities to formulate and solve all kinds of problems, simulate devices and technologies, and develop technology through remote collaboration. This is a new world in which information *forges* ahead in innovative ways, where sharing information collaboratively is completed through a variety of modalities, and experiencing reality can take place in alternative, remote ways independent of physically *being there*. Seeing, sensing, experiencing, and controlling things from afar produces an even more robust ecological system that can be appropriated for supply chains that involve natural gas production.

On account of the broad bandwidth of the CCRINGSS initiative, our goal is to focus-in on the areas that are most relevant to work that is resident in the College of Information Sciences and Technology, so that is one way we have set boundary constraints (as mentioned above, understanding distributed team cognition, human-centered design, and decision augmentation) for this chapter.

PREMISE

As we take a human-centered approach, the objectives we have in mind will be highly coupled with (1) information and cognitive sciences and (2) cognitive systems engineering as an applied framework. This represents a new perspective on problem solving that does not get the cart before the horse, as so often has been the case with technology-only or business-specific approaches in this area. The goal is to take a human-centered approach first, such that technologies and business practice fold into understanding the needs and requirements of human factors and cognitive systems, not vice versa as often has been the case in the past.

The MINDS Group in the College of IST at The Pennsylvania State University has researched a number of real-world domains that require the intricate integration of information, people, technology, and context wherein evolutionary patterns are generated from various source points, and where uncertainty is the norm not the exception. Examples of these kind of very challenging domains are emergency crisis management and response, fighter pilot operations and control, environmental protection planning and monitoring, police cognition, emergency medical operations, infectious disease control and operations, intelligent and image analyst work, automobile driver–vehicle dynamics, battle management command, control, communications, and intelligence C^3I, disaster management and response, and cybersecurity; are all domains in which we have applied a cognitive systems engineering approach (i.e., the *Living Laboratory Framework*, McNeese, 1996)—and as evidenced by the chapters resident in this volume—to increase understanding and develop innovative systems designs to increase performance and overcome barriers. Inherent in these complex domains are many different kind of challenges, problems, and issues that have to comprehended and resolved to insure safe and efficient operations. Many of the problems may have similar natures or can come about owing to similar causality (e.g., information overload can cause attention deficits, impoverished visualization can produce inept problem comprehension, cognitive bias can induce conflict among team members, cultural differences can influence decision making, to name a few). On the other hand, many domains have problems, issues, and constraints that come about because of the unusualness of the domain itself. An example might be uninhabited air vehicles where remotely distributed operators engage in interdisciplinary yet interdependent roles to fly a drone as well as integrate sensors, and control air weapons all simultaneously. This domain has its own set of unique problems that require a great amount of attention to address and overcome. Therein, the domain of the supply chain underlying natural gas exploration/exploitation is of similar stance. It too has a bevy of challenges, problems, constraints, and issues that have to be addressed in order to propose a human-centric approach to problem solving, and to develop innovative state-of-the-art designs that make a substantial difference in work.

The objective of this chapter is to expand on the challenges, problems, constraints, issues, and to generate research trajectories that jointly produce a research agenda, if carried out would make life better for managers, engineers, scientists, and workers all across the supply chain. We will begin with a top-down perspective of this work domain.

Viewing the supply chain in natural gas exploration/exploitation first from a top-down lens, we do see that it is similar to a number of other domains mentioned as it is *human-centered* (i.e., the domain is heavily ensconced with human perception, cognition, and decision making), is *work-focused* (activities result in work that achieves objectives and intentions), has work accomplished by humans operating in the mid of *complexity* (complex systems), has *emergent data streams* and *uncertainty* in various nodes, requires *human-systems integration* that often demands *naturalistic decision making-problem centered learning,* requires *teamwork* (within team and across team—team of teams—communications and coordination), displays *nonlinear dynamics* that can result in chaotic patterns, requires the *expertise of people* to make assessments and judgments, exhibits a *large number of interdependencies* that supply information-data at the right place at the right time thereby enabling *reliability, validity, resiliency,* and *redundancy* where needed, relies on *information, cognitive,* and *social technologies* that often amplify and augment human decision making and social interaction, and utilizes *human–computer interfaces* to facilitate the coupling of the environment with the actor. This collective set of domains may fall under what is termed sociotechnical systems, but each field of practice can have unique characteristics and perturbations. Indeed, our research has spanned across a number of sociotechnical systems wherein interdisciplinary study has progressed. When we look at natural gas supply chains then it is through this coloration that our exploration will precede.

If we view the supply chain operating within the natural gas exploration domain as a sociotechnical system, it is informative to break this down into three layers: (a) *individual operations and analysis,* (b) *teamwork and collaboration,* and (c) *organization and enterprise architecture.* Each layer must interconnect and exchange data, information, and knowledge with other layers in a meaningful, timely manner for success to be present. As we look at a purview of this area and generate a research agenda, these layers must be addressed as a composite whole but yet each layer emphasizes different perspective that is present in work. Also, each of these layers are feeding forward information to other layers in real time therein making each layer highly interdependent on the other layers for work processes to ensue and be complete. In these kinds of complex domains, no longer workers may assume a stance of being independent, isolated, and remote from other workers, and go about their business as if there is no coupling with man or machines. In fact, workers must share information, work on multiple teams, utilize sophisticated technological tools, and make handoffs (shift work). As humans and machines are not perfect—breakdowns, errors, mistakes, and even catastrophic consequences are bound to occur. However, by incorporating a viable, human-centric, and interdisciplinary purview of the work in natural gas supply chains—these consequences can be minimalized and planned for to insure safety, resilience, and learning across the system of systems. As we delve into this new domain for study, we will apply the LLF to understand challenges but also to promote a research agenda to enable a united front to increase performance in a variety of ways. The outcome is a whole new way to approach innovation and design in natural gas supply chains wherein human cognition is fully integrated with machines/computation to address the demands of the environment as they emerge with multiple requirements. We will now provide an overview of our research approach and breakdown requirements within the framework elements and processes.

APPROACH

As the goal of this chapter is to examine natural gas supply chains through the dual lens of information—cognitive sciences and cognitive systems engineering—it is appropriate to apply the *LLF* as our distinct approach. The living laboratory approach (McNeese, 1996; McNeese & Pfaff, 2012) is one kind of perspective to develop and plan a cognitive systems engineering program for a line of research in a specific field of practice (e.g., natural gas). By implementing a program framework like this, the result is a multimethodological approach that harnesses holistic knowledge to improve human performance within designated measures of success. The LLF has been utilized before to study numerous fields of practice that entail complex environments, distributed collaborative activities, emergent data and information flows, and embedded technologies (See McNeese & Forster, 2017 [Chapter 1 of this volume]). It is appropriate to apply it to areas cognate to the CCRINGSS initiative to discern research and development opportunities that are waiting to be fulfilled. Please refer to the figures used in the introductory chapter of this volume that demonstrate the full progression and approach of the LLF. The remainder of this chapter will look at each element with the purpose of providing a research agenda that fits within that element.

GENERAL PROGRESSIONS

As mentioned earlier in the chapter, much of the work in similar fields of practice involves dynamic complex systems that emerge with varying contingencies. Complex system dynamics come about when human(s) act within their environment based on information availabilities. Information may come from within human cognition (internal neurological states) or arise from the context itself (external cognition) and definitely may change over time, space, and people as different contingencies play out (enactments). When information is specified through the context, the role of technology may be very relevant (World Wide Web, mobile phones, and the Internet of things) in accessing, seeking, transforming, visualizing, and storing information to progressively enhance *use*. Utilizing the *LLF* to understand supply chains in natural gas exploitation/exploration can point to (1) individual interaction, (2) team interaction, and (3) enterprise–organizational interaction, as we indicated earlier as it invokes (a) information, (b) environment, and (c) cognition. The framework is setup to be flexible wherein information seeking in research areas can begin wherein availability of information is most prevalent. This may be different for different researchers depending on their situations and interests and practical constraints. We will setup two hypothetical cycles (top-down and bottom-up) through the living lab that emphasize different starting points, and propagate through interrelated progressions of the LLF elements. By no means are these the only pathways through, but for sake of example they will elicit research topics and agendas according to the pathway set forth.

The primary value of the LLF is to base multisource knowledge on mutual learning about problems that are indigenous to the field of practice under study. This ties learning to problem finding/problem solving that is the core value of the LLF. Learning is fueled by obtaining knowledge from each specific element of the framework as is possible—realizing that only in a perfect world would everything be fully available and accessed. As information is not fully available and there are uncertainties and constraints within each element studied, it is not possible to have all the knowledge to bear on problems. However, to the extent multiple sources of knowledge come to bear on understanding—continuous process improvement is still viable and is probably dependent on how much knowledge can be obtained and the timeline of problem solving that is needed. *Problem-based learning* then represents the core foundation of the entire LLF, and is usually where we start thinking about and exploring a field of practice–domain–context.

When we begin explorations for a new research area—the first thing that stands out is finding the salient problems that cause people the most difficulties and prevent optimal performance. Although optimal performance may not ever be possible, it is informative to find problems that reduce performance significantly and to address them in creative ways. Problems can then be defined in greater depth, and explored with various perspectives to get a handle on what issues/constraints are active in complex systems. Therein, the progression through the LLF begins by knowing what the specific problems are in a domain. The researcher becomes aware of problems either through exploring the field of practice itself or through prior literature that is published (literature/document reviews). The cycles we mentioned in the prior paragraphs are developed from these two pathways. So, on the one hand problems point upward to theories, whereas on the other hand they point downward to practice. For this particular chapter the approach will begin with the context of use and work through the elements with this particular starting point.

From the field of practice of supply chains in natural gas systems comes the idea of *use* (note Flach & Dominguez, 1995, indicate that use refers to an integrative term where use is integrated with the environment through mutual consideration of goal–instrument–user, wherein affordances, information, and effectivities are bound together as a holistic model of use-centered design). Use represents the work that occurs in practice for everyday operations, and involves users who have the skills, knowledge, and abilities to complete the work that is given to them through the information that specifies affordances that are present. When affordances or effectivities are blocked, invisible, or underspecified then problems–constrains–issues arise in the work itself. Many of the problems that occur may be due to perceptual issues or action potentials thwarted. As one can clearly see the LLF is tied to ecological psychology foundations (Gibson, 1979), and approaches the design of cognitive systems with a strong contextualistic bent. Therein, we begin the cycle through the LLF elements by beginning with the contextual environment where work may occur. One further note—contextual environment here is meant to include physical, social, and technological layers that surround users intending to do work in the natural gas operations. This may be referred to as the *sense surround* that formulates a complete workspace the user interacts with. In turn, the progression for this cycle begins with Element #1: Ethnography Fieldwork.

MIDWAY BREATHER—STOP AND REFLECT

The natural gas domain is a complex field of practice that is both social and informational, requiring novel approaches to understanding, conceptualizing, and developing sociotechnical systems in support of collaborative team decision making.

The living lab approach is a one kind of perspective to develop and plan a cognitive systems engineering program for a line of research in a specific field of practice (e.g., natural gas).

ELEMENT #1: ETHNOGRAPHIC FIELDWORK

Challenges: What do individuals do at the most basic operational level in natural gas exploitation (what effectivities are employed for accomplishing top intentions)? What problems cause the most pain and greatest lost from a performance/production perspective? What do workers seem to have trouble with as the supply chain emerges over time, space, and people? How can information specify the affordances that are present in the context to enable specific worker effectivities or the need for adaptive situations and sustainability? What work roles are required to accomplish intentions within the environment? How does the environment constrain worker intentions? How does a worker recover from errors? As the environment is coupled to teamwork or organizational processes that layer over individual and teamwork, one challenge will be to determine what the teamwork considerations are for enabling natural gas operations (both collocated and distributed work need considered). On account of information flows from and back to the organizational aspects of supply chain, another challenge relates to how information is acquired and shared through teamwork and organizational components involved with work? Yet another challenge to consider is what temporal contingencies are present across the teamwork and the organization, and how are they orchestrated in the environment; as this is very basic to supply chain production.

RESEARCH OVERVIEW

The role of research at this level is to conduct field studies at the natural gas work site to primarily look at how work processes are accomplished through the interaction of the user within the given environment. In addition, a secondary concern is the number of the roles required to foster accomplishment of work intentions and objectives. Workers often not only do perform one role but are involved in complex *context switching* where they perform multiple roles (each role may have an interconnected series of tasks, functions, and constraints attached to it). The role at this point of the exploration is for the researcher to observe a number of contexts, and workers who are imperative in carrying out major natural gas operations (this may be focused on a specific context of the supply chain operation owing to practical considerations). Therein, this element involves ethnographic and qualitative study of *use* as it emerges in the context of work. Ethnographic work consists of observing people's actions and accounts in everyday context, as opposed to researcher created contexts. "In other words, research takes place 'in the field'" (Hammersley & Atkinson, 2007, p. 4).

Data collected in an ethnography is multifaceted and includes interviews, document analysis, observations, and informal conversations (Hammersley & Atkinson, 2007). At the heart of ethnography methods is the desire to first and foremost engage in *observation learning*. Being able to make notes while watching the emerging context is very important to gain initial understanding and insights as to how the work is carried out given the constraints. This will also allow for the researcher to observe if there are mistakes, or errors, and what the unintended consequences are for individual, team, and organizational processes. The observation should also account for information flow during operations to derive how workers obtain information and then process/use it to gain advantages. Part of this also involves their use of tools, and how they communicate and coordinate their own work efforts, and how the culture of work develops for a given specific context. This will also mean that ergonomics will be salient as workers need to use tools/instruments to facilitate natural gas work. Having an eye for ergonomics can help identify weak points in human interaction with their environment that might result in better performance, safety, and efficiencies.

As this involves supply chain coordination it will be necessary to see how materials are present in the context or how they are acquired through logistics and just in time deliveries (e.g., the water truck has to arrive in time to provide water into the fracking process) and how the work roles are orchestrated and meld together into collaborative work to form the ecology. Frequently *team of teams* interaction underlines the necessity of enterprise and organizational considerations that impart strategies and tactics of how to operate in changing markets. At this level, safety, policy, sound engineering practice, and ethical treatments of the environment are all manifest.

Specific Research Agenda Items

- The fracking process itself (more locally bound natural gas extraction)—ergonomic considerations for individuals, coordination within and across fracking teams.
    - The role of affordances–effectivities–information when movement is of high consideration.
    - Understanding the culture of work related to fracking.
- The supply chain environment—organizational and enterprise factors—how the supply chain emerges within the various sectors.
    - Temporal processes (time limits, coordination, sequences, etc.).
    - Information sharing and flow within the environment (breakdowns, needs, and successes).
    - Contextual and situation awareness can be observed from a common operational picture perspective during this phase of activity.
- Collaborative systems/technologies present or absent in current environments (to what extent can it be amplified or forged differently).
- During this phase—a functional analysis can begin to be generated (and further developed and reinforced with other ongoing LLF elements) that begins to define affordances–functions–effectivities in related representational forms.
    - The methods used to begin this analysis should either be the Abstract Hierarchy (means/ends analysis), Rasmussen (1986), or via the IDEF2 functional decomposition method (see Zaff, McNeese, & Snyder, 1993).

Opportunities to Move Research Forward (via CCRINGSS or Otherwise)

- Study of specific environments with the Institute of Natural Gas Research (INGAR) at The Pennsylvania State University and/or GE in that they may have access to the aforementioned activities for specific locales (e.g., sites in PA).

- Observe the collaboration, technology, tasks, and contexts present within the natural gas supply chain domain.
- Review of security and cybersecurity aspects of the environment as it impacts human factors/cognitive system elements.

ELEMENT #2: KNOWLEDGE ELICITATION

RESEARCH OVERVIEW

Although the ethnographic fieldwork element provides an understanding of the environment where natural gas operations emerge and coalesce together, how *use* is enacted in that environment, and how this emerges over time while people are at a work setting; the knowledge elicitation element of the LLF represents a key lynchpin in the entire process. This is because the perspective switches from the researcher observing the actual work environment of natural gas to having the workers/users explain how they do their work, how they come to understand their work, how they collaborate with others, and how their work culture comes to pass. Knowledge elicitation is a lynchpin because it is often the place where verification of other elements occurs as a user provides their own perspective of use, the environment, the problems, the theory, and their ideas about what design should be to serve their intentionality and needs.

At this point, the onus is on acquiring or eliciting the knowledge that underlies problems, processes, and interaction whereby the user is the one that provides the basis of their experiences. This may occur during the time the researcher is at the work site (when they are not observing but actually switching to an active elicitation mode) or it could occur outside or beyond the work site (e.g., workers/operators are interviewed not during work but at an alternative site). There are many individual differences that influence how a person does their job, and within their own cognitive state they may perform a job with certain biases, may interpret the context they are in in a way that varies from others, and may be inclined to commit different errors based on their knowledge, skills, and abilities that they bring to the job. Therein, individuals can be probed about affordances–effectivities–information to help situate and extend what was collected from the actual environment of use. Many of the challenge questions put forth for ethnography can be enhanced and validated by communicating with the user. When teams formulate—individual cognition—has to be combined in ways that are productive, mutually compatible, and that accomplishes the intentions set forth. By eliciting knowledge from individuals and teams a better, more comprehensive view of problems–processes–constraints–issues–tradeoffs within natural gas supply chains is possible. Furthermore, the functional analysis can be refined, extended, and validated.

Knowledge elicitation is typically done in several ways: interviewing novices/experts, surveys and questionnaires, cognitive task/work analysis (see Hoffman & Militello, 2009; Vicente, 1999), concept mapping (McNeese & Ayoub, 2011). Our own work (AKADAM, McNeese, Zaff, Citera, Brown, & Whitaker, 1995) integrated concept mapping and functional analysis (IDEF2) with design storyboarding to infuse a *knowledge as design* perspective for utilizing expertise (see Perkins, 1986) within complex systems and cognitive systems methods (e.g., to define user awareness and cognitive tasks within the fighter pilot realm of expertise). These techniques may be appropriated at the individual or team level to gain further insights within the natural gas domain as well. Disagreements and conflicts in results can be treated in several ways (union or intersection of knowledge, frequently occurring themes or problems, and common ground). Users may not see eye-to-eye owing to their unique experiences, cases, and competence levels. In many cases, knowledge elicitation can expand into *Element #5: Design Prototyping of Technology* wherein designs are elicited directly from the user or functional means—ends are translated into specific design elements. We have done this with the AKADAM methodology by translating concept maps into interactive design storyboards centered on specific scenarios indicative of the contextual environment of the mission. Another way by which design has tied directly into knowledge acquisition and environmental analysis is through ecological interface design (Vicente, 1999).

There are many methods that may be utilized during the knowledge-elicitation phase to drill down deep into user knowledge resulting in detailed information that is pertinent to the entire LLF. Knowledge from the user can validate other LLF elements as well and inform content and processes necessary to invigorate a given element. As we reveal the application of LLF elements 3, 4, and 5 this will become more evident. The user becomes a central spectrum from which much light may flow to expand usefulness of the other elements.

Specific Challenges/Research Agenda Items

It is necessary to leverage and move forward items identified within the ethnographic element into the knowledge elicitation for mutual learning and enhancement of understanding. The following items are important to consider:

- Where do errors, slips, mistakes, and misunderstandings tend to occur within various states and processes of the actual natural gas operations for
  - Individual level?
  - Team level?
  - Organization–enterprise level?
- What impact does workload (high, medium, and low) have on worker outcomes for given situations?
- Is awareness (contextual, activity, and situation) present in individual work and teamwork? When does it tend to dissipate?
  - Are there occupational safety or hazardous work encounters that can cause serious harm?
- Does stress influence cognitive processes that in turn impact performance? How?
- How does automation play into the tasks and situations that need to be addressed on a daily basis and how could it play out in the future?
- Is fatigue a condition that impinges on work performance?
- How does the organization and enterprise facilitate optimum performance across the supply chain?
  - How does time pressure, material flow, and policy limit or facilitate successful performance?

- What other job position or roles are most interdependent in operations?
- How does leadership models fit with actual supply chain production and distributed work outcomes?
  - Are incentives a big issue in work satisfaction and performance?
- How salient is cybersecurity within the work setting? Has the enterprise been hacked or has information exploitation ever ensued in any known operations and business transactions?

Opportunities to Move Research Forward

- Convene with experts from GE/The Pennsylvania State University who actually are working in Natural Gas operations in Pennsylvania. This could be coupled to ethnographic operations but could also be considered independent:
  - Survey commissioned
  - Interview novices and experts
  - Interview trainers
  - Cognitive task/work analysis conducted at individual and team levels
    - Application of the AKADAM techniques
    - Use of the decision ladder (Rasmussen, 1986) activated as a cognitive model

ELEMENT #3: THEORETICAL FOUNDATIONS

Research Overview

As we move around the hub of the LLF, the next logical path sequence is considering as theoretical perspectives that are cogent in representing concepts, processes, and problems preeminent in NG supply chains. This path represents a more top-down path through the LLF. As we mentioned in the previous section, knowledge elicitation can be one basis (as a qualitative, bottom-up process) to reveal problems and expert's articulation as to why a problem is occurring. This directly moves problems as cues for certain theories and hypotheses as to *What is happening?* As we have taken a living ecological worldview of complexity, necessarily the concepts from ecological psychology, situated cognition, and distributed collaborative work are invoked. We have classified these research areas within Distributed Cognition Theory (Holland et al., 2001), which is highly contextualistic.

As members of the College of IST, the notion of *distributed information* (in time, place, and form) permeates the theoretical turbine that powers thinking about how NG supply chains play out in the sociotechnical systems fabric they exist in. It is our vision that this area will be very coupled and connected to the data–information–knowledge enveloped and that makes the system percolate to new levels. Distributed information necessarily in supply chains points to distributed teamwork as a means to make actions happen. Therein, cognition that exists both internally and externally is necessary to gain the advantage and power to produce success. Another level closely related is *information fusion* wherein layers of information fold into and complement each other in unique ways to provide the common operation picture of the supply chain orchestration. Information also fits into the temporal facet of context, as it can arrive at different times, can be interdependent, can emerge into the future therein showing trends, or be distributed across past experiences.

We have used different threads within the distributed cognition purview to make sense out of what experts and ethnographic analyses can etch out. So threads of theoretical positions taken up within emergency crisis management, emergency medicine, cybersecurity awareness, and UAVs—as applied—may be useful for the natural gas exploitation and production as well. Some of these specific theoretical concepts and, in turn, independent variables that could play out in the current domain are transactive memory, hidden knowledge, time pressure, information sharing and retrieval, learning and knowledge acquisition, and transfer, trust, and automation.

The theoretical foundations underlying natural gas supply chains should be highly reflective of the environment and information that leads to specific work outcomes (albeit in abstract forms). When experimental studies are conducted to test theory then the goal is to provide premium ecological validity within the experiment. As we mentioned earlier problems tie theory together with practice. Therein, the *problems experienced* set the boundary constraints on what theoretical positions are most relevant and related.

Just as theories about how things work and are accomplished can come directly from the user (from knowledge elicitation element) it may derive from the standard scientific method as well. That is, document analysis and literature review also provide a rich bandwidth of knowledge about a specific research area (in this case natural gas supply chains or related near neighbors). A document analysis is specified in other methods of cognitive systems engineering (Rasmussen, 1986; Vicente, 1999), but literature review remains the standard bearer for the scientific discovery process as it presents sound results from other researchers that inform problems associated with the theory or hypotheses one has in mind to study. Theories are tested and evaluated through the use of experiments (qualitative, quantitative, and design) in unique but sound means. In this way this element has natural vectors into the scaled world and the design prototype LLF elements.

Specific Challenges/Research Agenda Items

- What literature reviews and document analyses are relevant to conduct for natural gas supply chains?
- Given the prior challenge, what are the major concepts and independent/ dependent variables that are pertinent within the research literature that specifically relate to or refine knowledge obtained from the previous LLF elements?
- How does distributed information impact communications, coordination, collaboration, control, and intelligence within natural gas supply chains?
- How can teamwork and organization processes be improved in innovative ways?
- How does the degree of interdependencies across human team members produce or impact trust, ethics, conflict resolution, and loyalties?

Opportunities to Move Research Forward

- Utilize PhD dissertation research to study distributed cognition theory as related to natural gas supply chains (across all the colleges involved in CCRINGSS initiative).
- Conduct studies with real-world experts (and compare to novice participants) in how they process distributed information in emerging states of uncertainty.
- Utilize novices (students) to define how they acquire individual and team skills, knowledge and then transfer them to a new similar target problem (analogical problem solving)—this studies how novices become experts in given facets of the supply chain functions and processes necessary for resolving specific problem states.

At this point there is a natural transition from theoretical foundation into LLF element, the scaled-world simulation.

ELEMENT #4: SCALED-WORLD SIMULATION

RESEARCH OVERVIEW

Simulation represents another venue where much integration of data–information– knowledge can be reified, tested, and validated. Scaled worlds are simulations that take primary problems that exist within an environment and bring them more into a lab setting where control and focus can be implemented. The scaled world, in turn, represents an important element of an integrated LLF as it brings dynamics from the real world within the confines of a simulation where in-depth understanding *and convergence* of concepts can be adapted, adjusted, and refined even more so. A scaled world is much more than a simple lab test as employed in many psychological experiments. The context experienced by user(s) should be present to a threshold level of fidelity in the scaled world. It may not have the full sense surround intact but it should simulate a good portion of it to reflect actual practice. Therein, researchers will need to strategically look across the entire spectrum of how supply chains contribute to natural gas exploitation and exploration, and determine where *use* is (1) most important for the success in performance, safety and well being, and priority outcomes; (2) is troubled with an array of problems that create bottlenecks, errors, lack of situation awareness, overload, and other issues that demand human-systems integration; and (3) studied in a way to represent a variety of human factors present within the actual environment. Once target areas for simulation are established then the technological infrastructure, control-display surfaces, the data set/database utilized, and software architectures to implement the scaled world can be designed and created.

One way to think about scaled worlds is how they come into existence. This is where the kinetic cycling through different LLF elements results in productive outputs of knowledge. Practice should be evident as individuals and teams perform in the natural; gas supply chain scaled world. As we have seen—practice and use are collected from observations with ethnographic inquiry and knowledge elicitation.

So first and foremost—the scaled world emerges out of the findings obtained from these user-centered elements. Obviously, what novices and experts reveal about problems can come to reality in the scaled world and be presented, attuned, and adapted as needed. It is instructive to think of a scaled world as being able to produce the most difficult problems, situations, constraints, and issues within a controllable simulation to provide in-depth study of it under various conditions (e.g., the use of independent–control–dependent variables).

Second, scaled-world simulations typical commence via scenarios that represent sequences of decisions and actions that are possible when specific events and/or complex situations arise from one state to another. Decisions and actions are coupled to specific problem situations that emerge in time as the simulation plays out in depth and fullness. As the LLF is necessarily problem-centered, one could think of scaled-world simulation as providing ecologically valid situations that contain problem states and a potential number of solution states that come into place with correct elements of decision-action sequences. Simulations may also contain information-seeking elements where participants look-up information to further their awareness and action. In combination with the information architecture underlying the simulation, scenarios offer up the production of built-in affordances that are coupled with effectivities provided by both human team members and the requisite technology that is operable under any given condition. Affordances, decisions, and action are specified by information, which is requisite either within the interface provided or resident within data files that are activated from the information architecture under certain facile conditions. As there are many potential problems—issues that face natural gas exploitation and supply chains—the prioritization of the highest payoffs will need to be ascertained before a significant investment in simulation design is manifest. This creates prioritized plans for simulation that correspond to the challenges, and opportunities to move forward that have been accruing through each of the LLF elements. This last facet points to the last element of the LLF, the technology prototypes that we will get too shortly.

Third, scaled worlds become the vehicles on which specific theoretical hypotheses are tested. This is referred to usually as *human-in-the-loop testing*, and provides a robust engagement and enriched experience for participants as the simulation, scenarios, and conditions applied are *living* in that they are highly emulative of real-world problems-issues. Hence, the testable conditions under which a hypothesis is assessed and evaluated are deemed to have a higher ecological validity owing to the feed-forward of other LLF knowledge that has been collected/validated. Another aspect of validity is that novices and experts can be put in the simulator and corrections made accordingly. This improves the overall veridicality and realistic aspects of the experience. Also, placing novices or experts in the simulation and having them perform cognitive walkthroughs (Dix, Finlay, Abowd, & Beale, 2004) is an excellent extended form of knowledge elicitation that pays dividends.

Specific Challenges/Research Agenda Items
- Creating a scaled world with the appropriate fidelity to simulate the context, tasks, and decision-making processes of the real-world.
- Coupled with the first challenge, ensuring that the simulation is simplistic enough to be useful in experimentation but complex enough to not completely water down the inherent complexity within the natural gas domain.

- How do you identify the best context within the entire scope of natural gas supply chains to study, which can provide rich and salient data?
- Determining functions, tasks, and situations that are most salient from ethnography and knowledge elicitation that are most difficult for humans/ teams to perform, and setting them up appropriately within the information architecture present in the simulation.
- Ascertaining the relevant degree of interdependencies in natural gas supply chain teamwork/team performance scenarios that make sense (and are possible) to simulate within the scaled world.
- Developing a simulation for the participants that has the right mix of necessary systems, equipment, tools, and interfaces, that are present in the natural gas field of practice.

Opportunities to Move Research Forward
- Utilize PhD dissertation research to study routine and nonroutine activity and how it impacts responses to risk, uncertainty, and decision bias (for both novices and experts).
- Research the involvement of individuals and teams as a function of the depth of processing/depth of work involvement required for operations (i.e., simple versus involved versus complex conditions of work).
- Determine the extent of *team of teams* interaction and how the flow of handoffs incur.
- Related to last opportunity—study how the handoffs team make impact the effectiveness and efficiency of work activity, decision making, and performance.
- Research cooperative learning behaviors that arise during the collaborative decision-making processes within the natural gas supply chain domain (e.g., among control room operators).
- Discover the work habits of experts in complex domains of practice looking at how technology supports and/or hinders decision making in various impact situations (situations of information overload, time stress, hidden knowledge etc.).

ELEMENT #5: DESIGN PROTOTYPING AND INNOVATIONS

Research Overview

One powerful characteristic of the LLF is the application of output from one step in the process to serve as the input for one or more of the other processes. In the case of knowledge elicitation and ethnography, researchers can not only gain a deeper understanding of the collaborative naturalistic decision-making processes of workers in complex fields of practice, but can also recognize potential sociocognitive technologies that could be designed (envisioned) and developed to support these activities. Moreover, the scaled-world environment then presents a rich testbed for application and evaluation of solutions in a controlled context.

One of the primary aspects of the LLF is to approach knowledge as design (Perkins, 1986). The LLF merges knowledge from multiple elements to understand user experience and researcher experience to produce new ideas in the forms of prototypes that can be reconfigured as new knowledge is obtained. Prototypes are *envisioned designs* (Woods, 1998) that provide solutions to problems defined and explored within other elements of the LLF. As envisioned designs they often contain the knowledge fed forward from the other elements of the LLF. Subsequently, as LLF is a cyclic framework, once proposed solutions and innovations are tested, they can be moved from the scaled-world environment to that of the practice of the real world. This approach affords solutions to (1) be informed through ethnographic observation of teams (i.e., group-centric), (2) based upon individual knowledge of the domain and tasks through knowledge elicitation (i.e., individual-centric), whereby (3) scaled-world environments can evaluate and inform innovative solutions specific to the context (i.e., technology-centric). For example, a portion of our current research is interested in evaluating the effectiveness of intelligent agents in the knowledge acquisition and transfer process of workers within the natural gas supply chain. Using the scaled-world environment as a foundation, the prototype of an intelligent agent (or other sociocognitive technologies) can be designed and reconfigured as data is gathered and evaluated. This has the potential to inform the training programs in natural gas contexts with hopes of decreasing the amount of time required for newly hired or junior employees to become autonomous (i.e., the time that it takes for a novice to become an expert).

This same element of LLF, can be employed a multiplicity of times as research interests or prior evaluations dictate. Being a flexible LLF, solutions can be informed, designed, prototyped, evaluated, reconfigured, and then evaluated again cyclically. Much of the reconfiguration can be supported within the scaled-world infrastructure such that the underlying architecture remains stable while innovations are embedded within the simulation as necessary. Effectively, researchers can use informed simulations within the scaled world to study a plethora of problems addressed by potential solutions within varying fields of practice. For example, the NeoCITIES architecture (see McNeese & McNeese, 2016) has been useful for studying not only crisis management activities within a college campus environment, but has also served to evaluate cognitive and decision-making processes (e.g., hidden knowledge, cultural influences, time stress) within this same context, see McNeese, McNeese, Endsley, Reep, and Forster (2016). In addition, the underlying NeoCITIES architecture has also been adapted to understand the decision-making process of cyber analysts performing their daily tasks, which are typically ill-defined and difficult to explicitly learn and train (Tyworth, Giacobe, Mancuso, McNeese, & Hall, 2013).

Specific Challenges/Research Agenda Items

- Using all the knowledge gained within natural gas supply chains to discern what areas make sense for developing sociocognitive technologies and decision aids, and then thinking through what the necessary information architecture and requisite data/database is for a given prototype.
- Appropriating various cognitive decision aides and embedding them within a preexisting scaled-world environment can be difficult and time consuming.

- Figuring out the extent of function allocation between human and intelligent agents can be tricky and must evolve from other elements of the LLF.
- Determining when research *saturation* has occurred can be problematic given that LLF is cyclical and could potentially go on ad infinitum (i.e., at what point does the cycle end?).

Opportunities to Move Research Forward

- Evaluate the effectiveness of intelligent agents and other cognitive decision aiding techniques on the decision-making processes of novices versus those of experts.
- Compare various cognitive decision aides on knowledge acquisition and transfer among experts.
    - For instance, do these aides support or inhibit experts given their already preexisting knowledge of a given context?
    - Or, to what extent do these aides support or inhibit learning for novices?
    - In addition, had given these aides, which group (i.e., expert or novice) benefits the most?
- Testing human–autonomous interaction within the realm of distributed and collocated teams is an area that should provide meaningful results for the natural gas supply chain.

CONCLUSION

This chapter has reviewed elements of the LLF as they apply to the natural gas exploration and exploitation field of practice. Each element of the LLF has outlined various points of overview, requisite challenges, and potential research opportunities that are extant. In turn, an integrated overview and research agenda have been established for a new area. Natural gas exploration is becoming very prominent and should be approached from a human-centered design perspective. This chapter outlines how this perspective could become a reality based on the principles of knowledge as design (Perkins, 1986) and the LLF, which is highlighted in different ways throughout this volume.

REVIEW QUESTIONS

1. Define and describe the relationship of affordances and effectivities.
2. Describe the elements within the natural gas domain that make it similar to other socialtechnical system domain.
3. Briefly define the 6 elements of the LLF.
4. Understand how, when, and why to use the LLF.

REFERENCES

Akhondi, M., Talevski, A., Carlsen, S., & Petersen, S. (2010). Applications of wireless sensor networks in the oil, gas and resources industries. In *2010 24th IEEE international conference on advanced information networking and applications* (pp. 941–948). IEEE. doi:10.1109/AINA.2010.18

De Graaff, N. (2011). A global energy network? The expansion and integration of non-triad national oil companies. *Global Networks, 11*(2), 262–283. doi:10.1111/j. 1471-0374.2011.0030.x

Dix, A., Finlay, J., Abowd, G. D., & Beale, R. (2004). *Human-computer interaction* (3rd ed.). Harlow: Pearson Education Limited.

Flach, J. M., & Dominguez, C. (1995). Use-centered design: Integrating the user, instrument, and goal. *Ergonomics in Design, 3*(3), 19–24.

Gibson, J. J. (1979). *The ecological approach to visual perception*. Boston, MA: Houghton Mifflin.

Hammersley, M., & Atkinson, P. (2007). *Ethnography: Principles in practice* (3rd ed.). New York, NY: Taylor & Francis Group.

Hoffman, R. R., & Militello, L. G. (2008). *Perspectives on cognitive task analysis: Historical origins and modern communities of practice.* New York, NY: Taylor & Francis Group.

Holland, J., Hutchins, E., & Kirsh, D. (2000). Distributed cognition: Toward a new foundation for human-computer interaction research. *ACM Transactions on Computer-Human Interaction, 7*(2), 174–196.

Mathiesen, B. V., Lund, H., & Karlsson, K. (2011). 100% Renewable energy systems, climate mitigation and economic growth. *Applied Energy, 88*(2), 488–501. doi:10.1016/j. apenergy.2010.03.001

McNeese, M. D. (1996). An ecological perspective applied to multi-operator systems. In O. Brown & H. L. Hendrick (Eds.), *Human factors in organizational design and management - VI* (pp. 365–370). Amsterdam: Elsevier.

McNeese, M. D., & Ayoub, P. J. (2011). Concept mapping in the design and analysis of cognitive systems: A historical review. In B. M. Moon, R. R. Hoffman, J. D. Novak, & A. J. Cañas (Eds.), *Applied concept mapping: Capturing, analyzing and organizing knowledge* (pp. 47–66). Boca Raton, FL: CRC Press.

McNeese, M. D., & Forster, P. K. (2017). *Cognitive systems engineering: An integrative living laboratory framework.* Boca Raton, FL: CRC Press.

McNeese, M. D., & McNeese N. J. (2016, September, 19–23). *Intelligent teamwork: A history, framework, and lessons learned.* Proceedings of the 60th Annual Meeting of the Human Factors and Ergonomics Society, Washington, DC.

McNeese, M., McNeese, N. J., Endsley, T., Reep, J., & Forster, P. (2016). Simulating team cognition in complex systems: Practical considerations for researchers. Proceedings of the 7th international conference on applied human factors and ergonomics (AHFE 2016) and the affiliated conferences, Orlando, FL. In K. S. Hale & K. M. Stanney (Eds.), *Advances in neuroergonomics and cognitive engineering* (pp. 255–267). Springer International Publishing.

McNeese, M. & Pfaff, M. (2012). Looking at macrocognition through an interdisciplinary, emergent research nexus. In E. Salas, S. Fiore, & M. Letsky (Eds.), *Theories of team cognition: Cross disciplinary perspectives* (pp. 345–371). New York, NY: Taylor and Francis Group, LLC.

McNeese, M. D., Zaff, B. S., Citera, M., Brown, C. E., & Whitaker, R. (1995). AKADAM: Eliciting user knowledge to support participatory ergonomics. *The International Journal of Industrial Ergonomics, 15*(5), 345–363.

Perkins, D. N. (1986). *Knowledge as design.* Hillsdale, NJ: Erlbaum.

Rasmussen, J. (1986). *Information processing and human-machine interaction: An approach to cognitive engineering.* New York, NY: North Holland Publishers.

The Center for Collaborative Research on Intelligent Natural Gas Supply Chains (CCRINGSS): A White Paper. (2014). [January 24, 2014]. CCRINGSS. The Pennsylvania State University. University Park, PA.

Tyworth, M., Giacobe, N., Mancuso, V., McNeese, M., & Hall, D. (2013). A human-in-the-loop approach to understanding situation awareness in cyber defense analysis. *EAI Endorsed Transactions on Security and Safety, 13*(2), 1–10.

Vicente, K. J. (1999). *Cognitive work analysis.* Mahwah, NJ: Lawrence Erlbaum Associates.

Weber, J. G. (2012). The effects of a natural gas boom on employment and income in Colorado, Texas, and Wyoming. *Energy Economics, 34*(5), 1580–1588. doi:10.1016/j.eneco.2011.11.013

Wei, M., Patadia, S., & Kammen, D. M. (2010). Putting renewables and energy efficiency to work: How many jobs can the clean energy industry generate in the US? *Energy Policy, 38*(2), 919–931. doi:10.1016/j.enpol.2009.10.044

Woods, D. D. (1998). Commentary: Designs are hypotheses about how artifacts shape cognition and collaboration. *Ergonomics, 41*(2), 168–173.

Zaff, B. S., McNeese, M. D., & Snyder, D. E. (1993). Capturing multiple perspectives: A user-centered approach to knowledge acquisition. *Knowledge Acquisition, 5*(1), 79–116.

Section III

Theoretical Perspectives in Cognitive–Collaborative Systems

6 Methodological Techniques and Approaches to Developing Empirical Insights of Cognition during Collaborative Information Seeking

Nathan J. McNeese, Mustafa Demir, and Madhu C. Reddy

CONTENTS

The Need for Understanding Cognition during Collaborative Information
Seeking.. 107
What Is Collaborative Information Seeking? A Review of the Literature............. 108
    Conceptualizations and Definitions... 108
    How Has CIS Been Studied? Empirical Results of CIS 109
    Social and Technical Aspects of CIS: Cognition Is Missing........................... 111
Cognitive Systems Engineering Perspectives and Methods for Studying the
Cognitive Aspects of Collaborative Information Seeking.................................... 112
    Knowledge Elicitation... 113
        How Knowledge Elicitation Methods Can Be Applied to CIS 114
        Observations and Interviews 114
        Process Tracing 115
        Conceptual Techniques.. 115
        A Case Study of Knowledge Elicitation and Collaborative Information
        Seeking 116
    Cognitive Task Analysis Approach 117
        How Can a Cognitive Task Analysis Approach Be Applied to
        Collaborative Information Seeking?... 119
        A Case Study of the Cognitive Task Analysis Approach and
        Collaborative Information Seeking 120

The Living Lab Approach 121
    How Can the Living Lab Approach Be Applied to Collaborative
    Information Seeking? 122
    A Case Study of the Living Lab Approach and Collaborative
    Information Seeking 123
Conclusion 124
Review Questions 125
Acknowledgments 125
References 125

ADVANCED ORGANIZER

The chapter is organized as follows. First, we will review collaborative information seeking (CIS) literature with the purpose of explaining how CIS is conceptually viewed and studied, highlighting the lack of cognitive research. Next, we will present a set of cognitively oriented methodologies that can be used in CIS research. We present a metafamily of cognitive-elicitation techniques known as knowledge elicitation (KE). KE is a prominent set of methods and techniques for understanding both individual and team-level cognition. In the KE section, we will provide a high-level overview of the methodologies that fall under KE. We will then provide subsections outlining how to specifically use KE methods in CIS, concluding with the presentation of a case study. Next, we turn our attention to cognitive task analysis (CTA). CTA is a goal-specific approach (often employing KE methods) to understand cognition during the performance of a task. Similar to the KE section, we will provide an overview of CTA, present a section on how to apply CTA to CIS, and conclude with a case study. Finally, we present the living laboratory (LL) approach. The LL is a holistic cognitively based research approach that attempts to understand cognition in context, further understand it in the lab, and then develop technologies/recommendations to improve performance in the real-world context. Sections relating to how the LL can be applied to CIS, and a case study will be presented.

The need for collaboration in different contexts and settings define modern day work. In response to the growing utilization of teams, collaboration and methods to support collaborative efforts are becoming increasingly important. To better understand teamwork, and more specifically, team decision making, it is important to understand the role of collaboration during this process. A specific aspect of collaboration that we view as fundamentally important to team decision making is *CIS*. CIS has been studied through two streams of research: *social* and *technical* (Karunakaran et al. 2010). However, in both streams of CIS research, the role of cognition has been understudied. If the CIS research community is to fully understand and conceptualize CIS, it is important to take a broader perspective that accounts for the social, technical, and *cognitive* streams. Consequently, the purpose of this chapter is to provide researchers with multiple cognitively based research methods and approaches that can help them study cognition during CIS activities.

THE NEED FOR UNDERSTANDING COGNITION DURING COLLABORATIVE INFORMATION SEEKING

Over the years, the benefits of teamwork have been identified, especially its potential ability to positively increase efficiency and outcomes (Salas et al. 2008). For these reasons, teams are becoming embedded in almost every modern work domain. Domains such as education, homeland security, health care, and emergency crisis management are just a few that are dependent on high performing teamwork. The process of achieving effective teamwork is complex and complicated, relying on interactions among social, cognitive, and technical variables. Consequently, to better understand teams and the differences between high and low performing teams, researchers must seek to understand the many processes that are associated with teamwork. One process that is directly tied to teamwork is collaboration.

In response to growth of teams, collaboration and methods to support collaborative efforts are becoming increasingly important. Collaboration requires team members to communicate/coordinate and effectively work with each other. Historically, collaboration is conceptualized by variables of *time* and *space* (Schmidt and Bannon 1992). Collaborative efforts may occur instantaneously in real time (synchronous), or span over an extended time-period where communication is disparate and not instantaneous (asynchronous). Likewise, collaboration can occur within a physically isolated space (colocated), or via people communicating in multiple different spaces (distributed). Traditional means of collaboration are rooted in *synchronous* and *colocated* environments, but as technology continues to make advances, *asynchronous* and *distributed* collaboration is becoming more frequent. Modern day collaboration is dependent on varying levels of both synchronous/colocated *and* asynchronous/distributed efforts.

To better understand teamwork, and more specifically, team decision making, it is important to understand the role of collaboration during this process. A specific type of collaboration that we view as fundamentally important to team decision making is *CIS*. Foster (2006) defines CIS as "the study of the systems and practices that enable individuals to collaborate during the seeking, searching, and retrieval of information." In terms of team decision making, CIS is critically important. A team must first identify and articulate their problem set, and then seek to find information to achieve solving their problem. Therefore, if a team fails to collaboratively seek information, it is very possible that they will fail to solve their problem. CIS is a process that occurs early on during team decision making and continues throughout the team's lifespan.

When considering CIS during team decision making, there are two specific research gaps. First, some of the most prominent team decision making theories/models: Functional Theory of Group Decision-Making (Orlitzky and Hirokawa 2001), Multi-level Theory of Team Decision-Making (Hollenbeck et al. 1998), and Macrocognition in Teams Model (Letsky et al. 2007) fail to explicitly acknowledge the role and importance of CIS. Although information sharing is often identified as an overall step within the team decision-making process, there is little emphasis on information seeking activities. Second, and of particular importance to this chapter, the cognitive aspects of CIS have not been studied in depth.

CIS has been studied in both library/information sciences and computer-supported cooperative work (CSCW) mainly through two streams of research: *social* and *technical* (Karunakaran et al. 2010). The social stream focuses on examining how people perform CIS activities, seeking to understand the interactions that occur within their work domain. The technical stream focuses on translating many of the findings learned from the social stream and then developing systems or tools to support CIS-based activities or tasks. In both streams, the role of cognition during CIS is not widely discussed. When cognition has been studied within the context of CIS, it is a secondary goal and studied at the level of the individual and not the team (Shah and González-Ibáñez 2011). If the CIS research community is to fully understand and conceptualize CIS it is important to take a broader perspective that accounts for the social, technical, and *cognitive* streams. Searching, retrieving, and sharing information is dependent on both individual and team level cognition, so future research should account for understanding and articulating the role of cognition during CIS. Recent work by McNeese and Reddy (2013, 2015a, 2015b, 2015c) has sought to understand the impact that cognition has on CIS. Specifically, their work focuses on understanding how the concept of team cognition (see Mohammed et al. 2010, Cooke et al. 2013 for review) develops during CIS activities. In general, their research has indicated that team cognition does develop during CIS and that it greatly impacts to overall process of CIS. Further review of their work will be presented later in this chapter.

Although, it is important to acknowledge that future CIS research needs to focus on both individual and team cognition, it is also equally important that researchers know how to study cognition within the context of human behavior and performance. Without appropriate methods and theoretical understandings, it will be difficult to study cognition during CIS activities. Although eliciting and understanding cognition is complicated, understanding the specific methods to studying cognition is necessary to (1) appropriately capture cognitive activities and (2) understand these activities within the scope of context, tasks, and emerging demands. The purpose of this chapter is to provide researchers with multiple cognitively focused research methods and approaches to study cognition during CIS activities. Specifically, we look to the research domain of cognitive systems engineering (CSE) to identify different methodological techniques and approaches aligned with understanding cognition in the context of human behavior and performance.

WHAT IS COLLABORATIVE INFORMATION SEEKING? A REVIEW OF THE LITERATURE

CONCEPTUALIZATIONS AND DEFINITIONS

Traditionally, information seeking has been studied at the individual level (Shah 2010a). Yet, as work and organizations become more collaborative (Karsten 1999; Reddy and Dourish 2002), the importance of studying information seeking from a collaborative perspective has become apparent. This perspective on information seeking is referred to as CIS. CIS is an interdisciplinary research domain that spans multiple fields, such as information science (IS), human–computer interaction (HCI), and CSCW. CIS is also studied in varied variety of contexts including education, military, and medical settings.

In this chapter, we use the term CIS, yet there are many related and often interchangeable terms that also refer to the concept of CIS. Shah (2010a) has identified the following terms and research focuses as being strongly related to CIS: *collaborative information retrieval* (*CIR*) (Fidel et al. 2000), *collaborative search* (Morris 2013), *social searching* (Evans and Chi 2009), *collaborative exploratory search* (Pickens and Golovchinsky 2007), *co-browsing* (Gerosa et al. 2004), and *collaborative information behavior* (*CIB*) (Reddy and Jansen 2008). Even though the breadth and specificity of each of the terms is different, they all incorporate aspects of CIS.

The variety of terms presents challenges in defining specifically what CIS is for two reasons. First, there is not a clear understanding of what the breadth of the definition should include. For instance, some definitions include the act of seeking, searching, and retrieving information, whereas some only include one of those actions. Second, definitions of CIS greatly vary depending on the community. As Foster (2006) highlights, depending on the community, certain aspects of the definition "may emphasize information handling, search and retrieval, interaction, or the seeking and retrieving of information in support of collaborative work tasks." However, despite these challenges, there are a few widely acknowledged CIS definitions that are widely cited.

The most widely acknowledged definition comes from Foster (2006) who defines CIS as "the study of the systems and practices that enable individuals to collaborate during the seeking, searching, and retrieval of information." Similar to Foster (2006), Poltrock et al. (2003) have also defined CIS "as the activities that a group or team of people undertakes to identify and resolve a shared information need." Taking a more specific approach to CIS, Hansen and Järvelin (2005) use the following definition of CIS: "an information access activity related to a specific problem-solving activity that, implicitly or explicitly, involves human beings interacting with other human(s) directly and/or through texts (e.g., documents, notes, figures) as information sources in an work task related information seeking and retrieval process either in a specific workplace setting or in a more open community or environment." In addition, Karunakaran and Reddy (2012) define CIS in the simplest terms: "two or more individuals working together to seek needed information in order to satisfy a goal."

In addition to the definitional understanding of CIS, it is also important to note that CIS is a highly complex contextual and dynamic activity. The activity can involve both static and dynamic goals that are specific to the situation and context or there may be no goal at all. Also, depending on the situation and context, information may change throughout CIS. Finally, CIS requires multiple people with many individual differences to work together using many different means to seek information.

How Has CIS Been Studied? Empirical Results of CIS

There are two main streams of research which examine CIS: (1) the social stream in which CIS is examined in terms of its structures, processes, and multiple contexts and (2) the technical stream in which CIS is investigated in terms of the tools and systems that can aid and improve CIS activities (along with the growth, testing, and implementation of said tools/systems) (Reddy and Jansen 2008; Karunakaran and Reddy 2012).

In the social stream, the social aspects of CIS have been studied within multiple contexts, including education, military, and healthcare. In the educational contexts, CIS has received a great deal of attention. Work by Hyldegård (2006; 2009) extended the individually focused Information Search Process (ISP) (Kuhlthau 1991) model to encompass CIS. The goal of the study was to explore the differences between how individual and groups of students seek information. In addition, more specifically, if the ISP model would accurately explain group's behaviors and attributes of CIS. The researcher found that the ISP model was not accurate in explaining group's CIS, which led to her calling for the ISP model to be extended to support group collaboration. A group's contextual and social factors significantly change the process of collaborating to find information. In addition, researchers engage in CIS activities due to a lack of expertise in a certain area (Spence et al. 2005). In other contexts (e.g., the military or Command and Control (CC2), CIS is considered an important aspect of teamwork (Sonnenwald and Pierce 2000; Prekop 2002). Prekop (2002) investigated CIS behaviors in the Australian Defense Forces command and control unit. Specifically, the researcher analyzed how command and control workers collaboratively worked together during information seeking activities. This study focused on three specific aspects of CIS: *information seeking roles, information seeking patterns*, and *the contexts with which those roles and patterns are found*. In the study, multiple roles within CIS were found: *information gatherer, information verifier, information referrer*, and *information indexer*, and patterns of CIS behavior were identified: *information seeking by recommendation, direct questioning*, and *advertising information paths*. Finally, two different contexts of work were found identifying that CIS and organizational work activities are directly related. First, the CIS context was identified, which occupies the knowledge, skills, and known information within the group. Second, an organizational context was identified that encapsulates where CIS is occurring. Still within a CC2 context, another qualitative study (Sonnenwald and Pierce 2000) found three aspects related to CIB, including situation awareness, *dense social network*, and contested collaboration. In the medical field there have been multiple studies of CIS, including Reddy and Dourish's (2002) study that found that the members of a medical team would collaborate to seek required information *just in time* based on an understanding of each other's temporal rhythms. Other studies from the medical field focus on a variety of related topics, such as: how patient care teams collaborate to find information, identifying team information needs and CIS activity triggers, sensemaking during CIS, and the contextual factors of CIS and their impact on CIS activity (Spence and Reddy 2007; Reddy and Spence 2008; Paul and Reddy 2010; Spence and Reddy 2012, respectively).

In the technical stream, tools are being developed to support CIS activities. A number of tools and systems designed to support CIS have been introduced throughout the past twenty years. For the purposes of this chapter, we will not explicitly outline each system or tool in detail, but just highlight some of them here: *MUSE (Multi-User Search Engine)* (Krishnappa 2016), *Search Together* (Morris and Horvitz 2007), and *Cerchiamo* (Golovchinsky et al. 2008). One of the most recent and more advanced CIS-based tools is *Coagmento* (Shah 2010b), a system that enables groups of people to collaborate in order to find information through a web browser. It is a multileveled system, which records, first, the information that the groups find and, second, the processes used to find that information.

SOCIAL AND TECHNICAL ASPECTS OF CIS: COGNITION IS MISSING

CIS research focuses mostly on social and technical streams, but there has been little research examining CIS and cognition, specifically team cognition. In general, cognition is accounted for in individual information-seeking models, such as: the theory of sense-making (Dervin 1983), information search process (Kuhlthau 1988), the model of information-seeking behavior (Ellis 1989), and the model of information-seeking (Wilson 1997), but not in CIS based models.

When cognition is considered within the context of CIS, it is often in the technical stream, focusing on whether CIS-oriented tools and systems make sense to the individual using them. For example, Shah et al. (2009) acknowledged the importance of taking a cognitive perspective while developing *Coagmento*. The researchers examined how users worked with and understood the system by conducting individual cognitive walkthroughs. Another *Coagmento* study found that cognitive load is the same for users during CIS and individual information-seeking (Shah and González-Ibáñez 2011). In addition to the work done by Shah and colleagues, a cognitive work analysis was conducted on design researchers at Microsoft to further understand their motives for CIR. The study found that designers engaged in CIR when they were new to the team, when information was ambiguous, and when needed information was not documented (Fidel et al. 2004).

Although these studies are important and a step in the right direction, there is much more work that needs to be done before we understand the role of cognition during CIS. Specifically, we need to better understand the development of team cognition during CIS. Team cognition and CIS have been very rarely investigated together. In fact, until very recently, the only research study that has come close to investigating team cognition and CIS together was work by Paul and Reddy (2010) who explored the concept of sensemaking (Weick et al. 2005) during CIS. Sensemaking traditionally is a cognitive process that allows humans to *understand* their environment. In recent years, this process has been extended to the team cognition level leading to the concept of collaborative sensemaking. Paul and Reddy's research viewed sensemaking as an important aspect of CIS within the medical domain, and sought to conceptually understand how the two are related.

Yet, in recent years, researchers are beginning to acknowledge the importance of both individual and team cognition during CIS. Work by McNeese and Reddy (2013, 2015a, 2015b, 2015c) emphasizes the importance of team cognition in supporting CIS activities. Differences and similarities in the development of a CIS-based team mental model have also been recently described along with highlighting the importance of awareness mechanisms that help develop team cognition during CIS activities.

Although, cognitively focused work is increasing within the CIS research community, the social and technical streams still dominate the literature. To increase studies focused on cognition and CIS, the research community must know how to appropriately study cognition. In the next section, we outline specific methods and approaches that have long been successfully used to study cognition in multiple contexts, lending them useful for the CIS context.

COGNITIVE SYSTEMS ENGINEERING PERSPECTIVES AND METHODS FOR STUDYING THE COGNITIVE ASPECTS OF COLLABORATIVE INFORMATION SEEKING

Due to inadequate HCI and system design (SD) solutions in dynamic operational environments, a new interdisciplinary synthesis was introduced: CSE. Essentially, the CSE framework consists of several traditional fields (e.g., engineering, psychology, cognitive science, IS, and computer science) (Rasmussen 1983), and multiple intellectual threads, including control theory, information theory, ecological psychology, and Gestalt psychology (Flach 2016). CSE is an interdisciplinary engineering field concerned with several different factors, including: (1) the cognitive processing required for a human operator to perform their job; (2) the design of the technologies in the system(s) used by the operator to fulfill those cognitive requirements; and (3) the needs and goals of the human users as well as the systems/software engineers, and program managers (Hollnagel and Woods 1983). One of the core principles of CSE is that both human-human and human-automation system/technologies/tools need to be conceived, designed, analyzed, and evaluated in terms of cognition. This idea underlines how crucial each part of the system (i.e., human and automation) is for the overall outcome of the system (Hollnagel and Woods 1983).

Within CSE and to a greater degree the human factors community, knowledge has long been associated with cognition, with the two often being used interchangeably (Cooke et al. 2000). One of the main domain problems of interest to CSE is tacit knowledge (i.e., know-how) that provides the foundation of technological innovation within the perspective of human nature and human work (Winograd and Flores 1986). In 1968, Rasmussen introduced three core components in order to show how tacit understanding (i.e., what constitutes a *well-founded decision*) is directly associated with human nature and human work: (1) operator's beliefs and awareness, (2) interface representation, and (3) functional work and problem domain. These three components have a great impact on not only shaping the design of information technologies to support human performance, but also on cognitive science and the human-factors field (Flach 2016).

It is apparent that understanding cognition is of paramount importance in CSE. For this reason, the community has long utilized many research methods and approaches to both capture and understand cognition at the individual and team level. Below, we review and apply research methods (KE) and approaches (CTA, LL approach) to the CIS context. The first, KE, is the heart of most cognitively based methods and encompasses many different techniques. The last two, CTA and the LL approach, differ from KE in that they are specific goal or problem-directed approaches to studying cognition (while also using methods found in KE). We acknowledge that there are many methods and approaches that one could take to study cognition during CIS, but we view the three presented in this section as potentially the most effective ways of studying the cognitive aspects of CIS.

MIDWAY BREATHER—STOP AND REFLECT

- CIS is widely studied from two perspectives: social and technical, yet little of the research from either employs a team cognitive perspective.
- In order to increase the awareness of and interest in cognitive (individual and team) work in this area, cognitively oriented methods must be introduced and explained to the CIS community.

KNOWLEDGE ELICITATION

KE, is "the process of collecting from a human source of knowledge, information that is enough to be relevant to that knowledge" (Cooke 1994, 802). Over the years, many methods have been developed that fall within the overarching scope of KE. Although it is not possible to review every single KE method, we will provide a high level overview of the three main families of methods that compose KE. Much of our understanding pertaining to KE comes from a seminal article written by Cooke (1994), where she provides an extensive overview of KE methods and techniques. In this work, she classifies three broad areas of KE methods based on variables of specificity and formality: (1) observations and interviews, (2) process tracing techniques, and (3) conceptual techniques.

Observations and interviews are the most commonly used method to understanding individual and team level cognition. In particular, observations are often very useful in understanding cognition. Although cognition can be elicited via interviews, not all humans may be able to adequately describe their cognition before/during/after an activity. For this reason, it is useful to gather observational data as they provide direct insights into cognitive activities. As noted by multiple researchers, the link between cognition and interaction is quite strong (Van den Bossche et al. 2006; Cooke et al. 2013), so observations must be accounted for when attempting to understand cognition. In terms of specificity and formality, observations and interviews are informal and flexible. Depending on the scope and focus, the researcher may focus an observation on a specific task within the larger task (active, focused, or structured observation), or their interview may consist of varying levels of focus (unstructured, semi-structured, or structured).

Process-tracing methods refer to having humans describe their cognition, either in real time or after a task has concluded. More specifically, the methods of *thinking aloud* or *subsequent recall* protocols are used to capture the cognition associated with performing a specific task. Although these methods are useful and provide insights directly from the source (the human), as previously noted, humans cannot always articulate the entirety of their cognition. In addition, process-tracing methods may or may not work depending on the setting or task that they are employed in. For example, some tasks may require communication, meaning that the human is not able to *think-aloud* during their task. Likewise, employing process-tracing methods at a team level often does not work due to the need for team level communication. In comparison to observations and interviews, these methods provide more structure and specificity to capturing cognition.

Finally, conceptual techniques capture the relevant concepts in a domain and represent them in a structured and interrelated way. Specific methods that are often utilized within this family are concept mapping, paired comparison testing, and card sorting. These methods are very powerful because they have the ability to not only elicit cognitive context, but also explain how the human views that content in relation. Conceptual techniques are often used to elicit cognition at the team level. Cooke and colleagues explain the utilization of conceptual techniques at the team level by which "judgments of the proximity of domain-relevant concepts to one another are elicited from team members" (Cooke et al. 2007, 245). This approach explains "individual expert–novice differences and applies to team cognition in the elicitation of team knowledge, team member mental models, and the identification of interaction patterns in teams" (Cooke et al. 2007, 245).

KE have been traditionally used to elicit knowledge from individuals. However, through Cooke and other's work (McNeese and Ayoub 2011; Moon et al. 2011), these methods can be used for better understanding team level cognition through both individual aggregation and capturing team-level cognitive processing. In order to fully understand cognition during CIS, it is important to capture both individual and team-level cognition, which is why KE methods are of paramount importance.

How Knowledge Elicitation Methods Can Be Applied to CIS

There is great promise for using KE methods to better understand cognition during CIS. In fact, we posit that KE methods are the main way in which researchers can develop insights into how cognition develops and impacts the process of CIS. KE methods can allow researchers to elicit cognition both at the individual and the team level, as well as capture the relationships and linkages between interaction and cognition. The many specific methods that fall within KE allow for a comprehensive approach to studying cognition. Yet, because of the many methods, a researcher must fully understand which methods to use in accordance with the goals (and the limitations/practical constraints) of the research study, and the research context. Below, we detail how each of the previously noted families of KE methods can be applied to specifically studying CIS.

Observations and Interviews

First, we will discuss the role of observations during CIS. Many of the CIS studies referenced in this chapter are observational studies that provide conceptual and empirical findings. The reason for that is two-fold: first, much of the CIS work comes from the IS community, which often utilizes qualitative methodology, and second, and more importantly, CIS activities are often explicitly tied to observable interactions, making observations a key methodology for CIS research. Yet, not many researchers have used observations for the purposes of understanding cognition during CIS. The relationship between interaction and cognition is inextricably tied together, with Cooke postulating that team-level interaction is actually team cognition (Cooke et al. 2013; Cooke 2015). Team-level interactions, such as communication and coordination that can be captured during observations are representative of team cognition. Understanding this, a CIS researcher can develop insights into cognition during the process of CIS when team members are working on a task. However, one of the challenges in using observations to capture cognition during CIS is that the observer must be trained in understanding when the team is engaged in CIS and not just overall teamwork.

Interviews may be one of the most impactful and effective ways to elicit cognition within the CIS context. In most cases, the researcher will need to focus her interview on the CIS task being studied. In general, the researcher would want to use observations along with interviewing. This will allow the interviewer to orient and scope their interview to the task, and elicit task-relevant cognition. We recommend utilizing a semi-structured cognitive interview (Fisher and Geiselman 1992) that allows for a variety of probing techniques (Willis 1999). These interviews may be conducted both before and after task observation. In addition, interviews may be conducted individually and then aggregated to the team level and across all teams, or they can be conducted at the team level (focus group style), and aggregated across all teams. We recommend that researchers should conduct interviews at the individual level and then aggregate to the team and across all teams' level. This aggregation technique should also be coupled with real time observations (to gather team level cognitive processing). We recommend this primarily because more insight into cognitive activities can be elicited at the individual level, and it can also provide the researcher a method to measure for shared or overlapping cognition across individuals.

Process Tracing

Capturing cognition via process tracing during CIS activities is one of the most straight-forward and direct ways to elicit cognition. As CIS activities are inherently collaborative in nature, most require communication. This communication process represents a goldmine of cognitively relevant information. In fact, Cooke (2013, 2015) has postulated that interaction (via communication and coordination) is in fact representative of team-level cognition, and that communication is a specific way to capture concurrent cognition (Cooke et al. 2000). Therefore, it is essential that CIS researchers attempt to capture the communication that occurs during a CIS specific activity. Yet, it is also important to note that this method should be paired with another KE method because not all cognition is represented in team-level communication.

Conceptual Techniques

The utilization of conceptual techniques can be used to elicit both individual and team-level cognition during CIS. There are three main conceptual methods that we recommend for identifying cognition during CIS: individual and team concept mapping, card sorting, and paired comparison testing (McNeese and Reddy 2013). Concept mapping may occur at either the individual or team level, and may be focused on declarative or procedural knowledge. In order to administer concept mapping in the context of CIS, participants should complete a CIS task and then the researchers should provide a specific focus question for the participants to map. Card sorting is a method where participants are given cards with differing concepts on each, and then are asked to organize or group the cards based on similarities or relationships. After the cards are sorted, participants are asked to specifically name each group. In the instance of CIS, the researcher would provide concepts that are fundamental to a specific CIS task and then ask participants to sort and name each group. Paired comparison ratings are often utilized to understand team cognition, and can also be used in the context of CIS. Participants are given a series of two statements that may or may not be similar. The participant is then asked to rate the degree of similarity between the two statements. The researchers can then analyze them for similarity ratings. Once all team members have produced individual similarity ratings, they can then be compared across the team to understand shared knowledge. Paired comparisons can easily be adapted to CIS, with the researcher creating statements relating to cognitive aspects of the CIS task or activity.

A Case Study of Knowledge Elicitation and Collaborative Information Seeking

An example of using KE methods during CIS activities can be found in the dissertation work of McNeese (McNeese 2014). In this work, McNeese used all three families of KE to study how team cognition developed during a CIS activity. The dissertation consisted of two separate but related studies, below we outline each.

McNeese et al. (2014) study consisted of both dyad and triad teams working on a CIS specific activity in a colocated environment. Using these tasks, the researchers sought to understand the different processes the teams utilized to collaboratively seek information, and also the development of team cognition throughout the CIS tasks. When participants came to the lab, they were provided with a brief overview of each of the main topics that the study pertained to: CIS and team cognition. Then, the participants were given a set of three related CIS tasks in which they had a total of 30 minutes to complete. During these tasks, the participants were given freedom in how to complete the tasks, either through communication methods or individual work. As the participants were collaborating, their work was video recorded for the purposes of process tracing. Once the teams completed their tasks, they were separated to complete two different KE methods—interviewing and conceptual techniques. More specifically, each participant completed an individual interview, and an individual declarative concept map (focused on declarative knowledge) relating to the concept of CIS. Each of these methods provided the researchers insights into the individual and team cognition associated with CIS. After each participant completed both a cognitive interview and a concept map, the team came back together to produce a procedural concept map highlighting the process of CIS while also outlining the role of cognition during the process.

Results from this study indicated that team cognition specific to CIS develops in the form of a team mental model, that is often taskwork oriented (McNeese et al. 2014). The development of the team mental models aligns with previous work that has described how team mental models develop (McComb et al. 2010). During this study, individual team members begin the task with their own individual cognition and then shift their cognition to align with team-based goals. Accordingly, there is a need for a cognitive shift from the individual to the team level. So, individual team member's activities lead them to evolve from their individual mental models into a team mental model. This mental model convergence from individual to team is divided into three phases (McComb et al. 2010): (1) team members become familiar with the new domain; (2) each team member creates a unique view of the situation, which may differ from that of the other team members; and (3) they let their individual views progress into a team view. These three phases were also reflected in N.J McNeese's (2014) study. The concept mapping data from this study also indicated that the most central concepts of CIS were: *Research, Sharing Information, End Goal, the Internet,* and *Collaboration* (McNeese and Reddy 2015b). In addition, findings from this study also highlighted the important role of multiple awareness mechanisms that help to develop team cognition during CIS. More specifically, multiple specific awareness methods centered around the concepts of *search, information,* and *social* were all identified as being critical to team cognition development (McNeese and Reddy 2015c).

The second study of N.J. McNeese's dissertation work focused on understanding CIS processes and team cognition development in the context of distributed work. In this study, the same CIS tasks were given with the same time limitations for the purposes of comparing the distributed setting to the colocated setting. In this study, dyad teams communicated solely through the CIS system *Coagmento*. The chat logs that the participants communicated in were recorded as an outlet to process tracing. After the CIS tasks were completed, each individual team member took part in a cognitive interview (McNeese 2014). Results from this work indicated that even in the distributed setting, team cognition related specifically to CIS developed (McNeese and Reddy 2015a). Particularly, a CIS team mental model developed in the teams. Yet, in comparison to the colocated setting, these models were highly dependent on taskwork aspects of team cognition and not teamwork aspects found in the colocated teams. Aspects that were identified as helping to develop team cognition in this setting were: *team experience, communication and sharing information, awareness, trust and assumptions, leadership or lack of leadership,* and *the structured nature of distributed CIS.* Future research based on these studies will specifically highlight multiple CIS processes that the teams utilized during the CIS tasks.

COGNITIVE TASK ANALYSIS APPROACH

A specific research approach for measuring individual and team cognition within CSE is a CTA. This approach employs many of the previously identified KE methods in a systematic manner to understand the cognition necessary to *perform* a specific *task*. Schraagen et al. (2000, p. 3) define a CTA as "the extension of traditional task analysis techniques to yield information about the knowledge, thought processes, and goal structures that underlie observable task performance." Although a CTA is directly related to KE methods, it is a separate research approach. There is often confusion regarding KE methods and CTAs, with many publications using these terms interchangeably. From our perspective, while KE methods and CTAs share commonalties, a CTA is a goal-directed approach to understanding cognition of a specific task, whereas KE is a meta-family of methods. The relationship between the two is that a CTA's success is dependent on utilizing many KE specific methods.

In CTA, *cognitive* refers to the methods underlying reasoning and knowledge, such as: perceiving and paying attention, having the knowledge, skills, and strategies to adapt in dynamic situations, and knowing the purposes, goals, and motivations for cognitive work. *Task*, refers to the outcome people are working toward; not the activities they undertake on the way to achieving that outcome. Finally, *analysis*, refers to decomposing an item into its component parts in order to better understand those parts and how they relate to each other (i.e., the process from bottom-up to top-down) (Crandall et al. 2006). Overall, CTA is a process that researchers can use to determine the key drivers of cognition (i.e., determining accurate and complete descriptions of cognitive processes and decisions) in a variety of applications (Crandall et al. 2006; Clark et al. 2007). However, there are many methods that can be used during a CTA, and choosing the right one for the right task can be challenging for a researcher. Therefore, it is important to know which CTA methods will be successful in different situations. Fortunately, there are reviews outlining how to choose methods and techniques for conducting a CTA (Crandall et al. 2006; Clark et al. 2007). Crandall et al. (2006) outline what they delineate as three main *aspects* of a successful CTA: (1) KE, (2) data analysis, and (3) knowledge representations. Each of these aspects consists of their own specific considerations. For instance, KE methods are distinguished using two categories: (1) data collection and (2) method focus. Data collection simply refers to how the data were collected, which might be through interviews, self-reports (such as surveys, questionnaires, diaries, and logs), observations, or automated capture by computers. The second category, method focus, refers to aspects of time, realism, difficulty, or generality. These are explained in the following paragraph.

If the method focuses on *time*, then the researcher will want to know: when did the events happen? For instance, retrospective data allows researchers to find incidents based on particular or specific events, and also allows them to focus on certain kinds of events and aspects of cognitive performance. These types of data can be obtained via interview or self-report; guided questionnaires can be particularly effective because they can help people remember their experiences in more detail, ideally allowing them to provide a description of how they made their judgments and decisions. KE methods may also be focused on *realism*, that is, was the data gathered in real-world settings? In this case, researchers often use simulations, for instance: flight simulators can recreate real-world events, like accidents, and are used in individual and team trainings and to help develop new technologies. *Difficulty* is another possible focus—how challenging is the case and how tough is it to collect the data? Fields like healthcare, aviation, or nuclear power provide examples of difficult situations, such as accidents, where incident-based methods (observing and interviewing some specific accidents) can be used to understand how people make sense of and respond to these situations; hopefully providing insight into the sorts of errors people might make and how safety can be improved. Finally, the last focus is *generality*—how specific is the task? In this situation, researchers seek to elicit declarative knowledge in a given domain. Techniques such as concept mapping, in which core concepts and their relationships are depicted, can be used to survey general knowledge, specifically the goals people have when completing a task and how those goals are prioritized and related to each other (Crandall et al. 2006).

Stepping back from KE and moving to Crandall et al. (2006) second aspect of CTA, data analysis comprises: data preparation, data structure, discovering meaning, and representing key findings. Data preparation refers to assessing how complete and accurate the whole data set is (i.e., all data recorded should have a clear label and complete identifying information). The next piece of data analysis and data structure requires dissecting the data in order to rearrange it and find patterns inside it; this can be done, for instance, through coding, lists, and descriptive statistics. Discovering the meaning entails extracting key questions, issues, and emergent threads of meaning from the data (some procedures: categories, ratings, and statistical test). Finally, representing the key findings involves converting the data's discovered meaning into a visual form, bringing its story to the forefront (some techniques: charts, graphs, storyboards, and concept maps) (Crandall et al. 2006).

The third and final aspect of CTA is knowledge representation, which focuses on the critical tasks of data visualization, presenting findings, and communicating meaning. This aspect is also inherently linked to data analysis, and the combination of these two aspects provides different representations of the products from the analysis process than the representations produced from KE alone. In this case, there are four types of analytic products that are typically yielded: (1) textual description; (2) tables graphs and illustrations; (3) qualitative models (e.g., flowcharts); and (4) simulation, numerical, and symbolic models (e.g., computer models) (Crandall et al. 2006).

How Can a Cognitive Task Analysis Approach Be Applied to Collaborative Information Seeking?

A CTA can be a very useful research approach for understating the cognition associated with CIS activities. The nature of CIS is inherently task specific, meaning that teams of individuals are not going to take part in CIS activities unless they are presented with a task that requires gathering information in a collaborative manner. For example, patient handoffs in the healthcare domain are a specific task that is highly dependent on the effort of nurses working together to find needed information. Similarly, in the intelligence analysis sector, individual analysts must collaborate with other analysts to find information relevant to a specific intelligence problem or task.

As previously noted, CTAs and KE methods share a close relationship. The reason for specifically outlining CTAs as opposed to simply including them within the overall family of KE is that CTAs provide specific guidance for a CIS specific task. In the previous section, we highlighted the importance of KE methods, yet, we did not provide specific guidelines on when to use each method. The benefit of a CTA is that it provides the researcher a more focused plan for understanding the cognition in a specific task. The CTA approach also explains the importance of analyzing data, and then presenting it in a meaningful way for the purposes of improving the task. In the specific context of a CIS task, it will be extremely important in gathering data at the individual and team level to fully understand the task. Using the CTA structure outlined by Crandall, Klein, and Hoffman, we believe that there will be many opportunities to develop insights into cognition during CIS. Depending on the task and the context, a variety of the previously mentioned KE methods can also be employed.

A Case Study of the Cognitive Task Analysis Approach and Collaborative Information Seeking

To the best of our knowledge, there has not been a CTA focused on specific CIS related tasks. Therefore, for this case study, we will highlight similar work that focuses on the utilization of a cognitive *work* analysis (CWA) framework to better understand CIR (a context closely related to CIS). Once, we have reviewed this work, we will then present a specific CIS task and explain how the methodology presented by Crandall, Klein, and Hoffman can be applied.

A cognitive work analysis is extremely similar to a CTA in that the goals of each are to understand the cognition associated with adequately performing work. The cognitive work analysis framework is an approach that uses multiple methods to better understand human behaviors relating to work and cognition in context (Vicente 1999; Rasmussen et al. 1994). In general, a CTA is more focused than a CWA, which seeks to typically understand the larger scope of work, not just one specific task. A CWA also does not lend itself as well to specifically understanding the relationship of cognition to performance. A CTA directly seeks to understand how cognition impacts performance, whereas CWAs are typically aligned to understanding cognitive motivations, not examining how cognition actually affects task performance. This is not to discourage researchers from using CWAs to study the cognitive aspects of CIS. We feel that both CTAs and CWAs have value in further developing our understanding of cognition during CIS. The researcher will need to review the research context and the goals of the study to decide which approach to utilize.

CWAs are often employed within the information sciences community, where they are used to develop a better understanding into the design and evaluation of information systems (Fidel and Pejtersen 2004). For this reason, related areas of CIS, specifically CIR, have been studied using this methodological approach. CIS activities are becoming increasingly dependent on information systems and collaborative search platforms. If a CIS researcher is focused on understanding technical development, then a CWA might be helpful to develop new technical insights. Research by Bruce and colleagues (2003) explored CIR activities of two design teams through a CWA oriented field study. The study found that the concept of CIR is ambiguous—not lending well to a single definition that CIR is tied directly to context, and that CIR activities are not exclusive to collaborative activities. In addition, work by Fidel and colleagues (2004) employed a CWA to better understand why design engineers participate in CIR-based activities. Through a field study using observation and interviews, the researchers identified that engineers utilize CIR when they are new to an organization, when information is ambiguous, or when information is not documented (Fidel et al. 2004).

These two studies are beneficial in understanding the cognitive aspects of collaborative information behaviors that occur within the engineering context. Unfortunately, these types of studies are lacking, with very few CWAs taking place with the focus on understanding the cognitive aspects of collaborative information behaviors (with no CTAs taking place). Below we will briefly explain how to use a CTA during the specific task of intelligence analysts collaboratively seeking information. The intelligence context has long been defined by individual work, but collaborative work is becoming more apparent (McNeese et al. 2015).

First, when conducting any CTA, it is extremely important that the researchers understand the specific cognitively oriented task that they are studying. In the proposed setting, the researcher needs to focus their methods on the specific task of collaboratively seeking and sharing information during the analysis of an intelligence problem. This is difficult to achieve, as there are many other activities occurring within the greater scope of work. In order to develop this focus, we suggest the researcher actually observes all aspects of the environment. By doing this (possibly multiple times), the researcher will be able to identify specific instances of where and when CIS is occurring either directly or indirectly. With this knowledge, the researcher can then focus their KE methods to these task specific instances of CIS. This will help determine which of the many KE methods the researcher should employ. At a minimum, for this specific task and context, the researcher needs to observe the task (intelligence analysts collaborating to find and/or share information), conduct both individual and team level interviews aimed at understanding the cognitive activities that occur during the CIS task, and conduct at least one of the process tracing or conceptual techniques (at the individual and team level) previously outlined. The analysis of the data collected by these methods can be done using multiple different analytical techniques, ranging from thematic to statistical analysis. Finally, after the analysis, knowledge representation should occur to report data findings and to improve cognition associated with the task. We encourage researchers to not only view knowledge representation as a means for reporting data, but as a way to develop new recommendations/workflows/technologies to help improve cognition. For example, in this context, a technology that better allows analysts to share or view information in a collaborative manner may be developed based on the overall findings of the CTA.

THE LIVING LAB APPROACH

The experimental LL approach, first used by M.D. McNeese (1996), is considered as an integrative response (based on ecological psychology) to increased sociotechnical environments. As sociotechnical systems became more prevalent, technological error in team–technology interaction have also increased. Examples of such errors are the inadequacy of communication technology to meet human-systems engineering perspectives, jeopardized shared understandings across team members in highly heterogeneous teams (e.g., command-and-control teams, surgical teams), and a failure to be context-sensitive (McNeese et al. 2000). In order to address these issues, the LL approach was constructed as an ecological framework consisting of theory and practice, continued process improvement, and tool development in the human-factors field.

Being ecologically based, the LL framework focuses on transactions between team members and their environment (collaborative ecosystems) and, further, postulates that these transactions both determine and situate all dynamic and complex environments. In order to understand these transactions ecologically, the LL framework combines ethnographic observations of teams within their task environment with unique knowledge from each team member's perspective. These two sources are used to create simulated environments that provide a setting for new empirical research. There are four outcomes of the LL ecological framework (McNeese et al. 2000; Hall et al. 2008): (1) *ethnographic studies* (a field-based study of cognition), which focus on understanding the problems and issues related to a particular domain or application, and in which researchers work with end users to figure out what issues are related to decision making, human operation, kinds of data, the types of inferences people seek, and so on; (2) *knowledge discovery* in which KE methods (e.g., interviews, concept mapping, and observation) are used to understand the problem domain; (3) *scaled-world simulations* wherein researchers can create a high level fidelity environments in order to evaluate the evolving concepts, scenarios, and test sets; and (4) *design of support tools* are developed based on the results of scaled-world simulations, and these new tools can be applied to improve individual and team situation awareness and the decision-making process. If these *outcomes* produce satisfactory results in the lab, and then they can be introduced and applied in the real world.

How Can the Living Lab Approach Be Applied to Collaborative Information Seeking?

All four of the LL framework's outcomes are directly relevant to better understanding the cognitive aspects of CIS. Specifically, the first two outcomes—*ethnographic studies* and *KE* methods, were explained within previous sections. Studying CIS in context with a real-world environment is not only fundamental to articulating cognition, but it also provides insights into how to further study specific cognitive aspects in a controlled environment. Although ethnographic and KE are useful to identifying cognition and how it occurs within context, it is hard to delineate and study specific aspects of cognition within a real environment. The dynamic nature of the environment consists of multiple different variables, resulting in a great deal of empirical noise, and creating multiple confounding variables. These extraneous variables have the ability to impact the researcher's ability to capture data specific to a specific cognitive aspect. For example, if a study is interested in the concept of team situational awareness, it is best to study this not only in a real-world context but also in a controlled environment. The controlled environment (laboratory) allows the researchers to specifically study multiple independent and dependent variables and validate that the results of the study are not being confounded by other variables. The LL approach is unique because it emphasizes the need to understand the real-world environment and then design simulations based on the characteristics, capabilities, and limitations of that environment. This results in experiments consisting of high levels of fidelity, increasing validity and reliability (McNeese 1996). More importantly, because the results from the experiments are based on the principles of the real-world environment, it allows for technologies/tools to be developed in accordance with the real-world environment. In regards to CIS, a researcher can use the LL approach to understand how cognition occurs within particular contexts, and then based on that understanding build a simulated environment that is representative of the real environment. After the experiments, tools related to enhancing the cognitive aspects of CIS can be developed and then introduced back into the environment. In the next section, we briefly describe a potential study that employs the LL approach to study the cognitive aspects of CIS.

A Case Study of the Living Lab Approach and Collaborative Information Seeking

Multidisciplinary cancer teams (MCT) are a team-oriented approach to patient care where the needs of the patient dictate the personalized treatment they receive; the type of cancer, its stage, the patient's needs, and their overall health all factor into treatment decisions. The Department of Health (2004) defines an MCT as consisting of representatives from diverse healthcare areas along the care pathway—from diagnosis to follow-up and beyond—who have scheduled discussions about a given patient; each member gives their own input when making diagnostic or treatment decisions about the patient (Lamb et al. 2011a; Lamb et al. 2012). In order to ensure that each patient receives proper care (that might require several different kinds of therapies), each expert in the MCT needs to make sure that team interaction oriented care (i.e., communication and coordination within the team) is available to help patients through diagnosis, treatment, and recovery.

In some countries, MCTs are quite common. For example, in the United Kingdom, MCT meetings—the focal aspect of MCT—occur once a week (Lamb et al. 2011b). During these meetings, patients can be discussed at any point along the treatment pathway, although typically this occurs at diagnosis, following treatment, and sometimes if there is a recurrence or progression of the disease. Therefore, the main goal of these meetings is to collate and review information about the patient and their disease, discuss it, and make a decision for further investigation and treatment. This is a context that is highly dependent on CIS. The MCT must collaboratively determine how to treat a patient, and the process of finding relevant information to inform that a patient care plan is critical (Lamb et al. 2011a).

Within the context of the MCT, the LL approach could be implemented to better understand how cognition occurs during CIS. Cognition is clearly guiding the team decision-making processes associated with CIS. A better understanding of that cognition, and how it guides CIS processes may be beneficial in improving how MCTs collaboratively seek information, and ultimately make decisions.

Referring back to the four outcomes of the LL approach, researchers first need to conduct an ethnographic study of the context that MCT teams are working in. The MCT's work environment would be observed in order to understand how the teams work together, issues during MCT collaboration, and how the team interacts with the patient. During the ethnographic data collection, the researcher will need to learn about individual and CIS processes (specific to cognition) that occur within the MCT. Throughout this study, the researcher would apply KE methods (e.g., interviews, concept mapping, and observation) in order to understand the problem domain/cognitive aspects, and then, based on data from observing and interacting with the MCT and the patient, the researcher can create a simulated lab environment to evaluate the evolving concepts, scenarios, and test sets related specifically to CIS and cognition. Examples of this might be a simulation that focuses on a team cognitive concept, such as team mental models, transactive memory, and team situation

TABLE 6.1

Applying the Collaboratively Information Seeking Approach on the Multidisciplinary Cancer Teams within the Living Lab Framework

The LL Process	Applying the LL Framework on the MCTs
1. Ethnographic studies	In order to understanding the issues related to the MCT work environment, interaction within the MCT, and interaction of the MCT with the patient, the researchers work with the MCT experts to figure out what issues are related to collaboration and decision-making process of the MCT, by observing the field (examples: Lamb et al. 2011a,b; Ruiz-Casado et al. 2014; Dew et al. 2015).
2. Cognitive systems engineering	In order to understand the problem domain related to the collaboration within the MCT, and the MCT's interaction with the patient, the cognitive experts can use some of the methods in the MCT's work environment, including making observations of the MCT meetings (Lamb et al. 2011a; Ruiz-Casado et al. 2014), interviewing each individual expert in the MCT (Prades and Borràs 2011), conducting the questionnaires on the experts (Taylor et al. 2012), recording the MCT meetings (Dew et al. 2015), and using the previous MCT records (Ruiz-Casado et al. 2014).
3. Scaled-world simulations	Researchers are able to use the problems, situations, constraints, and other information from process elements 1 and 2 to formulate cases, stories, or scenarios that are then built into a simulation, and represent the core and foundation of the simulation. In this example, a simulation can be built that requires utilization of an MCT tool to evaluate its usage.
4. Design of support tools	The experts can develop new tools or improve the current ones to improve individual and team aspects (e.g., individual and team-situation awareness, decision-making process).

awareness during CIS processes in MCT teams. Finally, the researcher may develop solutions (recommendations or technologies) to improve both individual and team cognition during CIS processes of the MCT. The following approach to using the LL to study cognitive aspects of CIS is summarized in Table 6.1.

CONCLUSION

In order to fully understand CIS, an understanding of social, technical, and *cognitive* approaches is necessary. Although there is substantial understanding of the social and technical approaches of CIS, much less is known regarding cognition. Moving forward, it is critical that the research communities interested in CIS explore how cognition occurs during CIS and how it impacts the overall activity of CIS. Yet, in order to understand the role of cognition during CIS, the research community must first be aware of how to study cognition. In this chapter, we have outlined many traditional cognitive engineering research methods and approaches that are appropriate for studying cognition during CIS.

REVIEW QUESTIONS

1. Describe the need for more cognitive (both individual and team) oriented CIS research.
2. Describe the benefits and costs of using KE, CTA, and the LL approach for studying cognitive aspects of CIS.
3. Outline a study that examines cognition in CIS using the living lab approach.
4. Describe a potential CIS study that uses all three methods presented (knowledge elicitation KE, CTA, and the living lab) in conjunction with one another.

ACKNOWLEDGMENTS

The dissertation research of Dr. McNeese and Dr. Reddy's research into CIS was supported in part from a grant from the National Science Foundation (IIS#0844947).

REFERENCES

Bruce, H., R. Fidel, A. M. Pejtersen, S. Dumais, J. Grudin, and S. Poltrock. 2003. A comparison of the collaborative information retrieval behaviour of two design teams. *The New Review of Information Behaviour Research* 4(1): 139–153. doi:10.1080/147163103100 01631499.

Clark, R. E., D. Feldon, J. J. G. Van Merrienboer, K. A. Yates, and S. Early. 2007. Cognitive task analysis. In *Handbook of Research on Educational Communications and Technology*, edited by J. M. Spector, M. D. Merrill, J. van Merrienboer, and M. P. Driscoll, 3rd edition. New York: Routledge.

Cooke, N. J. 1994. Varieties of knowledge elicitation techniques. *International Journal of Human-Computer Studies* 41(6): 801–849. doi:10.1006/ijhc.1994.1083.

Cooke, N. J. 2015. Team cognition as interaction. *Current Directions in Psychological Science* 24(6): 415–419. doi:10.1177/0963721415602474.

Cooke, N. J., J. C. Gorman, C. W. Myers, and J. L. Duran. 2013. Interactive team cognition. *Cognitive Science* 37(2): 255–285. doi:10.1111/cogs.12009.

Cooke, N. J., J. C. Gorman, and J. L. Winner. 2007. Team cognition. In *Handbook of Applied Cognition*, edited by F. T. Durso, R. S. Nickerson, S. T. Dumais, S. Lewandowsky, and T. J. Perfect, 2nd edition. pp. 239–268. Chichester, UK: John Wiley & Sons. doi:10.1002/9780470713181.ch10.

Cooke, N. J., E. Salas, J. A. Cannon-Bowers, and R. J. Stout. 2000. Measuring team knowledge. *Human Factors: The Journal of the Human Factors and Ergonomics Society* 42(1): 151–173. doi:10.1518/001872000779656561.

Crandall, B., G. A. Klein, and R. R. Hoffman. 2006. *Working Minds: A Practitioner's Guide to Cognitive Task Analysis.* Cambridge, MA: MIT Press.

Department of Health. 2004. Manual for Cancer Services 2004. http://webarchive. nationalarchives.gov.uk/+/ww.dh.gov.uk/en/Healthcare/Cancer/DH_4135595. (Accessed on March 10, 2016).

Dervin, B. 1983. An Overview of Sense-Making Research: Concepts, Methods, and Results to Date. *International Communications Association Annual Meeting,* Dallas, TX.

Dew, K., M. Stubbe, L. Signal, J. Stairmand, E. Dennett, J. Koea, A. Simpson et al. 2015. Cancer care decision making in multidisciplinary meetings. *Qualitative Health Research* 25(3): 397–407. doi:10.1177/1049732314553010.

Ellis, D. 1989. A behavioural model for information retrieval system design. *Journal of Information Science* 15(4–5): 237–247. doi:10.1177/016555158901500406.

Evans, B. M. and E. H. Chi. 2009. Towards a model of understanding social search. arXiv:0908.0595 [cs], August. http://arxiv.org/abs/0908.0595.

Fidel, R., H. Bruce, A. M. Pejtersen, S. Dumais, J. Grudin, and S. Poltrock. 2000. Collaborative information retrieval (CIR). *The New Review of Information Behaviour Research* 1(January): 235–247.

Fidel, R. and A. M. Pejtersen. 2004. From information behaviour research to the design of information systems: The cognitive work analysis framework. *Information Research: An International Electronic Journal* 10(1): 210.

Fidel, R., A. M. Pejtersen, B. Cleal, and H. Bruce. 2004. A multidimensional approach to the study of human-information interaction: A case study of collaborative information retrieval. *Journal of the American Society for Information Science and Technology* 55(11): 939–953. doi:10.1002/asi.20041.

Fisher, R. P. and R. E. Geiselman. 1992. *Memory-Enhancing Techniques for Investigative Interviewing: The Cognitive Interview.* Springfield, IL: Charles C Thomas Publisher.

Flach, J. 2016. Supporting productive thinking: The semiotic context for cognitive systems engineering (CSE). *Applied Ergonomics.* doi:10.1016/j.apergo.2015.09.001.

Foster, J. 2006. Collaborative information seeking and retrieval. *Annual Review of Information Science and Technology* 40(1): 329–356. doi:10.1002/aris.1440400115.

Gerosa, L., R. Giordani, M. Ronchetti, A. Soller, and R. Stevens. 2004. *Symmetric Synchronous Collaborative Navigation.* Trento, Italy: University of Trento.

Golovchinsky, G., J. Adcock, J. Pickens, P. Qvarfordt, and M. Back. 2008. Cerchiamo: A collaborative exploratory search tool. *Proceedings of Computer Supported Cooperative Work (CSCW'08)*, pp. 8–12, San Diego, CA.

Hall, D. L., M. McNeese, J. Llinas, and T. Mullen. 2008. A framework for dynamic hard/soft fusion. In *2008 11th International Conference on Information Fusion*, pp. 1–8. Cologne, Germany.

Hansen, P. and K. Järvelin. 2005. Collaborative information retrieval in an information-intensive domain. *Information Processing & Management* 41(5): 1101–1119. doi:10.1016/j.ipm.2004.04.016.

Hollenbeck, J. R., D. R. Ilgen, J. A. LePine, J. A. Colquitt, and J. Hedlund. 1998. Extending the multilevel theory of team decision making: Effects of feedback and experience in hierarchical teams. *The Academy of Management Journal* 41(3): 269–282. doi:10.2307/256907.

Hollnagel, E. and D. D. Woods. 1983. Cognitive systems engineering: New wine in new bottles. *International Journal of Man-Machine Studies* 18(6): 583–600. doi:10.1016/S0020-7373(83)80034-0.

Hyldegård, J. 2006. Collaborative information behaviour—Exploring Kuhlthau's information search process model in a group-based educational setting. *Information Processing & Management*, Formal methods for information retrieval, 42(1): 276–298. doi:10.1016/j.ipm.2004.06.013.

Hyldegård, J. 2009. Beyond the search process – Exploring group members' information behavior in context. *Information Processing & Management* 45(1): 142–158. doi:10.1016/j.ipm.2008.05.007.

Karsten, H. 1999. Collaboration and collaborative information technologies: A review of the evidence. *SIGMIS Database* 30(2): 44–65. doi:10.1145/383371.383375.

Karunakaran, A., and M. Reddy. 2012. Barriers to collaborative information seeking in organizations. *Proceedings of the American Society for Information Science and Technology* 49(1): 1–10. doi:10.1002/meet.14504901169.

Karunakaran, A., M. C. Reddy, and P. R. Spence. 2010. Toward a model of collaborative information behavior in organizations. *Journal of the American Society for Information Science and Technology* 64(12): 2437–2451. doi:10.1002/asi.22943.

Krishnappa, R. 2016. Multi-user search engine (MUSE): Supporting collaborative information seeking and retrieval. Unpublished Master Thesis, Rolla, MO: University of Missouri.

Kuhlthau, C. C. 1988. Developing a model of the library search process: Cognitive and affective aspects. *RQ* 28(2): 232–242.

Kuhlthau, C. C. 1991. Inside the search process: Information seeking from the user's perspective. *Journal of the American Society for Information Science* 42(5): 361–371.

Lamb, B. W., K. F. Brown, K. Nagpal, C. Vincent, J. S. A. Green, and N. Sevdalis. 2011. Quality of care management decisions by multidisciplinary cancer teams: A systematic review. *Annals of Surgical Oncology* 18(8): 2116–2125. doi:10.1245/s10434-011-1675-6.

Lamb, B. W., N. Sevdalis, H. Mostafid, C. Vincent, and J. S. A. Green. 2011. Quality improvement in multidisciplinary cancer teams: An investigation of teamwork and clinical decision-making and cross-validation of assessments. *Annals of Surgical Oncology* 18(13): 3535–3543. doi:10.1245/s10434-011-1773-5.

Lamb, B. W., C. Taylor, J. N. Lamb, S. L. Strickland, C. Vincent, J. S. A. Green, and N. Sevdalis. 2012. Facilitators and barriers to teamworking and patient centeredness in multidisciplinary cancer teams: Findings of a national study. *Annals of Surgical Oncology* 20(5): 1408–1416. doi:10.1245/s10434-012-2676-9.

Letsky, M., N. Warner, S. M. Fiore, M. Rosen, and E. Salas. 2007. *Macrocognition in Complex Team Problem Solving.* London, UK: Ashgate.

McComb, S., D. Kennedy, R. Perryman, N. Warner, and M. Letsky. 2010. Temporal patterns of mental model convergence: Implications for distributed teams interacting in electronic collaboration spaces. *Human Factors: The Journal of the Human Factors and Ergonomics Society* 52(2): 264–281. doi:10.1177/0018720810370458.

McNeese, M. D. 1996 (October). Collaborative systems research: Establishing ecological approaches through the living laboratory. In *Proceedings of the Human Factors and Ergonomics Society Annual Meeting* (Vol. 40, No. 15, pp. 767–771). Los Angeles, CA: SAGE Publications.

McNeese, M. D. and P. J. Ayoub. 2011. Concept mapping in the analysis and design of cognitive systems: A historical review. In B. M. Moon, R. R. Hoffman, J. D. Novak, & A. J. Cañas (Eds.), *Applied Concept Mapping: Capturing, Analyzing, and Organizing Knowledge*, pp. 47–66. New York: CRC Press.

McNeese, M. D., K. Perusich, and J. R. Rentsch. 2000. Advancing socio-technical systems design via the living laboratory. *Proceedings of the Human Factors and Ergonomics Society Annual Meeting* 44(12): 2–610–2–613. doi:10.1177/154193120004401245.

McNeese, N. J. 2014. The role of team cognition in collaborative information seeking during team-decision-making. Unpublished Dissertation, The Pennsylvania State University.

McNeese, N. J., N. J. Cooke, and V. Buchanan. 2015. Human factors guidelines for developing collaborative intelligence analysis technologies. *Proceedings of the Human Factors and Ergonomics Society Annual Meeting* 59(1): 821–825. doi:10.1177/1541931215591249.

McNeese, N. J. and M. C. Reddy. 2013. Studying team cognition during collaborative information seeking: A position paper. In *Workshop on Collaborative Information Seeking: Consolidating the Past, Creating the Future.* San Antonio, TX: ACM.

McNeese, N. J., and M. C. Reddy. 2015a. Articulating and understanding the development of a team mental model in a distributed medium. *Proceedings of the Human Factors and Ergonomics Society Annual Meeting* 59(1): 240–244. doi:10.1177/1541931215591049.

McNeese, N. J. and M. C. Reddy. 2015b. Concept mapping as a methodology to develop insights on cognition during collaborative information seeking. *Proceedings of the Human Factors and Ergonomics Society Annual Meeting* 59(1): 245–249. doi:10.1177/1541931215591050.

McNeese, N. J. and M. C. Reddy. 2015c. The role of team cognition in collaborative information seeking. *Journal of the Association for Information Science and Technology.* doi:10.1002/asi.23614.

McNeese, N. J., M. C. Reddy, and E. M. Friedenberg. 2014. Towards a team mental model of collaborative information seeking during team decision-making. *Proceedings of the Human Factors and Ergonomics Society Annual Meeting* 58(1): 335–339. doi:10.1177/1541931214581069.

Mohammed, S., L. Ferzandi, and K. Hamilton. 2010. Metaphor no more: A 15-year review of the team mental model construct. *Journal of Management.* doi:10.1177/0149206309356804.

Moon, B., R. R. Hoffman, J. Novak, and A. Canas (Eds.). 2011. *Applied Concept Mapping: Capturing, Analyzing, and Organizing Knowledge.* Boca Raton, FL: CRC Press.

Morris, M. R. 2013. Collaborative search revisited. In *Proceedings of the 2013 Conference on Computer Supported Cooperative Work*, pp. 1181–1192. CSCW'13. New York: ACM. doi:10.1145/2441776.2441910.

Morris, M. R. and E. Horvitz. 2007. SearchTogether: An interface for collaborative web search. In *Proceedings of the 20th Annual ACM Symposium on User Interface Software and Technology*, pp. 3–12. UIST'07. New York: ACM. doi:10.1145/1294211.1294215.

Orlitzky, M. and R. Y. Hirokawa. 2001. To err is human, to correct for it divine a meta-analysis of research testing the functional theory of group decision-making effectiveness. *Small Group Research* 32(3): 313–341. doi:10.1177/104649640103200303.

Paul, S. A. and M. C. Reddy. 2010. Understanding together: Sensemaking in collaborative information seeking. In *Proceedings of the 2010 ACM Conference on Computer Supported Cooperative Work*, pp. 321–330. CSCW'10. New York: ACM. doi:10.1145/1718918.1718976.

Pickens, J. and G. Golovchinsky. 2007. Collaborative exploratory search. *Proceedings of Workshop on Human-Computer Interaction and Information Retrieval, Computer Science and Artificial Intelligence Laboratory (CSAIL)*, pp. 21–22. Cambridge, MA: Masschusetts Institute of Technology.

Poltrock, S., J. Grudin, S. Dumais, R. Fidel, H. Bruce, and A. M. Pejtersen. 2003. Information seeking and sharing in design teams. *SIGGROUP Bulletin* 24(1): 14. doi:10.1145/1027232.1027271.

Prades, J. and J. M. Borràs. 2011. Multidisciplinary cancer care in Spain, or when the function creates the organ: Qualitative interview study. *BMC Public Health* 11: 141. doi:10.1186/1471-2458-11-141.

Prekop, P. 2002. A qualitative study of collaborative information seeking. *Journal of Documentation* 58(5): 533–547. doi:10.1108/00220410210441000.

Rasmussen, J. 1983. Skills, rules, and knowledge; signals, signs, and symbols, and other distinctions in human performance models. *IEEE Transactions on Systems, Man and Cybernetics* SMC-13(3): 257–266. doi:10.1109/TSMC.1983.6313160.

Rasmussen, J., A. M. Pejtersen, and L. P. Goodstein. 1994. *Cognitive Systems Engineering*, p. xviii, p. 378. New York: John & Wiley Sons.

Reddy, M. and P. Dourish. 2002. A finger on the pulse: Temporal rhythms and information seeking in medical work. In *Proceedings of the 2002 ACM conference on Computer supported cooperative work*, pp. 344–353. New York: ACM.

Reddy, M. C. and B. J. Jansen. 2008. A model for understanding collaborative information behavior in context: A study of two healthcare teams. *Information Processing & Management* 44(1): 256–273. doi:10.1016/j.ipm.2006.12.010.

Reddy, M. C., B. J. Jansen, and P. R. Spence. 2010. Collaborative information behavior: Exploring collaboration and coordination during information seeking and retrieval activities. In, Foster, J. (Ed.) *Collaborative Information Behavior: User Engagement and Communication Sharing*, pp. 73–88. Hershey, PA: Information Science Reference.

Reddy, M. C. and P. R. Spence. 2008. Collaborative information seeking: A field study of a multidisciplinary patient care team. *Information Processing & Management*, Evaluation of Interactive Information Retrieval Systems, 44 (1): 242–255. doi:10.1016/j.ipm.2006.12.003.

Ruiz-Casado, A., M. J. Ortega, A. Soria, and H. Cebolla. 2014. Clinical audit of multidisciplinary care at a medium-sized hospital in Spain. *World Journal of Surgical Oncology* 12(1): 1–7. doi:10.1186/1477-7819-12-53.

Salas, E., N. J. Cooke, and M. A. Rosen. 2008. On teams, teamwork, and team performance: Discoveries and developments (cover Story). *Human Factors* 50(3): 540–547. doi:10.1518/001872008×288457.

Schmidt, K. and L. Bannon. 1992. Taking CSCW seriously. *Computer Supported Cooperative Work (CSCW)* 1(1–2): 7–40. doi:10.1007/BF00752449.

Schraagen, J. M. C., S. F. Chipman, and V. J. Shalin (Eds). 2000. *Cognitive Task Analysis*. Mahwah, NJ: Erlbaum.

Shah, C. 2010a. Collaborative information seeking: A literature review. In *Exploring the Digital Frontier*, edited by A. Woodsworth, Vol. 32. Bradford, UK: Emerald Group Publishing.

Shah, C. 2010b (February). Coagmento—A collaborative information seeking, synthesis and sense-making framework. In *Integrated Demo at CSCW*, Savannah, GA, February 6–10.

Shah, C. and R. González-Ibáñez. 2011. Evaluating the synergic effect of collaboration in information seeking. In *Proceedings of the 34th International ACM SIGIR Conference on Research and Development in Information Retrieval*, pp. 913–922. SIGIR'11. New York: ACM. doi:10.1145/2009916.2010038.

Shah, C., G. Marchionini, and D. Kelly. 2009. Learning design principles for a collaborative information seeking system. In *CHI'09 Extended Abstracts on Human Factors in Computing Systems* (pp. 3419–3424), ACM.

Sonnenwald, D. H. and L. G. Pierce. 2000. Information behavior in dynamic group work contexts: Interwoven situational awareness, dense social networks and contested collaboration in command and control. *Information Processing & Management* 36(3): 461–479. doi:10.1016/S0306-4573(99)00039-4.

Spence, P. R. and M. C. Reddy. 2007. The 'Active' gatekeeper in collaborative information seeking activities. In *Proceedings of the 2007 International ACM Conference on Supporting Group Work*, pp. 277–280. GROUP'07. New York: ACM. doi:10.1145/1316624.1316666.

Spence, P. R. and M. Reddy. 2012. Beyond practices: A field study of the contextual factors impacting collaborative information seeking. *Proceedings of the American Society for Information Science and Technology* 49(1): 1–10. doi:10.1002/meet.14504901131.

Spence, P. R., M. C. Reddy, and R. Hall. 2005. A survey of collaborative information seeking practices of academic researchers. In *Proceedings of the 2005 International ACM SIGGROUP Conference on Supporting Group Work*, pp. 85–88. GROUP'05. New York: ACM. doi:10.1145/1099203.1099216.

Taylor, C., K. Brown, B. Lamb, J. Harris, N. Sevdalis, and J. S. A. Green. 2012. Developing and testing team (team evaluation and assessment measure), a self-assessment tool to improve cancer multidisciplinary teamwork. *Annals of Surgical Oncology* 19(13): 4019–4027. doi:10.1245/s10434-012-2493-1.

Van den Bossche, P., W. H. Gijselaers, M. Segers, and P. A. Kirschner. 2006. Social and cognitive factors driving teamwork in collaborative learning environments team learning beliefs and behaviors. *Small Group Research* 37(5): 490–521. doi:10.1177/1046496406292938.

Vicente, K. J. 1999. *Cognitive Work Analysis: Toward Safe, Productive, and Healthy Computer Based Work*. Mahwah, NJ: Erlbaum.

Weick, K. E., K. M. Sutcliffe, and D. Obstfeld. 2005. Organizing and the process of sensemaking. *Organization Science* 16(4): 409–421.

Willis, G. B. 1999. *Cognitive interviewing: A 'How To' Guide*. Presented at the 1999 Meeting of the American Statistical Association, Research Triangle Park, NC: Research Triangle Institute. http://www.hkr.se/pagefiles/35002/gordonwillis.pdf.

Wilson, T. D. 1997. Information behaviour: An interdisciplinary perspective. *Information Processing & Management* 33(4): 551–572. doi:10.1016/S0306-4573(97)00028-9.

Winograd, T. and F. Flores. 1986. *Understanding Computers and Cognition: A New Foundation for Design*. Norwood, NJ: Intellect Books.

7 A Case Study of Crisis Management Readiness in Cross-National Contexts

Applying Team Cognition Theories to Crisis Response Teams

Tristan C. Endsley, Peter Kent Forster, and James A. Reep

CONTENTS

Introduction 133
Cognitive Systems Engineering in Crises and the Crisis Management
Environment 133
    What Are Crises? 133
    Cognitive Systems Engineering in Crisis Management 134
How Is It Impacted by Culture? 134
    The Crisis Management Context 135
    Crisis Cycle 136
    Teams in Crisis Management 138
    Team Cognition 139
    Challenges Teams Face in Crisis Management Activities 140
    Common Operational Picture 141
    Team Situation Awareness 141
    Information Sharing 142
    Hidden Knowledge 143
    Cultural Composition 144
A Case Study Examining Team Cognition and Culture 144
    NeoCITIES Simulation 146
    Manchester University Case-Study Context 146
    Communication 147
    Sharing Hidden Knowledge 149
    Discussion 150
    Penn State Case Study Context 151

Communication 151
Sharing Hidden Knowledge 152
Discussion 154
Comparison of Contexts... 154
Communications 154
Performance 155
Sharing Hidden Knowledge 155
Review Questions.. 155
Acknowledgments... 156
References... 156

ADVANCE ORGANIZER

Crisis response necessitates timely, effective responses in a chaotic, uncertain environment. This complexity challenges teamwork by straining resources (both physical and human cognitive). Overcoming these challenges demands that teams are able to effectively coordinate and collaborate across all levels of a response, meaning they must share information, develop a shared understanding of the context, and develop common ground. In this environment, the impact of cultural components on effective teamwork is often overlooked.

This chapter describes two case studies that evaluates the following:

- How cultural differences emerge in team cognition.
- The impact on team information sharing behaviors in emergency crisis management contexts.

The study uses the living lab framework (LLF)—NeoCITIES, a scaled world, human-in-the loop simulation to assess two culturally distinct groups to evaluate cognitive differences in emergency crisis response. The chapter:

- Defines crises, and discusses team activities and behaviors in crisis response.
- Provides important and relevant concepts to team cognition and details many of the challenges inherent in teamwork.
- Concludes with an analysis of the findings from two case studies.

The case studies presented in this chapter indicate that further research is needed to draw conclusions in regards to the impacts of cultural contexts on the ways humans engage in teamwork. In particular, exploring and understanding processes of information sharing (when, what, and how much) in distributed settings across cultural groups represents a much-needed area of expansion. Given the nature of international political settings, sharing of intelligence across national boundaries, sharing of response work to disaster events or international conflicts occurs more frequently.

Adequately understanding information sharing and comprehension of that information is pivotal to teamwork in many different settings. From a living lab perspective, this chapter provides a new approach. Although integrating research with innovation, it also breaks down the traditional territorial boundaries by conducting an international study.

INTRODUCTION

Catastrophic events such as the Fukushima nuclear disaster or the Malaysia Air 370 disappearance represent breakdowns in complex sociotechnical systems. These crisis responses require international collaboration and teamwork. Teams carry out tasks associated with complex problems, involving multiple goals and numerous actions all of which are overshadowed by uncertainty and a dynamic environment (Fischer, Greiff, & Funke, 2012). Cognitive systems engineering (CSE) methods utilizing the LLF seek to understand the cognitive processes and tasks of this environment and subsequently inform the design and development of sociotechnical systems used in these environments.

This chapter establishes theoretical perspectives of crisis management within the CSE context. It examines how case studies may be used to compare and contrast the theories of team cognition in two culturally diverse arenas and evaluates the criteria for effective response using cases derived from the living lab approach. Much of the derived information is viable for building cognitive systems that address ill-defined problem solving across multiple levels of collaborative activity. A common error in examining crisis management is assigning too much attention to the precrisis planning stage. From an emergency response perspective, much of the precrisis management planning might be considered part of a risk-management process. While this debate is beyond the scope of this chapter, the application of CSE helps inform future planning by capturing valuable information on the context or strategic environment of the crisis, situational awareness, or tactical knowledge, timeliness of response, and accuracy of response, which encompasses proper distribution of resources and the disclosure of hidden knowledge.

The chapter's foundational concepts are then applied to two case studies encompassing the living lab approach and using participants in the United States and the United Kingdom.

COGNITIVE SYSTEMS ENGINEERING IN CRISES AND THE CRISIS MANAGEMENT ENVIRONMENT

WHAT ARE CRISES?

Reminiscent of Von Clausewitz's (1997) *fog of war*, crises are opaque and dynamic. Crises are understood as incidents that emerge quickly and unexpectedly and present a significant threat to an organization's interests (Holsti, 1989). They are characterized by high risk and uncertainty that may erode commonly used systems, communication protocol, and infrastructure to some extent. (Carrithers, DeHart, & Geaneas, 1998; Shaluf, Ahmadun, & Said, 2003). Furthermore, crises may threaten public safety, financial stability, and/or result in damage to organizational reputation (Coombs, 2007). Crises demand decisions to be made often in an environment with less than optimal information (Pearson & Clair, 1998). Faulkner (2001) describes a crisis as "a situation where the root cause of an event is, to some extent, self-inflicted through such problems as inept management structures and practices or a failure to adapt to change" (p. 136).

Navigating this environment requires contextual knowledge, situational awareness, and a process to mitigate its impact, and coordinate effective and efficient response and recovery. Crisis complexities often require that responses be coordinated among various team members to effectively minimize damage and manage the risk. Given the convergence of a challenging environment and the reliance on team response, crises may create significant strain on team cognition and on overall system performance.

COGNITIVE SYSTEMS ENGINEERING IN CRISIS MANAGEMENT

Crises demand human responses. These responses are challenged by the diversity among individual responders' knowledge, both with regard to the event, and the context in which the event occurs. In a team environment, different levels of knowledge, how it is collated, and its dissemination may result in conflicting goals or diminished agility. To improve crisis response, information resources and data must be processed quickly, and efficiently disseminated to the decision makers (Ntuen, Balogun, Boyle, & Turner, 2006). CSE integrates both human knowledge and perspective of the environment, as well as cognition and behavioral artifacts (McBride, Adams, Ntuen, & MaZaeva, 2002). By capturing human actions that influence crisis management, CSE may be used to improve real-time response through a systematic analysis of processes and the integration of tools. If a sufficiently robust system exists or becomes an integral research framework, such as the NeoCITIES simulation discussed in this chapter, the improvement of crisis responses through the identification of strengths and weaknesses can facilitate better planning, situational awareness, and the practical application of lessons learned.

HOW IS IT IMPACTED BY CULTURE?

Cultural context combines both the concepts of culture (i.e., individuals as members of various groups) as well as recognizing that environment also affects people (Rapoport, 2008, p. 20). Broadly, the cases examine a British-cultural context versus an American one. Preliminary results from these cases may be used to gather further knowledge of how similarities and differences resulting from cultural context influence cognitive systems and improve understanding of potential inhibitors of successful crisis response.

Evidence suggests that cultural background can have a profound impact on information-sharing behaviors, and can affect the development of a common operational picture (COP) within a team. A COP emerges within a team for a particular context when all team members have a representative understanding of the environment. Cultural norms can dictate how and in what manner people view their roles within a team, the types of interactions they will engage in with their team members, and how they share information relevant to their task (Rasmussen, Sieck, & Smart, 2009). In culturally diverse teams, when these underlying cognitive perspectives determined by culture are not appropriately understood, small issues can snowball into larger issues. For example, cultural perspectives of leadership and subordinate roles can dramatically influence how and when people deem it appropriate to share information, despite the importance of that information to overall team goals (Strauch, 2010). For example, in 1990 a Colombian aircraft on approach to New York City crashed due to fuel exhaustion, when the level of their distress was not accurately communicated to air traffic controllers. This disaster was identified as culturally influenced (Helmreich, 1994) as the captain, who only spoke Spanish, issued a state of emergency for his first officer to translate into English; however, this was not clearly conveyed to traffic controllers. It is hypothesized that the first officer may have been reluctant to ask for clarification from the captain due to cultural expectations that subordinate members do not challenge those in leadership positions. In other aircraft incidents, the cultural perspectives of team members have been found to dictate how team members respond to uncertainty in information, and how they interact with their fellow team members. McHugh et al. (2008, p. 146) describe an overall trend toward harmony in Asian (described as collective in their study) teams as opposed to the United States, and European teams. This trend toward harmony in Asian teams was found in processes of collaborative decision making as an emphasis on the preservation of face for leadership, which meant that team members rarely voiced outward dissent during decision meetings; instead, the process of dissent, or voicing of opinions, occurred in private and typically prior to a decision being made so that the leader would not lose any face in the process (McHugh et al., 2008). Integration of cultural cognitive underpinnings of teamwork in CSE activities reflects a much-needed area for new research.

In crisis-management situations, properly identifying, classifying, and responding to events is vital to ensuring the safety of both emergency response personnel and those affected by the crisis event. This problem is further compounded when combined with cultural differences (Jones, 2003; Moore, 1989) and the distributed teams that often surround large-scale crisis situations (Rosenthal & Hart, 1991). Due to the criticality of team performance during a crisis, having a method of training team members and ensuring a high degree of efficiency is imperative. In this study, we seek to understand how cultural differences can impact team-cognitive processes when combined with geographically dispersed teams. Furthermore, we hope to be able to gather insight that might lend itself to foster efficiency within these culturally diverse teams regardless of their geographic location.

THE CRISIS MANAGEMENT CONTEXT

Within the context of this study, crisis management is explored in a local context. The testing environment provides for robust engagement among multiple local resources (e.g., police, fire, and hazardous materials) responsible for response but is not programmed to examine interjurisdictional or even international crises. As a result, the focus here is more tactical. Regardless of environment, crisis management is a process designed to prevent and mitigate a crisis' damage on the affected organization (Coombs, 2007). Crises share general characteristics but are impacted differently by time (Parsons, 1996), severity, and geodemographic factors.

Successful crisis management requires planning and execution across a variety of environments (i.e., prevention, response, mitigation, and recovery). Recognizing that total prevention is unachievable, planning needs to include response, mitigation, recovery, and resilience. As previously discussed, this study assesses response, which is highly reactive. When an event occurs, a response is initiated and the response's impact on the event is evaluated for effectiveness. NeoCITIES is a well-designed simulation for assessing key response criteria and tasks. Properly used, the data generated by NeoCITIES may be used to improve responses that require partnerships, improve information sharing among dispersed organizations, and improve communication and resource allocation while controlling for resource erosion (a likely occurrence) during a crisis.

CRISIS CYCLE

As previously mentioned, effective crisis management requires a broader understanding of the stages of a crisis and their interrelationships. Although the geographic reach, time of response, and severity of a crisis may vary, the processes of preparing for, responding to, mitigating, and recovering from a crisis are cyclical. The four-stage crisis cycle—preparation, response, stabilization or mitigation, and recovery—emerged from the 1978 report of the National Governors Association Emergency Preparedness Project (Altay & Green III, 2006).

The initial stage of crisis management actual occurs before an event takes and involves preparation or preparedness. *Preparedness* focuses on identifying, preventing, mitigating, and communicating threats and vulnerabilities. Green III (2012) describes *mitigation* as "the application of measures that will either prevent the onset of a disaster or reduce the impacts should one occur" (Altay & Green III, 2006; Green III, 2012). To be effective, the *planning* stage should account for and seek to mitigate potential biases that may exist in teams or among teams that are responsible for response. Recognizing and reducing potential areas of friction resulting from differing perspectives both improves team collaboration and overall system effectiveness. Third, a crisis plan should include training (Coombs, 2007). Although some training may take the form of community outreach, much of the training and exercises are dominated by those responsible for response. Ultimately, a well-crafted crisis management plan is similar to the scripting of plays at the beginning of a football game. However, as in a football game, the crisis' unique characteristics will influence response. Part of the pre-event planning is recognizing that the planned response will ultimately be discarded in order to react to events as they occur.

The *response* stage in the crisis cycle deals specifically with the activities following the onset of a crisis. Green III (2012) describes the response process as "the employment of resources and emergency procedures as guided by plans to preserve life, property, the environment, and the social, economic, and political structure of the community, during the onset, impact, and immediate restoration of critical services in the aftermath of a disaster" (see also Altay & Green III, 2006, p. 480). The response activities assessed in NeoCITIES reflect the intensity of this part of the cycle and are highly time sensitive. Typical response activities may include urban search and rescue (USAR), fire fighting, emergency rescue, fatality management, medical care, evacuation of affected populations, employment of emergency operations planning, and use of the emergency operations center (Altay & Green III, 2006; Coppola, 2011; Green III, 2012). However, in today's information networked environment, response in cyberspace is likely to be needed either in conjunction with the more traditional responses or independently.

Although outlining response processes is critical, creating a COP once an event occurs is also vital. According to the Department of Homeland Security, the COP is the core situation awareness (SA) capability with regard to decision making, rapid staff actions, and mission execution (Security, 2012). COP may be achieved through a series of questions. Situation awareness is improved by defining *what happened*. Sharing knowledge on the event is also critical to implementing response and mitigation strategies. Second, the *what happened* should be assessed to ensure that the event rises to the level of a crisis. Different parts of an organization have different perspectives on what constitutes a crisis. A third question, which can often be overlooked in the opaque environment surrounding a crisis, is what might happen? What cascading effects might exist and what are the potential unintended consequences of response or lack thereof? The Germanwings Flight 9525 crash in the French Alps in May 2015 is illustrative of failure to consider *what ifs*. Shortly after the crash, Germanwings notified people via Twitter that as more information became available they would make it available via their website (Luege, 2015). What they failed to consider was whether their website was sufficiently robust to handle the volume of traffic that they had knowingly directed to it. When the site crashed, Germanwings had no other communication plan, meaning they abrogated the communication part of their crisis response responsibility. The remaining stages of the crisis management cycle are *stabilization and recovery*. Although not germane to this study, they deserve a brief description. Stabilization examines culpability and contributes to the lessons-learned process. The recovery stage requires actions to help maintain an affected population and to simultaneously rebuild affected infrastructure, so that a sense of normalcy can return.

As discussed above in the convergence of CSE to crisis management, the human element is central. Furthermore, the distributive nature of knowledge that occurs in most emergencies requires teamwork to resolve them (Ntuen et al., 2006). To be successful, this element relies on effective planning and training (i.e., preparation), agile response that understands the crisis' domain, and an ability to collect and process information that results in lessons learned. Within each of these stages, establishing shared goals, understanding of roles and responsibilities, and sharing information that creates adaptability is essential.

Moreover, each stage of the crisis-management cycle is characterized by the response of multiple teams, and often teams of teams, collaboratively engaged in distributed tasks. When examining crisis environments, the decision-making process of teams cannot be separated from the task within which they are engaged. Consequently, research into this context must be examined through a lens suitable to evaluating teams and how they work together to make decisions and accomplish tasks. As such, in the following section we will examine teams within the crisis management domain and the activities in which they engage.

TEAMS IN CRISIS MANAGEMENT

Whether it is a natural (i.e., an earthquake) or man-made (i.e., terrorism) event, crises require effective team coordination. The following table outlines team responses to the stages of a crisis (e.g., preparedness, response, recovery, and mitigation) as well as discussing activities and providing examples of actions (Altay & Green III, 2006; Coppola, 2011). When reviewing the table's structure, it should be understood that mitigation serves a dual role in a crisis environment (Table 7.1).

Understanding that context plays an important part in cognition presents challenges to designers, developers, and researchers of cognitive systems being used in crises. This is especially relevant within the crisis management domain due

TABLE 7.1
Description of Team Activities in the Crisis Cycle

Team Activities within the Crisis Management Cycle

Stage	Activities	Example Actions
Preparedness	Activities aimed to prepare a community for response to a future crisis event	• Emergency planning • Conducting disaster exercises • Training for response personnel • Development of communications systems • Public education • Construction of emergency operations centers
Response	"Employment of resources and emergency procedures as guided by plans to preserve life, property, the environment, and the social, economic, and political structure of the community" (Altay & Green III, 2006)	• Urban search and rescue • Fire fighting • Emergency rescue • Fatality management • Medical care • Evacuation of affected populations • Situation assessment • Needs assessment • Employment of emergency operations planning • Use of the emergency operations center
Recovery	Long-term process of stabilizing a community and restoring normalcy following a crisis event	• Debris cleanup • Rebuilding of infrastructure • Roads • Hospitals • Airports • Bridges
Mitigation	Measures aimed to prevent or reduce the impacts of a crisis	• Barrier construction (i.e., sand barriers for floods, salt sheds) • Active preventative measures (i.e., avalanche prevention)

Source: Altay, N., and Green III, W. G., *European Journal of Operational Research*, 175, 475–493, 2006.

to the high levels of risk, uncertainty, and criticality of the domain. Rules-based action, such as that explained by Carraher et al. (1990), does not translate well into environments that are dynamic and highly volatile as is characteristic of the crisis management domain. As such, designers of cognitive systems supporting these environments must be fully cognizant of the context within which those systems will be deployed and utilized. Subsequently, the LLF (McNeese, Mancuso, McNeese, Endsley, & Forster, 2013) is a viable solution for developing scaled-world simulations for evaluation of these systems and for examination of real-world problems emergent within the domain.

TEAM COGNITION

Teams are often defined as "a distinguishable set of two or more people who interact dynamically, interdependently, and adaptively toward a common and valued goal/object/mission, who have each been assigned specific roles or functions to perform, and who have a limited life span of membership" (Salas, Dickinson, Converse, & Tannenbaum, 1992, p. 4). Teams are employed in complex environments, "when the task complexity exceeds the capacity of an individual; when the task environment is ill-defined, ambiguous, and stressful; when multiple and quick decisions are needed; and when the lives of others depend on the collective insight of individual members" (Salas, Cooke, & Rosen, 2008, p. 540). Teams working in the management of crisis events are employed to deal with high stakes, in *ill-defined, ambiguous, and stressful* environments. In order to overcome the challenges of the crisis environment, members of a team must be able to collaborate, cooperate, and communicate successfully through the negotiation of a shared team process, or through the establishment of team cognition. Oftentimes, teamwork is distributed across technological artifacts (interfaces, social networks, and mobile computing devices) resulting in complex, cognitive systems.

Team cognition is described as a state of shared team functioning, which requires the development of common ground among team members (Mancuso & McNeese, 2012), the negotiation of shared knowledge structures (Fiore, Smith-Jentsch, Salas, Warner, & Letsky, 2010), the development of a COP (McNeese, Pfaff, et al., 2006) shared situation awareness, and through the development of team mental models (or shared representation of the environment). Team cognition emerges as a construction of a team's negotiated understanding of their tasks and interdependencies within a specific context *through distributed, emergent activities (of actors) using various sources* (McNeese, 2003, p. 519). As such, communication, collaboration, and coordination are fundamental to the development of team cognitive structures that will facilitate interactions within a team and improve overall performance outcomes. These requirements are compounded in distributed teamwork environments, where teams are geographically, temporally, and virtually displaced.

Furthermore, about what and how one thinks cannot be separated from the context irrespective to one's *knowledge, memory, and memory capacity* (Oyserman, 2015, p. 8), but rather are shaped by how and when the information is stored, accessed, or interpreted (Fiske, 1992; Schwarz, 2007; Smith & Semin, 2004). This concept, called situated cognition, refers to the subconscious influence that social and environmental contexts have on cognitive processes. In their words, "Context-embedded problems were much more easily solved than ones without a context" (Carraher et al., 1990, p. 24).

Contemporary collaborative environments, such as those used within the crisis management domain, require software and systems to be designed to support the distributed nature of the cognitive processes (i.e., application of cognitive system engineering) involved in making life and death decisions, and enabling responders to see through the *fog of war* via complex technological systems (Von Clausewitz & Graham, 1997). Historically, cognition has been viewed as a process that exists and takes place within an individual's head, however, distributed cognition contends that cognition is a process that is also external and displaced through the immediate context and which takes into account artifacts of the environment as part of cognitive processing (Hutchins & Klausen, 1996; Hutchins, 1995). Traditionally, research in distributed decision making has been primarily aimed at military and battlefield engagement scenarios, which can present challenges for gathering necessary data. NeoCITIES, however, was designed for emergency crisis management as exemplified in planning and operations for a prototypical college town environment, allowing for more freedom in subject recruitment (Jones, McNeese, Connors, Jefferson, & Hall, 2004). Methodologically, emergency crisis personnel were interviewed and observed *in situ* in order to fully grasp the cognitive processes involved in the technologically and socially distributed nature of this context. Subsequently, scenarios were created in order to closely approximate situations that might exist requiring the use of technological tools and teamwork in order to properly respond to each event presented to the research participants. Using NeoCITIES, our lab has been able to study and measure a wide variety of cognitive processes (see Hamilton et al., 2010; McNeese, Jefferson, et al., 2006) and the impact that various changes in the interface and cognitive tools have on team performance and decision-making processes (Jones et al., 2004).

Many aspects of team cognition inform our understanding of teams in complex domains; therefore, it is important to understand what aspects of teamwork are relevant to work, and those which present potential pitfalls in crisis management contexts. The following section will examine some of the challenges that teams face within the crisis management environment and the theoretical lenses through which we gain insight into team cognition intricacies.

CHALLENGES TEAMS FACE IN CRISIS MANAGEMENT ACTIVITIES

These crisis events demonstrated that cognitively demanding environments leads to significant breakdowns in human performance caused by improper management of team cognitive behaviors. Creating situation awareness and communicating information among parties is the greatest challenge in a crisis (Resnick, 2014). Although these historical events brought to light the significant challenges teams face in these complex environments, it is also clear from several recent events. In 2008, four highly mobile teams of terrorist attacked Mumbai. One of the tragedies of this crisis was that preattack intelligence was available but was not effectively integrated.

The fundamental difficulty is the presence of gaps within India's intelligence network (Jenkins, 2009; Laraia & Walker, 2009). The lack of managing team cognition behaviors was again evident in the time it took the authorities to develop situational awareness such as how many attackers were involved. The Boston Marathon bombing in 2013 stands in stark contrast to the Mumbai attacks five years earlier. Integrated technologies such as the Web Emergency Operations Center (WebEOC) and the Health and Homeland Alert Network (HHAN), coordination with the media to control messaging, something that never occurred in Mumbai; as well as preparedness planning, a unified incident command, and sharing information on the suspects heightened situational awareness. The results were that every patient transported to a hospital from the attack survived and the suspects were either killed or detained within three days ("After Action Report for the Response to the 2013 Boston Marathon Bombings," 2014; Resnick, 2014). In the following sections, several pitfalls including the development of a COP. Information sharing, and hidden knowledge profiles, to the development of shared team cognition is discussed in the context of a crisis management simulation, NeoCITIES.

COMMON OPERATIONAL PICTURE

To develop COP, members of a crisis management team must establish common ground, or a shared understanding of the operational context. A COP describes "a visual representation of tactical, operational, and strategic information to support rapid assimilation and integration by team members" (McNeese, Pfaff, et al., 2006, p. 467). It is often used across multiple levels of command as an informational tool within command and control (C2) centers (McNeese, Pfaff, et al., 2006). Significantly, a COP is needed to essentially establish shared situation awareness among all members of the team, and has been linked to improvements in overall team performance (Hager, 1997). In the uncertainty of crisis events, it is important that a COP is established across the various roles and responsibilities of the organization/team responding. This contributes to overall information sharing, knowledge management, and collaborative sensemaking, thus enabling team members to consolidate knowledge and improve performance. In command and control settings, the presence or absence of a COP can significantly impact outcomes in crisis management contexts. A team's ability to develop shared situation awareness of their environment is pivotal for its success.

TEAM SITUATION AWARENESS

Successful crisis management requires both COP and SA, which are interdependent but not synonymous terms. As previously discussed, COP is more operational. It relies on and contributes to SA but its focus is on responding to the elements of the event and is composed of the circumstances surrounding the event. SA benefits from COP in that it provides information for future events but at the time of event SA is a broader concept. The United States Coast Guard (USCG) defines situational awareness as *knowing what is going on around you* (USCG, 1998); however, SA is really informed by the knowledge sources that individual team members contribute to its development. Ntuen et al. (2006) outlined five functional knowledge types: common sense, temporal, domain, explanatory, and maintenance. These knowledge types span the immediacy of the event such as temporal issues of time and space or why it happened but still only provide event knowledge. SA encompasses context as well as event knowledge. This might include the rookie responder who has a limited historical foundation on which to draw but is highly competent at using new tools to categorize data to the older responder who has the ability to put the event in a broader context and draws on lessons learned from similar events while offering insight into potential cascading effects, countervailing risks, and even ancillary benefits. Within a team, these knowledge sources collide. Effective response requires command and control that recognizes the value of functional and contextual knowledge and is able to integrate them and effectively manage the dissemination of information. On paper this response sounds simple, in the trauma and dynamism of a crisis the collision of these knowledge sources and their biases often create a barrier to team SA.

Situation awareness is fundamental for teams in crisis management context. SA reflects an understanding and comprehension of the environment that allows the team to use information now and in the future. At the individual level, Endsley's (1988) model of SA presents three stages to SA: (1) perception, (2) comprehension, and (3) projection. Perception involves orienting one to the surrounding environment; comprehension is the process gathering and understanding the current environment or situation; and projection is the ability to the future trajectory of the situation (Endsley & Jones, 2011). At the team level, team situation awareness is "the degree to which every team member possesses the SA required for his or her responsibilities" (Endsley, 1995, p. 39). As a result of teams' interdependent nature, some individual SA will overlap.

Moreover, through the process of team coordination, individual SA will become shared across team members, even though no two team members may have identical individual SA for a particular situation. The degree to which the requirements of each team member's individual SA are met contribute to the level of performance a team is able to achieve (Endsley, 1995; Endsley & Jones, 2011). Understanding team SA in this context is important due to "the urgency, complexity, and uncertainty of disaster environments [which] test the limits of human capacity for seeking, processing, and disseminating information to support coordinated action" (Comfort, 1993, p. 15). SA's three stages of perception, comprehension, and projection are critical to achieving shared goals, which are useful for coordinating, responding, and mitigating a crisis.

INFORMATION SHARING

Key to the development of a COP across members of a team or organization is the process of information sharing (Mesmer-Magnus & DeChurch, 2009). Information sharing and the processes by which knowledge is transmitted have significant impacts on the outcomes of team performance during crisis events. Formally, information sharing is defined as "the degree to which team members share information with each other" (Johnson et al., 2006, p. 106) or furthermore, "conscious and deliberate attempts on the part of team members to exchange work-related information, keep one another apprised of activities, and inform one another of key developments" (Bunderson & Sutcliffe, 2002, p. 881). How teams access and utilize distinctive knowledge can greatly influence outcomes, it can enhance socioemotional aspects of a team's functioning (Mesmer-Magnus & DeChurch, 2009) leading to better inter-actions, more openness between the team members as well as facilitate trust and cohesion. Conversely, improper management of the distinctive knowledge can create strife and limit successful outcomes.

A meta-analysis of 72 team studies by Mesmer-Magnus & DeChurch (2009) revealed that information sharing positively predicted team performance, team cohesion, member satisfaction, and knowledge integration (p. 539). Barriers to effec-tive information sharing among team members, within an organization and between organizations are numerous within the crisis management domain for a variety of reasons. In practice, information overload (leading to cognitive overload), hidden knowledge, inherent chaos, and inaccurate or outdated information can all con-tribute to human error. In addition, team homogeneity adversely impacts informa-tion sharing behaviors, in which more highly homogeneous teams will engage in more information sharing behaviors than more diverse teams (Mesmer-Magnus & DeChurch, 2009).

The inherent chaotic nature of crises creates a context in which there is both an overwhelming amount of information and simultaneously not enough information to solve the issues at hand. This feature of the environment can create significant challenges for crisis managers and responders who must sort through relevant data, outdated information, and identify inaccurate or purposeful misinformation leading to information overload and cognitive failures.

This can be compounded by the increasingly distributed nature of teamwork. In today's world, teams are dispersed across geographical, cultural, and temporal boundar-ies with information communication technologies (ICTs) as the facilitator for many of the traditional face-to-face meetings. As advantageous as they may be, ICTs also present challenges, such as, "creat[ing] gaps in work like time zones and geographical sepa-rations, differences in cultures," (Ahuja, 2010, p. 36) and impacting information shar-ing behaviors (Mesmer-Magnus, DeChurch, Jimenez-Rodriguez, Wildman, & Shuffler, 2011). Mesmer-Magnus et al. (2011) found that while the virtuality of these environments positively influences teams sharing unique information it also impedes the openness, or overt sharing of information whether unique or not, of information sharing.

Hidden Knowledge

Despite the obvious utility of a divide and conquer approach to information exchange and division of knowledge among team members, it is often the case that the pro-cesses of information sharing and discussion within a team will focus on infor-mation that is shared among team members, rather than that which is unique, and potentially significant, to individual team members. A phenomena described by Stasser & Titus (1985) as hidden knowledge profiles, is one in which the "infor-mation that supports the optimal decision alternative is largely unshared, whereas information that supports the less desirable option is mostly shared" (Bowman & Wittenbaum, 2012, p. 296). The presence of these hidden knowledge profiles can significantly restrict a team's ability to develop a COP, as important and relevant information key to fostering success goes unshared. In fact, as Stasser & Titus (1987) state, "the way in which unshared information was distributed among members could significantly bias their pre-discussion preferences and that discussion in their groups rarely countered or corrected this initial bias" (p. 82), leading to nonoptimal decision making, and failure to successfully complete tasks and goals. In a crisis setting, this can be disastrous. The impacts of hidden knowledge on information sharing in teams can be compounded further by the influence of other contextual factors. Distribution of teams, national culture, and time pressure, among other things can compound the impacts of hidden knowledge and influence the kind of information that team members share.

CULTURAL COMPOSITION

At a basic level, diversity within a team impacts the levels of information that is shared among team members (Mesmer-Magnus & DeChurch, 2009) with higher levels of homogeneity paving the way for more cohesive interactions; however, this can lead to a lack of richness in problem solving a decision making. In crisis management, appropriate consideration of the cultural perspectives of team members represents an area in need of significant growth; the high-stakes nature of crisis response is not forgiving to errors that might arise from cultural misunderstandings on multicultural teams or from ingrained perspectives that disallow important information to be conveyed in a timely and articulated manner.

MIDWAY BREATHER—STOP AND REFLECT

- Crises are complex events that require teamwork to resolve successfully. However, teamwork is strained by a variety of factors (e.g., team cognition, common operating picture, situation awareness, information sharing, and hidden knowledge).
- The conceptual strains to teamwork may be further exacerbated by the role of culture; however, creating an environment in which to study this effectively presents a research challenge best resolved by a multinational case-study approach.

A CASE STUDY EXAMINING TEAM COGNITION AND CULTURE

In order to examine the differences that emerge across cultural groups, two case studies were developed using a scaled-world simulation, NeoCITIES. NeoCITIES is a scaled-world, human-in-the-loop distributed team simulation that allows for the effective examination of team cognition within complex settings. It is centered on routine and nonroutine emergency crisis response events in which teams of three must take on the roles of fire, police, and hazardous materials to coordinate and collaborate on resource allocation to events throughout the simulation. Although the studies used the NeoCITIES simulation as a foundation for data collection, the challenges of controlling for all potential variables across multiple sites in different countries proved to be insurmountable. As a result, the researchers agreed that a case-study approach presents data that are collected via separate experiments in separate international venues. These studies, when viewed as individual events, provide insight into how cultural differences may influence key factors in crisis response outlined above. The results presented here reflect the premise that multiple factors influence crisis response, and one of those is cultural context.

Research in cross-cultural contexts has remained largely unexplored due to the complexities and complications that arise from conducting such studies. Through an existing interuniversity cooperation, The Pennsylvania State University and the University of Manchester (UK) tackled the challenges by designing a collaborative research project to test cultural differences. The project's goal was to determine how culture might influence crisis response with a particular focus on information sharing. By analyzing responses from the United States and the United Kingdom teams to a series of events that required communicating and sharing hidden knowledge, the researchers hoped to identify similarities and differences in the two culturally different environments. The realities of cross-cultural research challenges such as establishing a common research approach and environment resulted in gathering data from two unique cases rather than determining what similarities and differences existed between commonly controlled research environments. Although failing to extract significant data related to the intended results, the experiment offered case data that is presented.

Penn State and the University of Manchester (United Kingdom) have had a long-term relationship, built primarily around a shared credit-based curriculum that offers professional development education to multinational companies. This project in itself presents an interesting database for assessing cross-cultural education. With more than a decade of cooperation, the primary investigators (PIs) on the education project agreed that a collaborative research project would further strengthen the interinstitutional partnership. The long sought-after initiative came to fruition in the context of cultural influencers on crisis management response. Combined, the institutions provided a viable participant poll and environment to conduct the study, a shared simulation (i.e., NeoCITIES) from which to collect data, and an institutional commitment to moving the project ahead.

Successful collaboration supported by viable findings required an agreed upon research approach, establishment of a similar testing environment, and most critically that NeoCITIES could be adapted for the United Kingdom environment in terms of contextual relevance, technical installation, and training. The advantages of identifying the key areas of collaboration allowed the researchers to adjust the project's perspective when insurmountable challenges arose. The lessons learned through the process of setting up this study in itself offered insight into cultural differences that informed the team's understanding of what was needed to implement cross-cultural research.

Although there was conceptual agreement on the research approach, it ultimately became apparent that establishing a common research environment that was the basis of valid statistical research project was undoable. The available facilities at the two institutions were simply too different to create similar testing environments in the time allocated for the study. With a quantitative study unattainable, it was decided that instead the experiments should be conducted at the separate locations but compared using a collective case-study approach. Case studies offer in-depth analysis of a complex problems or events over a specific time. As a result, case studies are not characterized by the methods used to collect and analyze data, but rather focus on the unit of analysis (Willig, 2001, p. 74). Case-study boundaries result from specifics of the analysis—who, what, where, when, and how that yields the data collected via observation, simulation, document review, and interviews. A case-study approach is highly compatible with the LLF as it carries ethnography, knowledge elicitation, and scaled-world elements to interpret practice around certain boundary conditions and intended goals related to cognitive systems and culture. Through a collective or comparative case method, the results from similar, albeit not identical, events may be analyzed to understand similarities and differences, which would be compared with a particular focus on communication among team members including the sharing of hidden knowledge. Furthermore, the case-study approach employed is highly compatible with the LLF as it carries ethnography, knowledge elicitation, and scaled-world elements to interpret practice around certain boundary conditions and intended goals related to teams, cognitive systems, and culture.

NEOCITIES SIMULATION

NeoCITIES requires participants to coordinate response to a series of crisis management tasks that occur within an overall scenario. Participants are given two training scenarios to familiarize themselves with the design of the simulation, and the roles they will assume, and the resources they have at their disposal to respond to the presented tasks. For the actual experiment, participants receive two performance scenarios, which typically last for 15–20 minutes. During each performance scenario, participants receive an information briefing through the information briefing panel of the simulation interface (Figure 7.1).

This briefing provided each team member with unique intelligence that pertained to an upcoming event, such as the time when the event is likely to occur, how much time they have for an appropriate and successful response, and the correct ordering of resources to be used in response to the event. This information is described as the hidden knowledge profile for this study. As discussed above, hidden knowledge reflects the type of information about a scenario that it is not jointly held by all members of a team, instead it is often the information that only one member of the team has access to, but which is not often shared with other team members due to an inherent bias to discuss shared information (Stasser & Titus, 1987). The presence of hidden knowledge can have profound impacts on team performance in complex domains, as it can dramatically limit the effectiveness of a response, as teams operate without a full understanding of complexities within their environment.

Participants at both institutions were given exactly the same intelligence information for the same events. This information as well as general communications behavior formed the basis of the study and are collated and analyzed using the following criteria as shown in Table 7.2.

MANCHESTER UNIVERSITY CASE-STUDY CONTEXT

At Manchester, 12 teams of 3 individuals participated in the simulation. The participants were mainly second and third year undergraduate students, but did include two

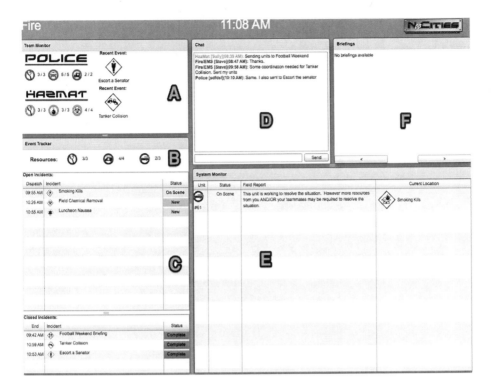

FIGURE 7.1   The simulation interface has the use of ways to keep track of one's own as well as team member's actions, the team monitor (A), the unit monitor (B), an event monitor (C), chat console (D), the dispatch panel (E), and a briefings panel (F).

graduate students. Overall age range was 18–26. Participants were recruited through a campus initiative for career assessments, which would give participant's experience with teamwork activities.

Specific adaptations were made to the NeoCITIES simulation so that it would be culturally and contextually relevant to participants in the United Kingdom. These included editing the language of the simulation so that it was reflective of British English, rather than American English, and updating the roles so that Environmental Services/Army was used in place of Hazardous Materials.* This case study, which contains a single scenario, presents an in-depth examination of team communications, sharing of hidden knowledge, and performance results.

COMMUNICATION

Communications were analyzed both quantitatively and qualitatively. The amount of communications per team, as defined by word count, compared to overall performance, as defined by percentage of successful completion of the scenario, provided initial data points. Observing completion rate as a factor of words used resulted in Table 7.3:

---

* For a further discussion of these adaptations, please view Endsley, Reep, McNeese, Forster, 2015.

TABLE 7.2
Communication Codes Used for Analyses

Code	Measure	Description
	Overall Rate of Communications	
	Total communications	Total number of communications in the session. the number of words in the experimental session
1	**Information Sharing**	
1.1	Information request	Request for information
1.2	Information transfer	Sharing of information
1.3	Unique information exchange	Exchange of information pertaining to the briefings in the hidden knowledge condition
2	**Action**	
2.1	Action request	Request for an action
2.2	Action intention	Statement of intended action
2.3	Action execution	Statement of action completion
3	**Coordination Behavior**	
3.1	Request for coordination	Asking for assistance
3.2	Coordinating action	Agreement to coordinate actions
4	Insignificant utterances	Utterances that do not convey information

Source: Adapted from Pfaff, M. S., Effects of mood and stress on group communication and performance in a simulated task environment. Doctor of Philosophy Dissertation, The Pennsylvania State University Electronic Theses and Dissertations Database, 2008; Entin, E. E., and Entin, E. B., *Measures for evaluation of team processes and performance in experiments and exercises.* Paper presented at the 6th International Command and Control Research and Technology Symposium, Annapolis, Maryland, 2001.

In the four teams that communicated with less than 50 words, there was an average of 90.25% completion. Five teams exchanged 100 or more words. Their average completion rate was 82% ranging from 67% to 89% completion. In the three teams with communications containing 50–100 words, however, had a much lower average completion rate, with a range of 66.7% to 83% completion. For example, one team evidenced this with the highest amount of communications (298 words). While this team did share response (e.g., 45 minutes) and event timing (e.g., event will occur at 6:45 p.m.) information, but not the ordering of responses for the hidden knowledge event.

TABLE 7.3
Communication and Completion Rates

Number of Teams	Range of Words	Average Completion Rate (%)
4	Less than 50 words	90.25
3	50–100 words	74.2
5	More than 100 words	82

It is interesting to note that they did not successfully complete the hidden knowledge event. Indeed, it seemed like this team was narrowly focused on timing aspects, and since the ordering information was never shared, they spent a significant amount of time trying to figure out how to respond and why their response was not working. This coordination effort occurred a minute and a half into receiving the event— so although they discussed timing aspects a bit more, they still did not seem to anticipate it well. Overall this team successfully completed about 89% of their total events, indicating that a strategy of a higher volume of communication, while not completely detrimental to their performance did not make them one of the more successful teams.

A second measure of communications was the quality of the communications. These were assessed by analyzing communications among the team members. Most teams at Manchester were engaged in coordinating activities within the communication chat window, very few of them did not actively coordinate actions, or at least discuss appropriate response activities. Generally, many of the coordinating actions took place in the form of questions to team members, as opposed to very specific directives. Following are some examples of the team communications:

> "Police [uom0337] (09:05 AM): do we have fire investigator on laptops?"
> "Fire/EMS [uom0352] (09:27 AM): might need police investigator for collapsed student"
> "Fire/EMS [uom0352] (09:52 AM): yap [sic] fire investigator sent"
> "Fire/EMS [uom0355] (11:48 AM): is police sent to stampede?"
> "Environmental Services/Army [uom0366] (09:37 AM): Investigator is on scene from me police on route?"

In some cases, these coordinating activities were requests for actions and resulted from cues from the simulation environment and thus were reactive. There was limited secondary discussion about the event itself or need resources. As a result, the researchers felt that situation awareness was improved reactively, which demonstrates an elementary level of team cognition.

> "Environmental Services/Army [uom0365] (09:52 AM): police"
> "Fire/EMS [uom0350] (09:56 AM): police"
> "Fire/EMS [uom0350] (10:41 AM): chemical"
> "Environmental Services/Army [uom0365] (11:14 AM): fire service"

SHARING HIDDEN KNOWLEDGE

In a crisis environment, some teams either have or receive unique information. Critical to improved response is the sharing of this hidden knowledge to inform decision making. This approach fundamentally involves the following two questions that were analyzed in the case studies:

- Was hidden knowledge shared between team members?
- Did the team use shared hidden knowledge?

In analyzing the sharing of hidden knowledge, it was evident that the Manchester teams were engaged in very little overt exchange of information from the briefings. Six of 12 teams failed the hidden knowledge performance indicating that either they did not effectively share the information they received in the briefings (a teamwork problem) or they did not use the knowledge they received from their teammates or the intelligence briefing to allocate appropriate resources (a situation-awareness problem). Communication data among four of the six teams that completed the hidden knowledge performance also indicated that they did not share or act on the hidden information. Three of these teams also had very limited communications, less than 50 words, but two completed 100% of the events and the third completed 83%. However, it is likely that the teams that successfully completed the hidden knowledge event used a brute force tactic of sending all resources. For the teams that shared information, one shared it late in the process and the communications on another did not make it clear the extent to which it had been shared. Both of these teams failed the performance measure.

One team shared the information early in the session, had a high level of communications and completed 94% of the events. Hence, one might deduce that early sharing and continuous communication enhances cognition. However, looking at another team it became clear that the tenor of the communications also is important. A second team that had a high level of communication but shared information more as a directive, with one team member ordering actions rather than eliciting input, successfully completed the hidden knowledge exercise but only completed 63% of the events. Thus, while teams may mention the hidden knowledge event generically (i.e., it takes place at the airport or involves luggage), they very rarely shared or discussed the unique information that they received in the briefings (i.e., the time and location of the event [6:45 p.m. at the airport], ordering and type of resources needed [SWAT, fire truck, and bomb squad], and amount of time to respond to the event [60 minutes]). Instead, team members who possessed the hidden information about the resources needed to successfully complete the event tended to ask others whether the resources were dispatched failing to realize that only they held that information. This is evidenced by vague references, which related to the resources but gave no indication that it was received in a briefing:

> "Fire/EMS [uom0357] (07:33 PM): has swat and bomb squad gone to the airport."

Overall, analysis of the communications shows that rather than using the information, the group simply completely allocated all resources.

DISCUSSION

No positive correlation existed between the amount of communications and their successful performance. In some instances, increased communication inhibited cognition leading to decreased performance. An optimum level of communications, in terms of number and quality exists; however, in this study there was insufficient data to determine that optimal level. Communication was deeply embedded in the simulation's context, which allowed team members to request actions and information without any additional effort. In some cases, however, communication strategies did not rise to the level of a collaboration that indicated the task had been completed. This may be due to a higher level of shared cognition, meaning that a lot of additional communication was not required for their coordination. Some teams were able to successfully complete the hidden knowledge event without sharing hidden knowledge information and completed the scenario. This reflects a strategic decision to simply allocate all resources to an event. While the scenario and the briefings were designed so that it would not be possible to complete the event without sharing that information, some teams were successful by using a brute force response, meaning they simply dispatched everything. This worked in the scenario environment, which does not account completely for the erosion of resource capability that occurs in real crises. Each team member was given specific information about one of the events. This information was unique to each role (e.g., Police received the timing information and fire-received information about the required resources). None of the teams actually shared the all of the unique information that they received in the briefing, and some teams shared none of the unique information they received in the briefings at all.

PENN STATE CASE STUDY CONTEXT

At Penn State, eight teams participated in the case study. Twenty-four students aged 20–24 were asked to complete the NeoCITIES scaled-world simulation. The demographics of the participants were 18 males, 6 females, 22 undergraduates and 2 graduate students. Participants were recruited through a single class in the College of Information Sciences and Technology, and were given course credit for their participation in the study. Students were part of a senior level class on team cognition.

In the U.S. case study, two scenarios were analyzed, but for the purposes of parallelism with the Manchester case study, only data from the same scenario will be presented below.

COMMUNICATION

Similar to the Manchester case study, communications in the U.S. teams were analyzed both quantitatively and qualitatively using the same methodology. The results are in the following (Table 7.4):

TABLE 7.4
Communication and Completion Rates

Number of Teams	Range of Words	Average Completion Rate (%)
1	Less than 50 words	94
7	More than 100 words	90

In general, teams in the United States seemed to have a high number of communications, with seven out of eight teams communicating with more than 100 words in the chat communication window throughout the scenario. Of the seven teams, four teams used over 150 words in their communications; these four teams had a dramatically lower average completion rate of events at 84.2%. Similar to the Manchester teams, it is clear that increased communication took up time and inhibited cognition leading to decreased performance.

Three U.S. teams used a medium level of communications (e.g., 101–114 words), shared information including hidden knowledge, and had an average completion rate of 98%. The fewest communications among the U.S. teams was 46 words but this team also completed the hidden knowledge event and 94% of all events.

Across all of the teams, the vast majority of communications pertained to requests for action. This involved directly requesting action from a teammate for a particular event or resources. For example, someone in the fire role may request for assistance from other teammates

"Police [psu0386] (11:46 AM): more resources at forum"
"Police [psu0386] (12:19 PM): more units at forum!!!!"
"HazMat [psu0387] (12:40 PM): I have two bomb squads says thats enough"
"Police [psu0386] (01:10 PM): more at sensitive matoerial [sic]"

Most, if not all, teams engaged in some level of coordination through their communications, typically asking for additional resources as in the example above, as well as through updates of their own resource allocations. In most cases, it seems as though there was an active approach to problem solving.

"Fire/EMS [psu0382] (04:47 PM): did you send an investigator?"
"Fire/EMS [psu0382] (04:12 PM): I sent it but it says i dont have the proper credentiaals [sic]"

Sharing Hidden Knowledge

A method similar to the Manchester case study was used to analyze hidden knowledge sharing in teams in the U.S. case study. As with Manchester, it became clear that the American teams did not overtly realize that the information they received in the briefings was unique to them. It was also apparent that some teams resorted to the *brute force* methods discussed in the case above.

In other cases, teams recognized that the information they received was unique but still did not discuss it in-depth. In analyzing the question of whether hidden knowledge was shared between the team members it became clear that for the most part, the U.S. teams actively shared some information with their teammates regarding the briefing. There also appeared to be improved cognition among a majority of the teams (i.e., five successfully completed the hidden knowledge event). Although many teams mention the event more generically (i.e., it takes place at the airport or involves luggage), there tended to be a lot of follow up, with a discussion of at least some details of the hidden knowledge.

"Fire/EMS [psu0385] (02:01 PM): say when you send units to airport"...
"HazMat [psu0387] (07:02 PM): Do we know what order for the airport?"...
"Fire/EMS [psu0385] (07:19 PM): swat, bomb squad, ambulance"

Three teams out of the eight failed the hidden knowledge event on the same scenario as the Manchester case study, which required that they share the unique information they received from their briefings.

In this scenario, only two teams overtly shared all of the uniquely held information that they received via the briefings during the NeoCITIES simulation. This either took the form of a copy/pasting of the briefing by each team member, or a more fragmented approach where important details were shared in a briefer manner.

"Fire/EMS [psu0400] (01:32 PM): Any events at the airport should be accompanied by all units in the following order: (1) Police (SWAT): to clear the crowds from the airport (2) HazMat (Bomb Squad): to search for and disarm any bombs on scene (3) Fire (Ambulance): to treat any victims."

"Police [psu0401] (01:46 PM): A mentally unstable student has decided that because she can't travel home for semester break she doesn't want anyone else to either. She is planning on detonating a suitcase bomb at the University Park Airport at around 6:45 today. All units should be on alert."

"HazMat [psu0402] (02:02 PM): nits [sic] should arrive on scene within 60 minutes (simulation time) of notification (i.e., 60 seconds actual time)."

However, of the eight teams, six alluded to the briefing though they did not share all of the information they received. Two did not share or discuss briefings at all. When information was shared, more often than not, it was only to convey ordering information given in the briefings. Notwithstanding, four of the six successfully completed the hidden knowledge event.

In many cases, participants would mention the briefing, as in the case of one team, "HazMat [psu0396] (02:03 PM): keeping a bomb squad in reserve for the airport threat." This however did not prompt other team members to share their uniquely held information, and in fact this team spent significant amount of communications trying to work out as an appropriate response to the luggage event at the airport, while also juggling responses to other events. Their communications illustrate the confusion:

"HazMat [psu0396] (07:10 PM): bomb sqaud [sic] and investigator to luggage"
"HazMat [psu0396] (07:29 PM): cleanuop [sic] to toxic gas"
"HazMat [psu0396] (09:09 PM): Firetruck maybe to luggage?"
"HazMat [psu0396] (09:23 PM): or ambulance…"
"Police [psu0395] (02:10 PM): on the way"
"HazMat [psu0396] (03:57 PM): sending a chem truck to help with swine flu decon"
"HazMat [psu0396] (08:05 PM): pulling back investigator from lugagge [sic]… probably need police"

Their inability to successfully understand the response requirements transmitted to them in the briefing suggests a breakdown in both individual and team situation awareness, particularly since they did not discuss any of the information from the intelligence briefing,[*] and did not act on the information they received. This may have been a form of perceptual blindness, in which the team was too focused on their ongoing tasks to attend to a new stimuli (briefing) to the detriment of their performance (Mack & Rock, 1998).

DISCUSSION

In the U.S. case, it was found that an optimum level of communications, both in terms of number and quality exists; however, a larger N is needed to further refine this conclusion. Overall it was found that communications need to be continuous and prompt particularly with regard to hidden knowledge to be effective. However, it appears that increased communications beyond a certain point took up time and inhibited cognition leading to decreased performance.

COMPARISON OF CONTEXTS

Two case studies of an emergency crisis management simulation deployed in two contexts revealed interesting variances between the different settings of Manchester and Penn State. The comparison of contexts focused on communication activity, overall performance, and sharing hidden knowledge. Although this is a preliminary study, two predominant questions emerged (1) What difference or variance was observed? (2) What were the explanations for the variance between contexts in the ways information was shared?

The majority of the American teams communicated more and performed better than the British teams. A great number of communications from the Americans pertained to requesting and verifying actions. In addition, Americans were more intent about communicating the actions that they had taken thus completing a feedback loop. In contrast, the Manchester teams devoted more time to actively coordinating actions than requesting information and assistance.

COMMUNICATIONS

It was observed that in their communications, Manchester teams were typically more polite in their requests for action by other team members. However, all in all there seemed to be a very limited exchange of information from the briefings, more directly they seemed to ask about whether the relevant resources for that event had been sent in the middle of the event, without reference to intelligence they had received via the briefings.

---

[*] It is important to note that when a briefing is released, participants are told via a pop up window. They must click *okay* in order to return to the simulation.

PERFORMANCE

Overall percentage completion of the events was much lower for the British (82.9% average completion rate with a low completion of 66.7%) teams than the Americans (90.6% with the low being 77%). Americans also seemed to communicate with one another more on average (1097 words) (137/team average) than the Brits (1328 words) (110 words/team average). In both cases, higher number of communications did not translate to improved performance. Suggesting that there may be multiple data points, supporting the assumption that an optimum communication level for teams exists, such that too much or too little communication erodes performance. Further research on this topic in distributed settings is certainly worthwhile.

SHARING HIDDEN KNOWLEDGE

Considering the number of teams, the difference between the British 50% completion and the U.S. 62.5% completion is insignificant. However, overall average completion for the teams who successfully completed the hidden knowledge event was remarkably close, with 95.4% for the U.S. and 94% for the British. Looking at the data, the best explanation for variance in sharing hidden knowledge is based on team composition rather than other factors. Regardless, continuous and prompt communication of hidden knowledge is apparently a key to success. Although this is not a startling reflection, it again substantiates the idea that sharing information is critical to successful response in crisis events.

In some ways these differences in styles of approach may have emerged due to there being generally more national diversity on teams at Manchester than at Penn state, which may have effectively required more overt coordination to achieve a COP during the simulation scenarios, due to limitations of teams not having inherent common ground. In addition, as students at Penn State were part of a larger class unit, with many interactions through the course of a semester, participants may have already developed a rapport with their teammates, while those at Manchester were much more ad hoc in nature.

Some of these differences may be attributed to cultural realities, which presents an opportunity for continued research. The British tend to approach issues from a much more analytic and data collection environment while Americans pursue action. This may partially explain some of the differences but more research is required. Overall, the way in which the same tasks were approached differed between the two contexts, indicating an interesting finding of culturally-based processes.

REVIEW QUESTIONS

1. Define and describe the stages of the crisis management cycle.
2. Recognize the challenges to cognition on teamwork in a crisis situation.
3. Describe how communication and information sharing influence effectiveness of team responses in crisis environments; considering both positive and negative outcomes.

4. Analyze how the level of COP and SA impact crisis response.
5. Understand that culture influences all aspects of crisis response, and offers extensive opportunity for future research.

ACKNOWLEDGMENTS

The authors thank the faculty, staff, and students at Manchester University, UK for their contributions for this work.

REFERENCES

After Action Report for the Response to the 2013 Boston Marathon Bombings. (2014). Retrieved March 3, 2016, from http://www.mass.gov/eopss/docs/mema/after-action-report-for-the-response-to-the-2013-boston-marathon-bombings.pdf

Ahuja, J. (2010). A study of virtuality impact on team performance. *IUP Journal of Management Research, IX*(5), 27–56. Retrieved from http://search.proquest.com/openview/4f2749ccc3d8617fa6b9cf4d9d10424a/1?pq-origsite=gscholar

Altay, N., & Green III, W. G. (2006). OR/MS research in disaster operations management. *European Journal of Operational Research, 175*(1), 475–493. doi:10.1016/j.ejor.2005.05.016

Bowman, J. M., & Wittenbaum, G. M. (2012). Time pressure affects process and performance in hidden-profile groups. *Small Group Research, 43*(3), 295–314. doi:10.1177/1046496412440055

Bunderson, J. S., & Sutcliffe, K. M. (2002). Comparing alternative conceptualizations of functional diversity in management teams: Process and performance effects. *Academy of Management Journal, 45*(5), 875–893. doi:10.2307/3069319

Carraher, T., Carraher, D., & Schliemann, A. (1990). Mathematics in the streets and in schools. *Psychology of Education: Major Themes, 26.* doi: 10.1111/j.2044–835X.1985.tb00951.x

Carrithers, J. R., DeHart, R. E., & Geaneas, P. Z. (1998). *Crisis management systems for emergency scenarios in international operations.* In SPE international conference on health, safety, and environment in oil and gas exploration and production. Caracas: Society of Petroleum Engineers.

Comfort, L. K. (1993). Integrating information technology into international crisis management and policy. *Journal of Contingencies and Crisis Management, 1*(1), 15–26. doi:10.1111/j.1468-5973.1993.tb00003.x

Coombs, W. T. (2007). Crisis management and communications. *Institute for Public Relations, 1*, 1–14. Retrieved from http://www.instituteforpr.org/crisis-management-and-communications/

Coppola, D. P. (2011). *Introduction to international disaster management.* Oxford, UK: Butterworth–Heinemann.

Department of Homeland Security. (2012). IT Program assessment. Retrieved from https://www.dhs.gov/xlibrary/assets/mgmt/itpa-ao-cop2012.pdf

Endsley, M. R. (1988). Design and evaluation for situation awareness enhancement. *Proceedings of the human factors and ergonomics society annual meeting, 32*(2), 97–101. Texas Tech University, TX: SAGE Publications. doi:10.1177/154193128803200221

Endsley, M. R. (1995). Toward a theory of situation awareness in dynamic systems. *Human Factors, 37*(1), 32–64. doi: 10.1518/001872095779049543

Endsley, M. R., & Jones, D. G. (2011). *Designing for situation awareness: An approach to user-centered design* (2nd ed.). Boca Raton, FL: CRC Press. Retrieved from http://books. google.com/books?hl=en&lr=&id=ufERJVeBHDcC&oi=fnd&pg= PP11&dq=Designing+for+situtation+awareness:+an+approach+to+user-centered+ design&ots=eJMJf0D0Tm&sig=FtY7w4PsOCezVS2r4pMcCqe4S-s

Endsley, T. C., Reep, J. A., McNeese, M. D., & Forster, P. K. (2015). *Conducting cross-national research: Lessons learned for the human factors practitioner.* Paper presented at the Proceedings of the Human Factors and Ergonomics Society Annual Meeting, Los Angeles, CA.

Entin, E. E., & Entin, E. B. (2001, June). *Measures for evaluation of team processes and performance in experiments and exercises.* Paper presented at the 6th International Command and Control Research and Technology Symposium, Annapolis, Maryland.

Faulkner, B. (2001). Towards a framework for tourism disaster management. *Tourism Management, 22*(2), 135–147. doi:10.1016/S0261-5177(00)00048-0

Fiore, S. M., Smith-Jentsch, K. A., Salas, E., Warner, N., & Letsky, M. (2010). Towards an understanding of macrocognition in teams: Developing and defining complex collaborative processes and products. *Theoretical Issues in Ergonomics Science, 11*(4), 250–271. doi:10.1080/14639221003729128

Fischer, A., Greiff, S., & Funke, J. (2012). The process of solving complex problems. *Journal of Problem Solving, 4*(1), 19–42. doi:10.7771/1932-6246.1118

Fiske, S. T. (1992). Thinking is for doing: Portraits of social cognition from daguerreotype to laserphoto. *Journal of Personality and Social Psychology, 63*(6), 877–889. doi:10.1037/0022-3514.63.6.877

Green III, W. G. (2012). Four phases of emergency management. Retrieved from http://www. richmond.edu/~wgreen/encyclopedia.htm

Hager, R. S. (1997). *Current and future efforts to vary the level of detail for the common operational picture.* Monterey, CA: Naval Postgraduate School.

Hamilton, K., Mancuso, V., Minotra, D., Hoult, R., Mohammed, S., Parr, A., . . . McNeese, M. (2010). Using the neocities 3.1 simulation to study and measure team cognition. *Proceedings of the human factors and ergonomics society annual meeting.* doi:10.1177/154193121005400434

Helmreich, R. L. (1994). Anatomy of a system accident: The crash of Avianca Flight 052. *The International Journal of Aviation Psychology, 4*(3), 265–284. doi:10.1207/ s15327108ijap0403_4

Holsti, O. R. (1989). Crisis decision making. In P. E. Tetlock, J. L. Husbands, R. Jervis, P. C. Stern, & C. Tilly (Eds.), *Behavior, society and nuclear war* (pp. 8–84). New York, NY: Oxford University.

Hutchins, E. (1995). *Cognition in the wild.* Cambridge, MA: MIT Press.

Hutchins, E., & Klausen, T. (1996). Distributed cognition in an airline cockpit. In D. Middleton & Y. Engeström (Eds.), *Communication and cognition at work* (pp. 15–54). Cambridge: Cambridge University Press.

Jenkins, B. M. (2009). *Terrorists can think strategically: Lessons learned from the mumbai attacks.* Santa Monica, CA: RAND Corporation. Retrieved from https://books.google. com/books?id=mI4NPwAACAAJ

Johnson, M. D., Hollenbeck, J. R., Humphrey, S. E., Ilgen, D. R., Jundt, D., & Meyer, C. J. (2006). Cutthroat cooperation: Asymmetrical adaptation to changes in team reward structures. *Academy of Management Journal, 49*(1), 103–119. doi:10.5465/AMJ.2006.20785533

Jones, R. E. T., McNeese, M. D., Connors, E. S., Jefferson, T., & Hall, D. L. (2004). A distributed cognition simulation involving homeland security and defense: The development of neocities. *Proceedings of the human factors and ergonomics society annual meeting, 48*(3), 631–634. doi:10.1177/154193120404800376

Jones, R. K. (2003). Miscommunication between pilots and air traffic control. *Language Problems & Language Planning, 27,* 233–248. doi:10.1075/lplp.27.3.03jon

Laraia, W., & Walker, M. C. (2009). The siege in Mumbai: A conventional terrorist attack aided by modern technology. In M. R. Haberfel & A. Hassell (Eds.), *A new understanding of terrorism: Case studies, trajectories and lessons learned* (pp. 309–340). New York, NY: Springer.

Luege, T. (2015). How Germanwings Failed at Crisis Communications. http://www.sm4good.com/2015/03/26/germanwings-failed-crisis-communication

Mack, A., & Rock, I. (1998). *Inattentional blindness.* MIT Press. Cambridge, MA. doi:10.1016/j.aorn.2010.03.011

Mancuso, V. F., & McNeese, M. D. (2012). Effects of integrated and differentiated team knowledge structures on distributed team cognition. *Proceedings of the human factors and ergonomics society annual meeting, 56*(1), 388–392. doi:10.1177/1071181312561088

McBride, M. E., Adams, K. A., Ntuen, C. A., & MaZaeva, N. (2002). Application of cognitive systems engineering to decision aiding design. In *Proceedings of institute for industrial engineering research conference* (pp. 1–5). Atlanta, GA: IIE Management Press.

McHugh, A. P., Smith, J., & Sieck, W. R. (2008). Cultural variations in mental models of collaborative decision making. In J.-M. Schraagen, L. Militello, T. Ormerod, & R. Lipshitz (Eds.), *Naturalistic decision making and macrocognition* (pp. 141–158). Burlington, VT: Aldershot.

McNeese, M. D. (2003). Metaphors and paradigms of team cognition: A twenty year perspective. *Proceedings of the human factors and ergonomics society annual meeting, 47*(3), 518–522. doi:10.1177/154193120304700356

McNeese, M. D., Jefferson, T., Bains, P., Brewer, I., Brown, C., Connors, E. S., … Terrell, I. (2006). Assessing the impact of hidden knowledge profiles on distributed cognition and team decision-making: Recounting the development of the NeoCITIES simulation. In *Proceedings of the human factors and ergonomics society 49th annual meeting human factors and ergonomics society* (Vol. 1124), Thousand Oaks, CA: SAGE Publications.

McNeese, M. D., Mancuso, V., McNeese, N., Endsley, T., & Forster, P. (2013). Using the living laboratory framework as a basis for understanding next-generation analyst work. In B. D. Broome, D. L. Hall, & J. Llinas (Eds.), *SPIE 8758, next-generation analyst* (p. 87580F). Bellingham, WA: SPIE.

McNeese, M. D., Pfaff, M. S., Connors, E. S., Obieta, J. F., Terrell, I. S., & Friedenberg, M. A. (2006). Multiple vantage points of the common operational picture: Supporting international teamwork. *Proceedings of the human factors and ergonomics society annual meeting, 50*(3), 467–471. doi:10.1177/154193120605000354

Mesmer-Magnus, J. R., & DeChurch, L. A. (2009). Information sharing and team performance: A meta-analysis. *Journal of Applied Psychology, 94*(2), 535–546. doi:10.1037/a0013773

Mesmer-Magnus, J. R., DeChurch, L. A., Jimenez-Rodriguez, M., Wildman, J., & Shuffler, M. (2011). A meta-analytic investigation of virtuality and information sharing in teams. *Organizational Behavior and Human Decision Processes, 115*(2), 214–225. doi:10.1016/j.obhdp.2011.03.002

Moore, L. (1989). Chan v. Korean Air Lines, Ltd.: The United States Supreme Court eliminates the American rule to the Warsaw Convention. *Hastings International and Comparative Law Review, 13,* 229.

Ntuen, C. A., Balogun, O., Boyle, E., & Turner, A. (2006). Supporting command and control training functions in the emergency management domain using cognitive systems engineering. *Ergonomics, 49*(October), 1415–1436. doi:10.1080/00140130600613083

Oyserman, D. (2015). Culture as situated cognition. In R. Scott & S. Kosslyn (Eds.), *Emerging trends in the social and behavioral sciences* (pp. 1–20). Hoboken, NJ: John Wiley & Sons.

Parsons, W. (1996). Crisis management. *Career Development International, 1*(5), 26–28. doi:10.1108/13620439610130614

Pearson, C. M., & Clair, J. A. (1998). Reframing crisis management. *Academy of Management Review, 23*(1), 59–76. doi:10.5465/AMR.1998.192960

Pfaff, M. S. (2008). Effects of mood and stress on group communication and performance in a simulated task environment. Doctor of Philosophy Dissertation, The Pennsylvania State University Electronic Theses and Dissertations Database.

Rapoport, A. (2008). Some further thoughts on culture and environment. *Archnet-IJAR, 2*(1), 16–39.

Rasmussen, L. J., Sieck, W. R., & Smart, P. R. (2009). What is a good plan? Cultural variations in expert planners' concepts of plan quality. *Journal of Cognitive Engineering and Decision Making, 3*(3), 228–252. doi:10.1518/155534309×474479

Resnick, L. (2014). *It Takes a Team - The 2013 Boston marathon: Preparing for and recovering from a mass-casualty event.* Retrieved from http://sites.jbjs.org/ittakesateam/2014/report.pdf

Rosenthal, U., & Hart, P. (1991). Experts and decision makers in crisis situations. *Science Communication, 12*(4), 350–372.

Salas, E., Cooke, N. J., & Rosen, M. A. (2008). On teams, teamwork, and team performance: Discoveries and developments. *Human Factors, 50*(3), 540–547. doi:10.1518/001872008×288457

Salas, E., Dickinson, T. L., Converse, S. A., & Tannenbaum, S. I. (1992). Toward an understanding of team performance and training. In R. W. Swezey & E. Salas (Eds.), *Teams: Their training and performance* (pp. 3–29). Norwood, NJ: Ablex Publishing.

Schwarz, N. (2007). Attitude construction: Evaluation in context. *Social Cognition, 25*(5), 638–656. doi:10.1521/soco.2007.25.5.638

Shaluf, I. M., Ahmadun, F., & Said, A. M. (2003). A review of disaster and crisis. *Disaster Prevention and Management, 12*(1), 24–32. doi:10.1108/09653560310463829

Smith, E. R., & Semin, G. R. (2004). Socially situated cognition: Cognition in its social context. *Advances in Experimental Social Psychology, 36*, 53–117. doi:10.1016/S0065-2601(04)36002-8

Stasser, G., & Titus, W. (1985). Pooling of unshared information in group decision making: Biased information sampling during discussion. *Journal of Personality & Social Psychology, 48*(6), 1467–1478. doi:10.1037/0022-3514.48.6.1467

Stasser, G., & Titus, W. (1987). Effects of information load and percentage of shared information on the dissemination of unshared information during group discussion. *Journal of Personality and Social Psychology, 53*(1), 81–93. doi:10.1037/0022-3514.53.1.81

Strauch, B. (2010). Can cultural differences lead to accidents? Team cultural differences and sociotechnical system operations. *Human Factors, 52*(2), 246–263. doi:10.1177/0018720810362238

United States Coast Guard (USCG). (1998). Team coordination student training guide. Retrieved from https://www.uscg.mil/hq/cg5/cg534/nsarc/TCT.doc

Von Clausewitz, C. (1832). *On War*, Michael Howard & Peter Paret (Eds. and Trans. 1976). Princeton, NJ: Princeton University Press.

Von Clausewitz, C., & Graham, J. J. (1997). *On war.* [COL James John Graham, Trans. (London: N. Trübner 1873)].

Willig, C. (2001). Introducing qualitative research in psychology: Adventures in theory and method. *Qualitative Research, 2*(7), 217. doi:10.1177/1468794106058877

8 Emotion, Stress, and Collaborative Systems

Mark S. Pfaff

CONTENTS

Introduction 162
Theoretical Perspectives 164
    Emotion, Stress, and Individual Cognition 164
    Emotion, Stress, and Team Cognition 166
Midway Breather 167
Research Approaches 167
    Manipulations and Measures 168
    Team-Level Measures 170
    Applications of Team-Level Emotion and Stress Research 172
Conclusion 173
Review Questions 173
Discussion and Future Works Topics 173
Additional Activities and Explorations 174
References 174

ADVANCED ORGANIZERS

This chapter discusses the issues of emotion and stress in terms of how they can be addressed at the team level to support collaboration in technologically complex environments. The chapter begins with a discussion of the applicable bodies of theory at both the individual and team level, as one cannot understand a team also without understanding the individuals within that team. This is followed by a description of the diverse range of manipulations and measures used to perform research on emotion and stress in collaborating teams. It concludes with an example of how this knowledge has been applied in a problem-based approach to connect theory with practice, as described by the living laboratory framework, and how it leads to actionable interventions to address problems faced by teams in high-stress and emotionally-charged work contexts.

INTRODUCTION

Emotion and stress are conceptually distinct and have traditionally been considered as separate fields of study. Stress is frequently investigated in practical terms in applied environments, whereas emotions are often examined in theoretical terms in experimental environments. Emotion and stress are examples of concepts that have different meanings and manifestations, depending on the discipline of the person studying it, with what methods, in what setting, and with what goals. Both are plagued with terminological ambiguity, such as the interchangeable use of terms like *affect, mood, emotion,* and *feeling.* In this chapter, the term *emotion* is used to broadly encompass experiences ranging from the thrill of success to a frightened reaction to a prolonged depression. Readers are referred to Russell (2003) for a detailed decomposition of emotional concepts.

Similarly, the concept of stress may be found as an independent variable (a cause), a dependent variable (an effect), or a process (Cooper, Dewe, & O'Driscoll, 2001). Stress can come from the external environment (like high temperature or noise) and affect how someone works within it, or stress can arise internally from an individual's interactions with the environment (such as working with a system that frequently fails). A stressed individual, in turn, can affect the actors or systems around him. For example, he may communicate poorly with teammates or commit errors with higher frequency. Poor communication can lead to stress among teammates and yet more errors, leading to dangerous conditions, and even more stress.

As these examples illustrate, stress can be studied at different levels of analysis (e.g., individual, group, and organizational) and observed through different theoretical lenses (e.g., physiological, psychological, operational, and industrial). Hence, the study of emotion and stress in collaborative systems is the area of research well suited to the multitheoretic and multimethodological approach defined in the living laboratory framework.

The topic of stress is increasingly of interest to human factors research, often in response to high-profile accidents. One of the classical examples is the USS *Vincennes* accidentally destroying an Iranian civilian airbus in 1988, which led to the U.S. Navy initiating the Tactical Decision Making under Stress (TADMUS) project (Salas, Cannon-Bowers, & Johnston, 1998). More recent cases where operator stress has been studied include the September 11, 2001 attacks on the World Trade Center and Pentagon (Comfort & Kapucu, 2006) and the 2011 tsunami in Japan leading to the disaster at the Fukushima Daiichi nuclear power facility (Dauer, Zanzonico, Tuttle, Quinn, & Strauss, 2011). Retrospective analyses of these types of events often point to operator error as the primary failure, with official reports offering explanations that stress caused the error. The Fogarty report following the Vincennes tragedy concluded that, overall, the officers-in-charge had performed properly and acted within the rules of engagement, yet "stress, task fixation, and unconscious distortion of data may have played a major role in this incident" (Fogarty, 1988, p. 63). The simple view is that stress causes mistakes, but it is often unclear which specific types, characteristics, or consequences of stress are to blame. If stress is left as a *black box* then it is very difficult to figure out the most effective ways to prevent such accidents in the future. Instead, one should emphasize the *living* in the living laboratory framework to highlight how teams in complex work environments will exhibit dynamic and emergent behaviors modulated and constrained in different ways by different types of individuals, doing different types of tasks, under different types of stressors.

Stress and emotion must therefore be disentangled to understand the many ways they can occur and the multidimensional ways in which it is experienced. Research emerging from fields such as cognitive systems engineering (CSE) and human–computer interaction (HCI) provides evidence that accidents like these might not be solely attributable to operator error due to a stressful situation (like a surprise attack), but also to errors made in the design of systems operators have to use to deal with that situation. In addition, systems often fail to account for emotional factors that accompany stress, such as frustration, anxiety, or fear, which are known to have specific impacts on cognitive performance (e.g., Ahn & Picard, 2006; Lisetti & Hudlicka, 2015; Pfaff, 2012). Hence, any discussion of the effects of stress on human performance needs to also include the emotional dimension.

Although stress is being more widely addressed in many work domains, emotion is only recently gaining traction in the field of CSE. In complex work environments, the effects of stress and emotion are both factors worth investigating, though the emotional components tend to be subsumed by the more popular target of stress. As a result, emotions are less frequently discussed in detail due to the overwhelmingly negative emotional connotation of stress in the research literature. Interestingly, stress is not consistently detrimental to all aspects of performance. Indeed, in certain contexts (such as military training and sports) some stressors are not only expected, but serve to motivate operators to high levels of performance (Morris, Hancock, & Shirkey, 2004). However, the task or environmental characteristics that make a stressful situation beneficial or detrimental to performance are unclear.

To explore these issues, it is necessary to conceptually differentiate stress and emotion while unpacking their relationship to each other. This is an ambitious task. Both stress and emotion consist of multiple complex dimensions. *High stress* and *low mood* are phrases that may work colloquially but are otherwise unquantifiable statements. There are ongoing debates in the literatures of emotion and stress regarding which models best captures how each construct truly operates in the context of human cognition and behavior. These individual battles have spilled into each other, in some cases to debate whether stress includes emotional dimensions or if stress itself is subsumed among the emotions (Lazarus, 1993).

Research into these problems must explore both individual and team cognition in context in order to understand the specific interrelationships of multiple actors in technologically complex settings. To do so requires an interdisciplinary problem-based approach to holistically examine these issues in the context of real-world use as well as through theory-driven experimentation (Cooke, Gorman, Myers, & Duran, 2013; Cooke, Salas, Kiekel, & Bell, 2004). Many bodies of research investigate the effects of specific emotional conditions or stressors on individual cognitive activities, but research at the group level is far less common. Therefore, it is valuable to review the theoretical perspectives that can bridge the gap between individual cognition and team cognition, and the methodological approaches that can be used to derive actionable insights to improve how people perform complex work.

THEORETICAL PERSPECTIVES

There is no single overarching theory to explain how emotion and stress affect, and are affected by, team cognition. Rather, an integrative multitheoretic approach is necessary to study these problems and generate actionable findings. Theories of how emotion and stress influence individual cognition are an appropriate starting point. However, these must be addressed alongside theories of how multiple individuals work together, and distinctions must be drawn between conceptualizations of emotion and stress at the individual and team levels.

EMOTION, STRESS, AND INDIVIDUAL COGNITION

Until fairly recently, emotions and stress have been considered separate fields of study (Lazarus, 2006). Stress was (and still is) a substantial concern to many fields of study, both experimental and applied, whereas study of emotions has been more of an abstract puzzle with uncertain practical applications. Multiple vectors of research intersect the phenomena related to teams working under stress. Stress is an ongoing process between an individual and his or her environment that has multiple interacting psychological and physiological results, including changes in emotional state, cognitive ability, and physical performance. Complicating matters is the vague and overlapping usage of concepts like *stress, stressor, emotion,* and *mood.* Some studies refer to *emotional stress* or *stressful moods* (e.g., Mikolajczak, Roy, Verstrynge, & Luminet, 2009; Sheppard-Sawyer, McNally, & Fischer, 2000). A careful review of emotion- and stress-related research helps disambiguate the terminology and avoid confounding them in experimental or applied research. Without clearly understanding these cognitive factors, researchers are likely to misattribute causes to effects, which can lead to ineffective interventions when applying such findings to real-world problems.

The literature presents emotion and stress as having complex relationships with each other. Lazarus (2006, p. 36) argues that "we cannot sensibly treat stress and emotion as if they were separate fields without doing a great disservice to both." It is true that the literature on the relationship between stress and cognition shows many similarities with the literature on emotion and cognition. In some cases, when looking at the manipulations and measures used in both areas of research, one may suspect that emotion researchers and stress researchers are examining similar phenomena but choosing different terminology to describe it. For example, Bootzin and Max (1981) explain that while theorists disagree over the causes of anxiety, there is general agreement over the constitution of the anxiety response, which happens to be a list of verbal reports, behaviors, and physiological responses very similar to those attributed to stress. Moreover, in Mogg et al. (1990), the authors admit that it is unclear whether their manipulation (difficult task with false negative feedback or easy task with false positive feedback) was truly a manipulation of mood, stress, anxiety, or performance feedback. They chose *stress* as the most all-encompassing term, though the desired outcome of the manipulation was actually *anxiety*—a mood state. Furthermore, some conceptual frameworks of stress show stress leading to negative emotions, which in turn causes multiple physiological and psychological outcomes, though some also describe a direct effect of stress in addition to one mediated by a negative affect (Aldwin, 1994).

In organizational stress research, emotions are included among the individual differences in reacting to stress, and generally with very negative connotations: "When emotions are directly involved in action, they tend to overwhelm or subvert rational mental processes, not to supplement them" (Elster, 1985, p. 379). Dysfunctional reasoning is blamed on *emotional stress* (Leung, Chan, & Yu, 2012). This perspective holds that stress may be inescapable in high-stakes circumstances, but emotions (positive and negative ones) are something to be avoided at all costs. However, many researchers have shown this broad generalization to be untrue.

Some research has shown stress and emotions to be conceptually distinct and independent of each other, at least to some degree. One can feel happy or sad, but not stressed; one could be stressed and feel exhilarated, depressed, or neither. Many studies have shown how people may experience both positive and negative moods in the presence of stress (Lazarus, 2003). For example, observations of caregivers of terminal patients demonstrated that a small positive event amid a majority of negative events can prompt a positive affective response despite ongoing stress (Folkman & Moskowitz, 2000). This shows that positive moods may coexist with stressful conditions and such positive responses have a meaningful coping function.

Lazarus and Folkman (1984) emphasized the primacy of the cognitive process of appraisal preceding any reaction to stress, emotional or otherwise. Zajonc (1984), on the other hand, responded that the emotional reaction precedes cognitive processing of the threat, and these two reactions may sometimes be in conflict with each other. Sun and Zhang's (2006) model of individuals interacting with objects suggests that both are right. Their model proposes that while *traits* are positioned as the launching point for determining interaction, neither *emotional* nor *cognitive* reactions are given precedence in determining how the individual intends to act. In fact, these reactions influence each other, a position that aligns with the *parallel-processing* view that cognition and emotion are both interactive information-processing systems (Norman, Ortony, & Russell, 2003). Minsky integrates emotion and cognition even further with the position that each emotional state corresponds to a different manner of cognition (Minsky, 2006).

Classical views of decision making maintain that emotions and stress derail otherwise optimal rationalistic thought. Poor decisions are frequently criticized for having been made with emotion rather than reason (Hammond, 2000). However, contemporary views of naturalistic decision making are interested in how severely stressed decision makers can perform as well as they do in spite of a demonstrable departure from rationalistic decision processes. It appears that the challenges posed by stressors (such as noise, pain, or time pressure) to attention, memory, and information gathering are more disruptive to the slow structured decision processes than to fast recognition-primed naturalistic decision processes (Klein, 1998).

The complex effects of stress on individuals—particularly the individual differences in the appraisal of stressors—amplify the need to better understand the similarities, differences, and interactions between stress and emotion in collaborative task settings. From a CSE perspective, it is essential to understand how stress and emotion not only change how individuals perform tasks, but also change how they communicate and collaborate to perform tasks *together*. The perspectives regarding individual stress, mood, and cognitive performance detailed above may imply several effects relevant to the conditions individuals encounter in team's problem-solving contexts. However, they do not specifically explain how individual moods or reactions to stress will impact team cognition and the outcome of team-based activities.

EMOTION, STRESS, AND TEAM COGNITION

Team performance research simultaneously considers the individual and group level of analysis, and it is important not to conflate the two. Although individuals perform actions and experience emotion and stress, it can be misleading to define team performance, emotion, or stress as simply the aggregation of those variables from the individual level. A *team* cannot feel sad, for example. However, the regulating processes and collective outcomes of the interactions among individuals, including the influences of emotion and stress, are clearly in the domain of team activity. Therefore, it can be a challenge to distinguish which aspects of a research problem belong to the individual and which belong to the team.

The concept of *team* requires further definition, particularly in comparison to the term *group*, as the two are often used interchangeably in the literature. Of the two, *group* is the more general term. Groups tend to consist of a less-differentiated membership, coalesce for a given goal, and then disband. Teams generally have a shared history, a projected future, and distinct functions and roles. These are not rigid distinctions, as flexible teams may often interchange roles or redistribute tasks (Brannick & Prince, 1997). The term *team* refers to a collection of individuals organized in such a manner to serve a function (e.g., police department) for a host organization (e.g., municipal emergency center). To manage a complex event, multiple teams (e.g., the police and hazardous materials services) may have to share information and coordinate activities to address a compound threat (such as an overturned chemical truck) as a team of teams.

The adaptive team model (Serfaty, Entin, & Johnston, 1998) explains how stress is both an input and an output of team processes. Their model demonstrates the interaction of the effects of stress on individual cognition with those stressors found in team contexts, such as role uncertainty or miscommunication. It shows that not only does individual stress impact team processes, but difficulties in the team processes themselves can generate stress. Performance feedback also modulates stress—poor performance informs the team that there are failures at one or more layers in the process. Successful teams are responsive to these feedback structures, rather than adhering rigidly to some fixed approach to teamwork.

Emotion research at the team and organizational level has focused on the effects of emotional contagion, defined as a social influence involving the conscious or unconscious transfer of emotions to other group members, which happens not only in face-to-face interactions, but even during computer-mediated communication such as text-based chat (Pirzadeh, 2014; Pirzadeh & Pfaff, 2012). Therefore, there should be value in research focusing on team performance under stress, assessing not only how team members share information with each other, but also how they share stress and emotions with each other through verbal and nonverbal cues.

The emotional contagion concept of group or collective emotion has been shown to influence team performance not just as a result of the average emotional state of a group, but also the team's emotional diversity (Brief & Weiss, 2002). The model of group emotional contagion (Barsade, 2002) describes a process beginning with the emotional state of a group member, expressed in terms of valence (positive or negative) and intensity (high or low). This emotional state is perceived by group members, but with that perception moderated by attentional processes; greater attentional allocation leads to greater contagion. Attention may be affected by environmental or task-related factors, but also by type and intensity of the emotion being expressed. Contagion may occur by one of two paths, the first being primitive emotional contagion, characterized by subconscious reactions to someone's emotion expression, usually in the form of changes of facial expression, body language, or speech behaviors. This type is most strongly associated with in-person emotional contagion. The second path is more cognitively effortful, in which an individual perceives another's mood and adjusts his or her own according to the situation. In this case, emotion is perceived as information about how one should be feeling, and can be transmitted and received by other channels, such as text-based chat (Pirzadeh, 2014). Awareness of this emotion can provide feedback about how the team is performing, similar to the *affect as information* perspective (Schwarz & Clore, 1996), in which happy moods provide the information that the task is going well, but sad moods suggest that there may be problems present. Positive moods appear to improve group cohesion and cooperativeness, which in turn benefits team performance (Barsade, 2002).

MIDWAY BREATHER

Decades of research provide fundamental knowledge of the roles of emotion and stress in individual human cognition. However, trying to apply this to the team level is not as simple as aggregating individual effects and behaviors. At a team level, new interpersonal issues emerge about emotion and stress that are not addressed at the individual level of analysis. Many research domains and traditions address different aspects of the sociotechnical problems found in stressful technologically complex collaborative environments. Armed with knowledge of these theories, as well as their intersections, overlaps, and even contradictions, a researcher can more effectively address problems facing teams working in complex conditions.

RESEARCH APPROACHES

As described in the preceding section, multiple theories exist to explain the effects of various stressors and emotions on attention, problem solving, and decision making in individuals. These have been tested experimentally across a broad range of psychological perspectives: cognitive, behavioral, social, and clinical. However, they are less frequently studied within the context of team activity. Team-level research has additional challenges of experimentally controlling and measuring several independent and interdependent actions leading toward a group outcome. Other complications with emotion and stress include the ethical and technical difficulties in manipulating people's emotional states or recreating realistic crisis environments to examine team decision making under stress. Research situated in team contexts takes many forms, and often requires the use of qualitative and quantitative methods practiced in multiple disciplines. Hence, it is appropriate to apply the living laboratory framework to address these problems by integrating observational research, interviews with domain experts, controlled experimentation, and development and evaluation of prototype solutions. This section provides a sampling of methodological approaches for studying the effects of emotion and stress on team cognition.

Manipulations and Measures

Organizational stress research proposes many physical (e.g., heat, noise, fatigue) and nonphysical stressors. Cannon-Bowers and Salas (1998) identified the following list of stressors observed among naval crews, which have been used as realistic stress manipulations in many team contexts: multiple information sources, incomplete or conflicting information, rapidly changing and evolving scenarios, requirements for team coordination, adverse physical conditions (e.g., heat, noise, and sleep deprivation), performance pressure, time pressure, high work/information load, and threat (p. 19). From an organizational viewpoint, stress may consist of simultaneous reactions to external conditions (e.g., time pressure), as well as internal states (e.g., feelings of social inadequacy). For example, Driskell and Salas (1991) operationalized stress as both a threat to the individual's well being as well as an increase in personal responsibility for team task outcome.

By stimulating an individual with a story, song, or a memory associated with a particular emotional state, it is possible to put an individual into a suggested mood temporarily. The intensity and persistence of these mood-induction procedures varies depending on the target mood desired, yet they have proven to be very successful when properly implemented. According to Westermann, Spies, Stahl, and Hesse (1996), presenting a short film or story is generally the most effective method to manipulate a subject's mood, either positively or negatively.

These emotion and stress manipulations could be applied uniformly across a team under study. Alternatively, they could be selectively applied to specific team members to examine how a particular mood or stress state affects a team member in a different state (e.g., emotion contagion), or whether team members recognize and respond to another team member in stress.

Emotion and stress can be measured in multiple ways, including measures of physiological responses, as well as self-report survey measures. Physiological techniques provide real-time measurement over the duration of the experimental task, whereas survey measures are limited to snapshots in time, usually before and after the task. Physiological measurements such as heart rate, skin conductance, and skin temperature provide highly sensitive and real-time measures of stress and emotional responses, which are difficult for the subject to *fake* or consciously control. However, the intrusiveness of physiological measurement equipment needs to be considered carefully with respect to the tasks and setting of the study. For a comprehensive review of psychophysiological measurement techniques, see Cacioppo, Tassinary, and Berntson (2007).

Lazarus' transactional perspective (Lazarus & Folkman, 1984) focuses on the individual's perception of stress, which can be assessed via surveys. The short stress state questionnaire (SSSQ), provides a rapid and reliable assessment of three primary dimensions of stress (engagement, distress, and worry) with only 24 questions (Helton & Garland, 2006). Several context-specific stress assessments exist, such as the combat exposure scale (CES) that quantifies combat-related stressors (Keane et al., 1989). In cases where workload is used to manipulate stress, the NASA–TLX (Hart & Staveland, 1988) is a popular workload assessment tool for high-demand settings that provides a self-reported score from 0 to 100 computed from the weighted average of six subscales. However, the relationship between task load and stress is often an indirect one. Shorter, more contextually targeted surveys such as the CES have shown high construct validity within their intended domain when care is taken to ensure that the experimental context is aligned with the assumptions of the survey instrument.

Mood assessments are also self-reported using a range of survey instruments calibrated for specific emotional conditions. These may be used to assess feelings at the present moment (state measurements), or over longer periods of time (trait measurements). One such instrument is the Spielberger State–Trait Anxiety Inventory (STAI) (Spielberger, 1983), a popular and reliable measure of both state and trait anxiety. Respondents indicate their agreement with short phrases like *I feel pleasant*, *I tire quickly*, and *I feel like crying*. The widely used positive and negative affect schedule (PANAS) (Watson, Clark, & Tellegen, 1988) comprises two brief 10-item surveys that provide two independent measures of positive-(PA) and negative affectivity (NA). This instrument collects agreement with terms such as *interested*, *irritable*, *excited*, *ashamed*, *strong*, and *nervous*. Studies on emotion contagion often jointly use self-report and observational methods to capture perceived emotion as well as observable events or behaviors (e.g., facial expressions and posture) that reveal changes in emotional states, as neither alone is sufficient to measure effects that may be subtle and subconscious.

The generally negative connotation of stress makes it difficult to experimentally discriminate stress from negative emotion, as similar effects are sometimes seen in the presence of either one. As an example, manipulations of mood (Gasper & Clore, 2002), stress (Braunstein-Bercovitz, Dimentman-Ashkenazi, & Lubow, 2001), and anxiety (Derryberry & Reed, 1997) all demonstrated similar cognitive consequences. Therefore, a concern in this research is that measures of stress and measures of mood must not confound each other. There is some occasional overlap in the self-report instruments used for stress and mood measurement. In experimental trials, the SSSQ distress factor has appeared to be exclusively measuring negative effect, whereas the engagement factor measures motivation, and the worry factor measures cognitive effects (Helton & Garland, 2006). One could argue that there is little difference between *I feel sad* on the SSSQ distress factor and *I feel upset* that appears on both the positive affect scale of the PANAS and short form STAI. However, the SSSQ engagement factor shares two items with the PANAS: *I feel alert* and *I feel active*. This suggests that there is the potential for these instruments to confound stress and mood measurements, which recommends techniques such as factor analysis when multiple self-report instruments are used in conjunction with each other.

TEAM-LEVEL MEASURES

There are many widely used emotion and stress measures that have been well-validated at the individual level. However, to what extent are these applicable to team-level research? On one hand, emotion and stress are individual responses to internal and external stimuli. A team has no faculty to experience either one, so it could be argued that there is no such thing as *team emotion* or *team stress*. Rather, team-level expressions of emotion or stress can be discovered through patterns of interpersonal behavior, which may be observed directly or derived from artifacts such as activity logs or communications records. Often researchers looking at either factor in a team setting compute a team measure by averaging each team member's score on a mood or stress scale and using this as a variable in the analysis of the team's performance. For example, Pfaff (2012) averaged self-reported mood and stress scores from pairs of participants working on a shared task as a manipulation check. This was necessary because the task had a single shared outcome that provided only a team-level measure of performance, making correlation with individual mood or stress inappropriate. This could be described as a *bottom-up* approach to team emotion and stress.

On the other hand, the *top-down* approach asserts that when many people are gathered together, individual emotions are governed more by the emotional dynamics of the group. Much of this perspective comes from early research on emotion and crowd dynamics (McDougal, 1923). It argues that individuals imitate the emotional characteristics of those around them as a result of interpersonal relationships and social norms. Barsade and Gibson (1998) argue that a blend of top-down and bottom-up approaches can be valuable when studying group emotion, by considering both the mean *and* the variance of emotion measures within the team, as well as the influence of the most emotionally extreme members of the team. By extension, this approach equally applies to measuring team stress.

Once emotion and stress are measured, measuring their impact on team cognition and performance is a more complex problem. Aggregation of individual performance measures is a less valid measure of team performance, given that true team tasks involve interdependent actions and reactions among team members. Therefore, an individual's score that does not take into account the dependency of that individual's performance on another individual will misrepresent its contribution to team performance. On the other hand, when team tasks have a single measurable outcome, such as three individuals in different roles collaborating on a targeting task, the outcome of that task could be considered an appropriate measure of team performance. Unfortunately, a single performance score only reveals the *what* in terms of task outcome, but obscures the more important *why* of the specific individual and team cognitive activities that produced that outcome. An actionable understanding of team cognition must address the processes and behaviors that allow teams to become more adaptive and resilient under challenging conditions (Cooke et al., 2013).

Teams perform complex tasks via distributed cognition, in which cognitive functions are distributed among team members, and direct interaction between team functions occurs socially via communication channels. Indirect interaction happens through systems that allow one team member to monitor another, without requiring explicit communication between the two, such as with *over-the-shoulder* displays that allow one user to remotely observe a teammate's interaction with an interface (Gutwin & Greenberg, 2001). As communications can be captured in both colocated and geographically distributed work, it becomes a valuable data source that has been applied to performance, coordination, and more (Cooke & Gorman, 2009). However, though the data are relatively easy to gather (via recordings or chat logs), the researcher faces difficult choices about the most appropriate ways to analyze the data. Methods to analyze communication data range from fast and automatable quantitative methods, which are efficient but risk oversimplifying the data, to complex and effortful qualitative methods, which are labor-intensive but can produce richer and more detailed results.

Many researchers choose quantitative methods to analyze communication data, especially when there are large volumes of data to process that would be impractical to analyze manually. Measures that can be drawn from recordings or communications logs include frequency and duration of messages as well as turn-taking behaviors, from which a researcher could derive patterns of dominance or other conversational dynamics that could predict team outcomes.

Another popular quantitative approach employs dictionary-based word count software like the widely used Linguistic Inquiry and Word Count (LIWC; Tausczik & Pennebaker, 2010) that analyzes text for characteristics such as attentional focus, social relationships, cognitive styles, individual differences, and even emotions. This analysis can provide broad characterizations of the relative proportions of different types of verbal expressions, such as social words (indicating interactions or relationships between team members) or cognitive words (references to causation or insight). However, such analysis pays little attention to context that could change the implications of words under circumstances, and also can require substantial preprocessing of data to remove extraneous artifacts (e.g., time stamps) or correct misspellings. Boonthanom (2004) developed a taxonomy of verbal and nonverbal cues of emotion expression unique to text-based communication that go beyond literal emotion words (e.g., sad, happy) to other cues that are analogs to nonverbal physical behaviors, like gesture or tone of voice. These include things like vocal spelling (mimicking vocal inflections by distorting the spelling of a word, like *weeeellll* or *soooooo*), lexical surrogates (verbal expressions that normally would not be written out, like *uh huh* or *haha*), emoticons, and others. Unfortunately, these cues must be detected and counted manually by the researcher as they do not match the standard dictionary of terms. Using these methods, the researcher can analyze whether the effects of different emotional or stress conditions are reflected in counts or proportions of many different types of communications cues collected during a team task. For example, Pirzadeh and Pfaff (2012) analyzed chat logs from a team emergency management exercise to find that stressed participants used significantly more negative emotion words, but fewer vocal spellings than nonstressed participants. Such a finding demonstrates stress influencing mood, as well as influencing how individuals choose to express that mood to other team members. In this case, if people are less likely to express emotion using vocal spellings when under stress than during routine communications, other members of the team are less likely to know how their teammates are feeling, and consequently how that might affect their ability to perform their tasks.

Qualitative communications analysis methods such as content analysis and conversation analysis require manual inspection of the series of utterances between individuals to identify themes, categories, and patterns that reflect the cognitive activities of the team members. Sequences of utterances are the means by which individuals coordinate and accomplish team activities (Mazeland, 2006). These could take the form of a request followed by a decision, a question followed by an answer, or comment followed by an acknowledgment. Taken in context with the unfolding team activity, these sequences are examples of ways the researcher can uncover the complex cognitive activities underlying the team's success or failure at the current task.

A third option is to take a hybrid approach blending the strengths of both qualitative and quantitative approaches. When the task domain is well known, and there are boundaries to the types of things the team will discuss, a well-defined coding scheme can capture both quantity and meaning of utterances between team members. Adaptive Architectures for Command and Control (A2C2; Entin & Entin, 2001) is a coding scheme that classifies utterances into three categories: information (the status of events), action (ground-level activities), coordination (delegating or accepting responsibility to act), and acknowledgments. Each of the first three categories is further divided into requests and transfers. For example, an information request could be a question asking about how long before a resource will arrive at its destination, and an information transfer would be the response to that question. The ratio of transfers to requests produces an anticipation ratio, where high numbers reveal teams that anticipate the need to transfer information, action, or coordination without waiting for a corresponding request. This can be used as a measure of communication efficiency.

APPLICATIONS OF TEAM-LEVEL EMOTION AND STRESS RESEARCH

Pfaff (2012) adapted the A2C2 (Entin & Entin, 2001) coding scheme to examine how different mood and stress states would influence team cognition, as reflected in changes in these categories of messages between team members. The results showed that teams under stress became more proactive about communication, with an increased overall anticipation ratio. Specifically, it was requests and transfers of information (as opposed to action or coordination) that increased under stress, indicating an impact of stress on memory and attention that interfered with sensemaking. For example, if a message updating the status of an event was missed while the team member was attending to another task, it would require an information request to find out what happened to that event. Similarly, as the activity was designed to interleave tasks (i.e., new tasks were initiated before the preceding task could be resolved), team members would have difficulty retaining the status of several ongoing events in memory while figuring out to address the task currently at hand, again requiring a request (and subsequent transfer) of information within the team. Action anticipation was higher in the happy mood than the sad mood, paralleling results of individual-level emotion research showing that people turn their attention inward in sad moods. In addition, team members in sad moods more often replied with an acknowledgment ("OK") than an explicit action transfer ("I'll send a police car").

Applied at a team level, these two results showed that a negative mood can diminish team awareness (Gutwin & Greenberg, 2001). In fact, this finding was one of the motivators for Hellar (2010) to design a widget as part of a team command and control interface that monitored resources and task loads of teammates. This widget automatically cued available teammates that another team member was overloaded (and therefore subjected to the cognitive effects of mood and stress noted previously). Anyone who was available could then take the initiative to help the stressed teammate, as research has shown that stressed teammate is less likely to initiate such a request (Pfaff, 2012).

This example shows how theory drawn from knowledge about complex and stressful work domains can inform controlled research to test theory, design potential interventions, and evaluate their potential impact on real-world practice. In addition, this exemplifies how CSE and the living laboratory framework employ multitheoretic perspectives and multimethodological approaches to provide valid and reliable solutions to complex problems.

CONCLUSION

The interactions between emotion, stress, and cognition provide countless interesting problems for researchers to solve. The goal of this chapter was to describe the conceptual underpinnings and research methods associated with studying these issues at the team level. Of course, much of this knowledge has its foundations at the individual level of analysis. However, in order to generate knowledge that can make practical changes to the tools, procedures, training, and environments in which real work gets done, researchers are encouraged to shift from the microcognitive perspective of individual cognitive functions to the macrocognitive perspective of many individuals utilizing all of their cognitive functions in realistic work settings (McNeese & Pfaff, 2011; Schraagen, Klein, & Hoffman, 2008).

REVIEW QUESTIONS

1. What are some of the conceptual and operational ways in which a researcher could distinguish a stress reaction from an emotional reaction?
2. Contrast qualitative and quantitative approaches to analyzing team communication records for evidence of the effects of stress and emotion.
3. How might a researcher address the validity of using well-established individual measures of stress and emotion when studying the work of collaborative teams?

DISCUSSION AND FUTURE WORKS TOPICS

Think about the use of emotion as a variable in an experiment. For example, a researcher might want to manipulate emotional states as a categorical independent variable (e.g., sad versus happy), in order to see what the effects might be on certain types of cognitive tasks. On the other hand, a researcher may be interested in how different types of tasks affect a person's emotions, making it a dependent variable.

Researchers disagree about the most appropriate way to conceptualize and operationalize emotions. Some schools of thought look at emotions as a set of discrete states (e.g., fear, sadness, anger, disgust, and happiness), whereas others assign emotional states to points in a multidimensional space. For example, Russell (2003) defines emotional states in a two-dimensional space with valence on the X-axis (ranging from happy to sad) and activation on the Y-axis (ranging from low to high).

The choices a researcher makes about operationalizing emotion can have substantial effects on the results of a study. However, all of the approaches just described are based on the study of individual emotions. As the concept of *team emotion* is not well defined, it is unclear what the most valid ways are to conceptualize and operationalize it. To what extent can we map individual-level models of emotion to team-level ones? Do we have to start from scratch with completely new frameworks to understand team emotion? Further efforts to better understand and measure emotion at the team level could have significant implications on the design and use of collaborative systems.

ADDITIONAL ACTIVITIES AND EXPLORATIONS

Think about the most recent project that you worked on as part of a team, and where much of the work was performed and coordinated through technology (e.g., e-mail, chat, content management system). Where there any aspects that created stress for you personally, or for other members of the team? If so, what do you think the effects may have been on your team's ability to work effectively together and achieve the team's goals? Similarly, how would you describe the emotional climate within the team? Where there are any emotional events (such as one team member getting upset with another) that in turn affected the emotions of the entire team? On reflection, what role do you think the technology could have had in causing or contributing toward those problems, and what possible things could system designers do that might address those problems?

Note: The author's affiliation with The MITRE Corporation is provided for identification purposes only, and is not intended to convey or imply MITRE's concurrence with, or support for, the positions, opinions, or viewpoints expressed by the author. Approved for Public Release; Distribution Unlimited. Case Number 16-2894.

REFERENCES

Ahn, H., & Picard, R. W. (2006). Affective cognitive learning and decision making: The role of emotions. *Proceedings of the 18th European Meeting on Cybernetics and Systems Research (EMCSR 2006)*.

Aldwin, C. M. (1994). *Stress, coping, and development: An integrative perspective*. New York, NY: The Guilford Press.

Barsade, S. G. (2002). The ripple effect: Emotional contagion and its influence on group behavior. *Administrative Science Quarterly, 47*(4), 644–675. doi:10.2307/3094912

Barsade, S. G., & Gibson, D. E. (1998). Group emotion: A view from top and bottom. *Research on managing groups and teams, 1*(82), 81–102.

Boonthanom, R. (2004). *Computer-mediated communication of emotions: A lens-model approach* (Unpublished PhD dissertation). Florida State University.

Bootzin, R. B., & Max, D. (1981). Learning and behavioral theories. In I. L. Kutasch & L. B. Schlesinger (Eds.), *Handbook of stress and anxiety*. San Francisco, CA: Jossey-Bass Publishers.

Brannick, M. T., & Prince, C. (1997). An overview of team performance and measurement. In M. T. Brannick, E. Salas, & C. Prince (Eds.), *Team performance assessment and measurement: Theory, methods, and applications* (pp. 3–16). Mahwah, NJ: Lawrence Erlbaum Associates.

Braunstein-Bercovitz, H., Dimentman-Ashkenazi, I., & Lubow, R. E. (2001). Stress affects the selection of relevant from irrelevant stimuli. *Emotion, 1*(2), 182–192.

Brief, A. P., & Weiss, H. M. (2002). Organizational behavior: Affect in the workplace. *Annual Review of Psychology, 53*(1), 279–307. doi:10.1146/annurev.psych.53.100901.135156

Cacioppo, J., Tassinary, L. G., & Berntson, G. G. (2007). *The handbook of psychophysiology* (3rd ed.). Cambridge: Cambridge University Press.

Cannon-Bowers, J. A., & Salas, E. (1998). Individual and team decision making under stress: Theoretical underpinnings. In J. A. Cannon-Bowers & E. Salas (Eds.), *Making decisions under stress* (pp. 17–38). Washington, DC: American Psychological Association.

Comfort, L. K., & Kapucu, N. (2006). Inter-organizational coordination in extreme events: The World Trade Center attacks, September 11, 2001. *Natural Hazards, 39*(2), 309–327. doi:10.1007/s11069-006-0030-x

Cooke, N. J., & Gorman, J. C. (2009). Interaction-based measures of cognitive systems. *Journal of Cognitive Engineering and Decision Making, 3*(1), 27–46. doi:10.1518/155534309×433302

Cooke, N. J., Gorman, J. C., Myers, C. W., & Duran, J. L. (2013). Interactive team cognition. *Cognitive Science, 37*(2), 255–285. doi:10.1111/cogs.12009

Cooke, N. J., Salas, E., Kiekel, P. A., & Bell, B. (2004). Advances in measuring team cognition. In E. Salas & S. M. Fiore (Eds.), *Team cognition: Understanding the factors that drive process and performance* (pp. 83–106). Washington, DC: American Psychological Association.

Cooper, C. L., Dewe, P. J., & O'Driscoll, M. P. (2001). *Organizational stress: A review and critique of theory, research, and applications*. Thousand Oaks, CA: Sage Publications.

Dauer, L. T., Zanzonico, P., Tuttle, R. M., Quinn, D. M., & Strauss, H. W. (2011). The Japanese tsunami and resulting nuclear emergency at the Fukushima Daiichi Power Facility: Technical, radiologic, and response perspectives. *Journal of Nuclear Medicine, 52*(9), 1423–1432. doi:10.2967/jnumed.111.091413

Derryberry, D., & Reed, M. A. (1997). Anxiety and attentional focusing: trait, state, and hemispheric influences. *Personality and Individual Differences, 25*, 745–761.

Driskell, J. E., & Salas, E. (1991). Group decision making under stress. *Journal of Applied Psychology, 76*(3), 473–478.

Elster, J. (1985). Sadder but wiser? Rationality and the emotions. *Social Science Information, 24*(2), 375.

Entin, E. E., & Entin, E. B. (2001, June). *Measures for evaluation of team processes and performance in experiments and exercises*. Paper presented at the 6th International Command and Control Research and Technology Symposium, Annapolis, MD.

Fogarty, W. M. (1988). *Investigation report: Formal investigation into circumstances surrounding the downing of Iran air flight 655 on 3 July 1998*. Washington, DC: U.S. Department of Defense.

Folkman, S., & Moskowitz, J. T. (2000). Positive affect and the other side of coping. *American Psychologist, 55*(6), 647–654.

Gasper, K., & Clore, G. L. (2002). Attending to the big picture: Mood and global versus local processing of visual information. *Psychological Science, 13*(1), 34–40.

Gutwin, C., & Greenberg, S. (2001). *The importance of awareness for team cognition in distributed collaboration* (Report 2001-696-19). Calgary: Department of Computer Science, University of Calgary.

Hammond, K. R. (2000). *Judgments under stress.* New York, NY: Oxford University Press.

Hart, S. G., & Staveland, L. E. (1988). Development of NASA-TLX (Task Load Index): Results of empirical and theoretical research. In P. A. Hancock & N. Meshkati (Eds.), *Human mental workload* (Vol. 1, pp. 239–350). Amsterdam: North Holland Press.

Hellar, D. B. (2010). *An investigation of data overload in team-based distributed cognition systems* (Ph. D. dissertation). University Park, PA: The Pennsylvania State University. Dissertations & Theses @ CIC Institutions. database. (Publication No. AAT 3380915).

Helton, W. S., & Garland, G. (2006). Short stress state questionnaire: Relationships with reading comprehension and land navigation. *Proceedings of the Human Factors and Ergonomics Society, 50*, 1731–1735.

Keane, T. M., Fairbank, J. A., Caddell, J. M., Zimmering, R. T., Taylor, K. L., & Mora, C. A. (1989). Clinical evaluation of a measure to assess combat exposure. *Psychological Journal of Consulting and Clinical Psychology, 1*(1), 53–55.

Klein, G. A. (1998). *Sources of power: How people make decisions.* Cambridge, MA: MIT Press.

Lazarus, R. S. (1993). From psychological stress to the emotions: A history of changing outlooks. *Annual Review of Psychology, 44*, 1–21.

Lazarus, R. S. (2003). Does the positive psychology movement have legs? *Psychological Inquiry, 14*(2), 93.

Lazarus, R. S. (2006). *Stress and emotion: A new synthesis.* New York, NY: Springer.

Lazarus, R. S., & Folkman, S. (1984). *Stress, appraisal, and coping.* New York, NY: Springer.

Leung, M.-Y., Chan, I. Y. S., & Yu, J. (2012). Preventing construction worker injury incidents through the management of personal stress and organizational stressors. *Accident Analysis & Prevention, 48*, 156–166. doi:10.1016/j.aap.2011.03.017

Lisetti, C., & Hudlicka, E. (2015). Why and how to build emotion-based agent architectures. In R. A. Calvo, S. D'Mello, J. Gratch, & A. Kappas (Eds.), *The Oxford handbook of affective computing* (pp. 94–109). Oxford: Oxford University Press.

Mazeland, H. (2006). Conversation analysis. In K. Brown (Ed.), *Encyclopedia of language and linguistics* (2nd ed., Vol. 3, pp. 153–162). Oxford: Elsevier.

McDougal, W. (1923). *Outline of psychology.* New York, NY: Scribner.

McNeese, M. D., & Pfaff, M. S. (2011). Looking at macrocognition through a multi-methodological lens. In E. Salas, S. M. Fiore, & M. P. Letsky (Eds.), *Theories of team cognition: Cross-disciplinary perspectives* (pp. 345–371). New York, NY: Routledge Academic.

Mikolajczak, M., Roy, E., Verstrynge, V., & Luminet, O. (2009). An exploration of the moderating effect of trait emotional intelligence on memory and attention in neutral and stressful conditions. *British Journal of Psychology, 100*(4), 699–715. doi:10.1348/000712608x395522

Minsky, M. (2006). *The emotion machine: Commonsense thinking, artificial intelligence, and the future of the human mind.* New York, NY: Simon and Schuster.

Mogg, K. K., Mathews, A. A., Bird, C. C., & MacGregor-Morris, R. R. (1990). Effects of stress and anxiety on the processing of threat stimuli. *Journal of Personality and Social Psychology, 59*(6), 1230–1237.

Morris, C. S., Hancock, P. A., & Shirkey, E. C. (2004). Motivational effects of adding context relevant stress in PC-based game training. *Military Psychology, 16*(2), 135–147.

Norman, D. A., Ortony, A., & Russell, D. M. (2003). Affect and machine design: Lessons for the development of autonomous machines. *IBM Systems Journal, 42*(1), 38–44.

Pfaff, M. S. (2012). Negative affect reduces team awareness: The effects of mood and stress on computer-mediated team communication. *Human Factors, 54*(4), 560–571. doi:10.1177/0018720811432307

Pirzadeh, A. (2014). Emotional communication in instant messaging. In *Proceedings of the 18th international conference on supporting group work* (pp. 272–274). Sanibel Island, FL: ACM.

Pirzadeh, A., & Pfaff, M. S. (2012). Emotion expression under stress in instant messaging. *Proceedings of the Human Factors and Ergonomics Society, 56*, 493–497. doi:10.1177/1071181312561051

Russell, J. A. (2003). Core affect and the psychological construction of emotion. *Psychological Review, 110*(1), 145–172. doi:10.1037/0033-295x.110.1.145

Salas, E., Cannon-Bowers, J. A., & Johnston, J. H. (1998). Lessons learned from conducting the TADMUS program: Balancing science, practice, and more. In J. A. Cannon-Bowers & E. Salas (Eds.), *Making decisions under stress* (pp. 409–413). Washington, DC: American Psychological Association.

Schraagen, J. M., Klein, G. A., & Hoffman, R. R. (2008). The macrocognition framework of naturalistic decision making. In J. M. Schraagen, L. G. Militello, T. Ormerod, & R. Lipshitz (Eds.), *Naturalistic decision making and macrocognition* (pp. 3–25). Hampshire: Ashgate Publishing Ltd.

Schwarz, N., & Clore, G. L. (1996). Feelings and phenomenal experiences. In E. T. Higgins (Ed.), *Social psychology: A handbook of basic principles* (pp. 433–465). New York, NY: Guilford Press.

Serfaty, D., Entin, E. E., & Johnston, J. H. (1998). Team coordination training. In J. A. Cannon-Bowers & E. Salas (Eds.), *Making decisions under stress* (pp. 409–413). Washington, DC: American Psychological Association.

Sheppard-Sawyer, C. L., McNally, R. J., & Fischer, J. H. (2000). Film-induced sadness as a trigger for disinhibited eating. *International Journal of Eating Disorders, 28*(2), 215–220.

Spielberger, C. D. (1983). *Manual of the State-Trait Anxiety Inventory, form Y.* Palo Alto, CA: Consulting Psychologists Press.

Sun, H., & Zhang, P. (2006). The role of affect in IS research: A critical survey and a research model. In P. Zhang & D. Galletta (Eds.), *HCI and MIS (I): Foundations* (Vol. 5). Armonk, NY: M. E. Sharpe.

Tausczik, Y. R., & Pennebaker, J. W. (2010). The psychological meaning of words: LIWC and computerized text analysis methods. *Journal of Language and Social Psychology, 29*(1), 24–54. doi:10.1177/0261927x09351676

Watson, D., Clark, L. A., & Tellegen, A. (1988). Development and validation of brief measures of positive and negative affect: The PANAS scales. *Journal of Personality and Social Psychology, 54*(6), 1063–1070.

Westermann, R., Spies, K., Stahl, G., & Hesse, F. W. (1996). Relative effectiveness and validity of mood induction procedures: A meta-analysis. *European Journal of Social Psychology, 26*(4), 557–580.

Zajonc, R. B. (1984). On the primacy of affect. *The American Psychologist, 39*(2), 117.

Section IV

---

*Models and Measures
of Cognitive Work*

9 Measuring Team Cognition in Collaborative Systems
Integrative and Interdisciplinary Perspectives

Susan Mohammed, Katherine Hamilton,
Vincent Mancuso, Rachel Tesler, and
Michael D. McNeese

CONTENTS

Introduction 182
    Team Cognition 184
    Temporality and Team Cognition 185
    Storytelling and Reflexivity 185
Experimental Methodology 186
    The NeoCITIES Team Simulation 186
        Rationale for Selecting NeoCITIES 186
        Scenario Development in NeoCITIES 187
    General Experimental Procedure 188
        Dependent Measures 188
Independent Variable Manipulations.. 189
    Storytelling 190
        Story Content: Collaboration and Timing 190
        Story Context: Metaphorical versus Analog....................................... 190
    Reflexivity 191
        Guided Team Reflexivity 191
        Guided Individual Reflexivity 191
Summary of Key Findings and across the Four Experiments........................ 191
    Research Objective 1: To Understand the Differential Effects of Multiple
    Types of Team Cognition 191
    Research Objective 2: To Infuse a Temporal Focus into the Study of Team
    Cognition... 192
    Research Objective 3: To Investigate the Effectiveness of Interventions
    Designed to Improve Team Cognition 192

Lessons Learned.. 193
   The Value of an Integrative Perspective 193
   The Value of an Interdisciplinary Perspective............................... 193
Conclusion 194
Review Questions... 194
Acknowledgments.. 195
References.. 195

ADVANCED ORGANIZER

This chapter describes a program of research using the living laboratory framework that extended team cognition research into previously unexplored avenues. First, we simultaneously investigated multiple types of team cognition (information sharing, shared situation awareness (SA), and team mental models). Second, we infused a temporal focus (deadlines, pacing, and sequencing) into the study of team cognition and outcomes to more realistically mirror, organizational, and military team contexts. Third, we investigated the effectiveness of various interventions (storytelling, team reflexivity, and individual reflexivity) designed to enhance team cognition in distributed teams. Across four programmatic experimental studies, research questions were tested via three-person student teams performing NeoCITIES, a scaled-world simulation. Incorporating temporality into the study of team mental models (TMMs) proved fruitful in predicting unique variance beyond traditional teamwork- and taskwork-content domains.

Results also provided encouraging evidence that storytelling and guided reflexivity interventions may help overcome the collaborative obstacles faced by team members in distributed environments, particularly when administered at the group level.

INTRODUCTION

Increasingly, work is carried out by teams consisting of diverse members who have never interacted previously, but are required to perform complex, dynamic tasks in novel and time-pressurized environments (e.g., Tannenbaum, Mathieu, Salas, & Cohen, 2012). Given the difficulty of this context, the issue of how to enhance team collaboration and performance is increasingly salient in organizational and military settings. Indeed much work has been done in cognitive systems engineering and human-centered design (Lee & Kirlik, 2013) to reflect the active integration of cognition, teamwork, technologies, and context to enable building tangible, resilient, and reliable systems that meet both individual and team demands. Although already a considerable challenge in colocated teams, getting members *on the same cognitive page* is even more problematic when team members are geographically separated (e.g., Gilson, Maynard, Young, Vartiainen, & Hakonen, 2015).

Addressing these research challenges, our interdisciplinary research team investigated antecedents and consequences of multiple types of team cognition toward the goal of improving team coordination and performance in distributed decision-making teams. Contemporary work in cognitive systems engineering interweaves and is often coupled to naturalistic decision making and team cognition (e.g., see Glantz, this volume, Chapter 15). The living laboratory framework (McNeese, 1996; McNeese, Perusich, & Rentsch, 2000) provides a foundation for integrating and aggregating multiple methodological approaches to gain holistic knowledge for designing cognitive systems that are practically effective and useful. The purpose of this chapter is to provide demonstration and use of the living laboratory framework especially as relevant for problems involving team cognition where (1) theory, modeling, and measurement; and (2) scaled-world simulation are of high priority. Within the living laboratory framework using NeoCITIES as a scaled-world simulation (McNeese, Mancuso, McNeese, Endsley, & Forster, 2013), we conducted four programmatic experiments investigating the influence of several interventions on team cognition and subsequent team outcomes. This work was supported by the Office of Naval Research and had the following three overall research objectives:

1. To integrate and empirically investigate multiple types of team cognition (information sharing, team SA, and TMMs) that occur in complex systems to understand their differential effects on team outcomes.
2. To infuse a temporal focus (deadlines, pacing, and sequencing) into the study of team cognition and team outcomes, to more accurately reflect the context of organizational and military teams.
3. To investigate the effectiveness of various interventions (storytelling, team reflexivity, and individual reflexivity) designed to improve team cognition and collaboration in distributed teams.

These research objectives contributed to the team literature in several ways. First, learning how various forms of team cognition integratively influence team outcomes has been identified as a critical research need (e.g., Mohammed, Ferzandi, & Hamilton, 2010; Salas & Wildman, 2009). As such, we empirically investigated information sharing, SA, and TMMs in the same set of studies, allowing for a deeper understanding of their differential effects. Second, as team cognition measures have been deemed temporally deficient (Mohammed, Tesler, & Hamilton, 2012) and time has been identified as "perhaps the most neglected critical issue" in team research (Kozlowski & Bell, 2003, p. 364), we infused time into assessments of information sharing, shared SA, and TMMs. In doing so, we answered numerous calls to sharpen the temporal lens used in conducting team studies (e.g., Ancona, Goodman, Lawrence, & Tushman, 2001; Mohammed, Hamilton, & Lim, 2009).

Third, we were the first (to our knowledge) to empirically investigate how the powerful benefits of ad hoc stories could be leveraged as a team-training tool to allow team members to achieve higher levels of shared understanding, and thereby perform more effectively. Although several authors have proposed that storytelling can improve team learning and performance (e.g., Bartel & Garud, 2009; Denning, 2001; Fiore, McDaniel, & Jentsch, 2009), their claims have gone untested because much of the work on storytelling is conceptual (e.g., Denning, 2001), and empirical work has been largely focused at the individual level of analysis (e.g., Ang & Rao, 2008). In combination with storytelling, we also examined reflexivity (a team's reflection on its performance and future strategies) as a tool to enhance team cognition and collaboration among members.

Fourth, through the integration of team cognition, temporality, and storytelling, we expanded these literatures into previously unexplored avenues identified as potentially theoretically fruitful (e.g., Fiore et al., 2009), but not investigated empirically. For example, we answered previous research calls to broaden the list of empirically verified antecedents of TMMs (e.g., Mohammed et al., 2010) by examining the effects of storytelling on team cognition. Fifth, beyond these theoretical and empirical contributions, NeoCITIES methodologically offered a novel, but ideal, interface to investigate multiple aspects of team cognition because it is easily adaptable and designed to test team cognition and collaborative-system processes.

In this chapter, we review the four studies we conducted as an exemplar of a living laboratory framework integrating cognitive systems, communication technologies, and human-in-the-loop simulation in a virtual setting (McNeese et al., 2013). We begin by providing a brief introduction to the team cognition literature and the need for a temporal focus as well as storytelling and reflexivity as interventions. Next, we describe the methodology of the four experiments, including the NeoCITIES simulation and team cognition measures. After providing an overview of study results, we end with lessons learned, highlighting the benefits of adopting an integrative and interdisciplinary perspective.

TEAM COGNITION

Team cognition is a general term referencing how knowledge is collectively represented in a team (Cannon-Bowers & Salas, 2001). Over the past decade, the amount of team cognition research has increased substantially across disciplinary boundaries (e.g., Badke-Schaub, Neumann, Lauche, & Mohammed, 2007; Fiore & Salas, 2006; Undre, Sevdalis, Healey, Darzi, & Vincent, 2006). In addition to being identified as one of the hallmarks of expert teams (Salas, Rosen, Burke, Goodwin, & Fiore, 2006), a recent meta-analysis found that team cognition positively predicted team motivation, processes, and performance (DeChurch & Mesmer-Magnus, 2010).

As specific forms of team cognition, we assessed information sharing, shared SA, and TMMs. The sharing of (especially unique) information is recognized as a critical cognitive process and the basis by which other forms of team cognition develop (McComb, 2008). Shared SA refers to a shared perspective regarding the perception, comprehension, and future projection of current environmental events (Wellens, 1993). TMMs are mental representations of key taskwork and teamwork elements within a team's relevant environment that are shared across team members (Mohammed & Dumville, 2001). Whereas SA captures fleeting knowledge that is immediately relevant for specific contexts, TMMs characterize more durable knowledge acquired through experience (Espinosa & Clark, 2012). In selecting these three types of team cognition, we measured cognitive processes (information sharing) as well as dynamic (shared SA) and more enduring (TMMs), emergent cognitive structures.

TEMPORALITY AND TEAM COGNITION

Establishing and maintaining congruence in team members' temporal perceptions is a nontrivial task because members may enter the group context with different orientations toward time (Mohammed & Harrison, 2013). When team members disagree about deadlines and the pace by which tasks should be completed, coordination and performance problems can result (Mohammed & Nadkarni, 2014). However, the role of time has been largely downplayed in team-cognition research, which has emphasized task procedures (*what*) and team interaction patterns (*how* and *who*) without addressing timing (*when*) (Mohammed et al., 2012). It is especially important to address the development of cognitive systems (jointly consisting of information—technology—people) that need to adapt and respond in consistent ways when contextual variation occurs in a given field of practice. When changes emerge in an environment, time may become a major issue in team decision making/problem solving. Addressing this research need, we added a temporal referent to many of our measures. For example, temporal mental models were defined and operationalized as agreement among group members concerning deadlines for task completion, the pacing or speed at which activities take place, and the sequencing of tasks (Mohammed, Hamilton, Tesler, Mancuso, & McNeese, 2015). In addition, the scenarios within the NeoCITIES simulation were designed to emphasize temporal sequencing and pacing.

STORYTELLING AND REFLEXIVITY

As one of the oldest methods of transmitting knowledge across people (Denning, 2001), a story's purpose is to clearly structure and convey complicated ideas in a simpler way (Klein, 1998; Schank, 1995). Given that storytelling may be especially valuable when exploring complex, novel, and poorly understood phenomena, it may prove especially useful in aiding geographically distributed members to achieve shared understanding. Indeed, Fiore and colleagues (2009, p. 34) proposed that when information is presented *through the lens of...narrative perspective...[this] may strengthen a team's shared mental model associated with their task and teammates*. However, empirical work was needed to test this proposition. We hypothesized that teams whose members received a planned story intervention would have more similar TMMs than teams whose members did not receive a planned story intervention.

Reflexivity involves an analysis of what the group has accomplished, what it needs to accomplish, and how it can do so. Most studies have measured team reflexivity in correlational studies, rather than investigating the causal effects of *an intervention to induce reflection in groups*, which has been termed guided reflexivity (Gurtner,

Tschan, Semmer, & Nagele, 2007, p. 128). Guided individual reflexivity occurs when an intervention calls for individual members to reflect on their performance and how they can improve. Guided team reflexivity occurs when the intervention is at the group level, requiring all team members to collectively reflect on their performance and develop strategies for improvement. Although both processes could theoretically occur in the same team, no study (to our knowledge) has compared the use of both interventions to the use of just one.

EXPERIMENTAL METHODOLOGY

THE NeoCITIES TEAM SIMULATION

Research questions and hypotheses were tested via student teams performing a three-person group task called NeoCITIES, a human-in-the-loop scaled-world simulation emulating a command and control computer communication (C4) environment. NeoCITIES was designed to allow researchers to closely examine team behaviors and monitor the performance outcomes of distributed decision-making teams (McNeese et al., 2005). By trimming away the complexities and confounds of a more complex task, NeoCITIES permitted the testing of various aspects of team cognition and collaborative decision making. Methodologically, NeoCITIES offered a novel, but ideal, interface to investigate multiple aspects of team cognition due to its adaptability, flexibility, and controllability.

NeoCITIES features team resource allocation problems designed to emulate crisis management of a city's emergency services in a virtual environment. Within the simulation, teams were made up of three players, who played the roles of fire/EMS, police, or hazardous materials (HazMat). Each role was assigned three unique resources. For example, the fire unit had ambulances, fire investigators, and trucks at their disposal. Participants had to decide the severity of incoming emergency management events, what resources were needed, how many resources were needed, and whether assistance was required of other team members (Hellar & McNeese, 2010). Participants communicated with each other solely through instant chat technology. A small routine event (e.g., a trash can fire) could escalate into a more severe event (e.g., a building fire) if not responded to in a timely manner. NeoCITIES allowed for both qualitative (chat logs) and quantitative (performance scores and other behavioral metrics) data collection (Hamilton et al., 2010).

Rationale for Selecting NeoCITIES

There are several reasons why NeoCITIES was well suited to test our research objectives. First, it is designed to test team collaborative decision-making process, knowledge acquisition, and knowledge management within a command and control computer-mediated communication environment. As such, it was the ideal interface to investigate information sharing, SA, and TMMs. Second, emergency crisis management offered an excellent real-world domain that met the requirements of a complex system articulated by Rouse, Cannon-Bowers, and Salas (1992), including a highly emergent, dynamic, and complex environment, ambiguous and uncertain information, as well as individuals with differing roles needing to interact and share information as a team. Third, as an adaptable interface, NeoCITIES could be scripted to represent routine (e.g., law enforcement) and nonroutine (e.g., terrorism) scenarios ranging from individual events to highly interdependent occurrences that escalate when resources are not properly allocated. Therefore, the simulation was scalable and flexible enough to accommodate the new scenario development that was needed to test hypotheses involving team cognition, temporality, and storytelling. Fourth, objective-dependent variables were easily collected, including team performance (quantitative score indicating overall progress), time-to-decision, decision errors, and number of failed events. Fifth, participants found NeoCITIES to be an engaging and realistic exercise, which maintained interest and high levels of student participation.

Scenario Development in NeoCITIES

Emergency events were scripted around the university campus to increase realism and relevance to student participants (e.g., student rioting, bomb threats during exams, intoxication, apartment fires, and fumes from the chemistry building). Events varied in their severity (e.g., trash can fire versus a chemical tanker–truck collision). NeoCITIES scenarios were written to require a dynamic environment, high information load, the need to distinguish between relevant and irrelevant information, different roles with unique knowledge within the team, and team member interdependence. Some events were independent in that they only required resources from one unit (e.g., only EMS was required to revive a collapsed student suffering from dehydration). However, other events were interdependent in that they required resources from multiple roles (e.g., police, fire, and HazMat units were needed to respond to a large fire that may have been chemically induced, and involved the possibility of arson). Some interdependent events required a response from two units and others required all three. In addition, temporal requirements were infused in scenarios by prescribing that teammates respond on the scene of events within a particular time frame or in a certain sequence (e.g., police first, fire/EMS second, and HazMat third).

MIDWAY BREATHER—STOP AND REFLECT

- We measured three types of team cognition (information sharing, shared SA, and TMMs).
- We infused a temporal focus (deadlines, pacing, and sequencing) into the study of team cognition to more realistically mirror organizational and military teams.
- The NeoCITIES simulation offered an ideal real-world domain to investigate team cognition and allowed for a modern technological infrastructure and high fidelity scoring model.

GENERAL EXPERIMENTAL PROCEDURE

Extensive piloting preceded each study with undergraduate research assistants as well as subject pool students. Piloting involved checking the technical features of NeoCITIES, the clarity of the training materials, the difficulty of scripted events, the timing of the study, and the variability in responses to measures. In addition, several adjustments were needed to ensure that manipulation checks and measures were effective, and modifications were repiloted before formal data collection commenced.

Student participants were obtained from undergraduate courses and received course or extra credit for their participation. The sample size ranged from 71 to 185 3-person teams across the 4 experimental studies (1,686 students). After random assignment to a three-person team and a member role (police, HazMat, or fire/EMS), students were first trained on playing the NeoCITIES simulation. This training covered NeoCITIES mechanics as well as team dynamics within the simulation. After two practice scenarios, participants received a computer-generated team performance score and had the opportunity to review the solutions to each event.

Following the training, students completed two NeoCITIES performance scenarios, each of which was 14–15 minutes in length. During each performance session, participants received one or two information briefings containing data that was useful in resolving key events in the scenario. Each participant was given unique information pertaining to the same event (e.g., the time at which the event would occur or in what order units needed to respond), and only by every teammate sharing his/her briefing's content with the other teammates could that event be solved.

Throughout the study, online as well as paper and pencil measures were collected. The NeoCITIES simulation was paused once during each performance scenario to allow participants to complete a measure of SA. At the end of each scenario, students were provided with computer-generated team performance scores. Experiments were two to two and a half hours in length.

Dependent Measures

Information sharing was measured quantitatively as well as qualitatively. The quantitative measure consisted of phrase counts. Qualitatively, communication chat logs recorded during the simulation were coded (e.g., resources required, order, and pacing), allowing for a rich and more objective indicator of information sharing than self-report measures. At least two research assistants coded each set of chat log transcripts, and interrater reliability was typically above .70.

Shared SA was assessed using three different metrics, one of which was an objective indicator (e.g., situation awareness global awareness technique [SAGAT]) and two of which were subjective indicators (adaptations of the mission awareness rating scale [MARS; Matthew & Beal, 2002] and the situational awareness rating technique [SART, Taylor, 1990]). As the most popular and validated measure of SA (Endsley, 1995), SAGAT is a freeze probe technique that requires the task to be randomly stopped in order to ask participants a series of questions regarding their perceptions of the current situation. NeoCITIES and scenario sequencing had to be modified to integrate the SAGAT freeze into the simulation, so that participants would not be able to predict and prepare for its occurrence. The SAGAT measure consisted of ten multiple choice and short answer questions based on simulation events (e.g., based on the event description, what would MOST likely happen if units did not arrive on scene within 60 seconds in the case of tanker collision?). There was only one correct answer for each question that could be verified based on scenario scripting and a detailed action history of participant responses. In contrast to SAGAT, which was measured during NeoCITIES scenarios, MARS and SART were administered after participants had played the simulation. In addition, whereas the SAGAT questions had to be tailored to NeoCITIES, MARS and SART were generalized measures that could be used in a variety of contexts.

Team Mental Models

Three different types of TMM content were measured: team (how and with whom work gets done), task (what work gets done), and temporal (when work gets done). All three types were assessed using similarity ratings, one of the most popular measurement techniques (Mohammed et al., 2010). Each mental model contained a list of six dimensions that were generated through careful analysis of the task requirements, collaborative processes, and temporal dynamics involved in NeoCITIES. The six dimensions were briefly defined and arranged into a 6 × 5 grid (Hamilton et al., 2010). Participants were asked to rate the similarity of each dimensional pair on a scale of 1 (extremely unrelated) to 5 (extremely related). Sharedness was assessed using QAP correlations (e.g., Mohammed et al., 2015).

In addition to similarity ratings, temporal TMMs were also assessed via concept mapping, another common assessment technique used in the literature (e.g., Mohammed et al., 2012). After each performance session, participants were given a list of three events and asked who should arrive to the event first, second, and third. The similarity among team member maps and the number of direct links shared in Studies 1 and 2 were assessed (Mancuso, Hamilton, et al., 2011).

Team performance was operationalized objectively through the scores generated by the NeoCITIES simulation. The overall team performance score was calculated based on an event growth formula that accounted for the number of events, severity of events, and the time taken to successfully resolve events. As such, task performance in NeoCITIES rewarded players that quickly and correctly allocated the correct number and type of resources to an event. Players received lower scores when opportunities were lost through inaction or slow and/or incorrect responses (Hellar & McNeese, 2010). In addition to the overall team performance score, NeoCITIES was also modified to provide more circumscribed performance scores for temporally-infused events, independent events, and interdependent events.

INDEPENDENT VARIABLE MANIPULATIONS

Given that team cognition has been shown to facilitate team effectiveness (e.g., DeChurch & Mesmer-Magnus, 2010), we developed interventions designed to enhance the development of shared understanding in teams. Each is reviewed below.

Storytelling

Story Content: Collaboration and Timing

We directed our efforts toward story creation in two areas that were relevant within an emergency management context: collaboration and timing. The collaboration theme highlighted the breakdowns that occur when teams fail to communicate critical information, and the timing theme highlighted the breakdowns that occur when teams fail to attend to the pacing and ordering of member actions.

Story Context: Metaphorical versus Analog

The term *metaphorical storytelling* was used to refer to stories that help to connect what listeners are less familiar with to what they are more familiar with. As stories should be relatable (Denning, 2001), and popular TV shows have popularized this setting, the metaphorical story was based on a true story that occurred at a hospital in New York City. In the story, the protagonist breaks his leg during a winter vacation. The patient's postsurgery recovery encountered complications that stemmed from collaboration and timing errors of the medical team.

In contrast to metaphorical storytelling, analog storytelling maintains a closer connection between the story and the target domain. Derived from analogical problem solving (Bransford & Stein, 1984), the goal of analog storytelling is to apply the principles of narrative to a domain to a similar situation that the story represents. The NeoCITIES/analog story focused on a graduate student working in a laboratory environment during a severe snowstorm. During an unforeseen power outage, the protagonist has an accident where a beaker containing acid that he was carrying splashed its contents on his left arm. The protagonist has severe chemical burns and requires the assistance of police, fire, and HazMat. Complications stem from collaboration and timing errors of the response team.

Our stories were designed to embed the deep structure about the NeoCITIES simulation into an engaging and applicable story format using the principles of narrative. Each of the four stories went through several drafts, informed by internal discussions among the research team as well as sources with medical expertise (for the medical stories) and emergency crisis management experience (for the NeoCITIES stories). We had to balance the need to incorporate the deep level structure of collaboration and timing requirements in the NeoCITIES simulation with the need for the story to be realistic, credible, and engaging. We consulted with a professor and writer from the English department to edit our stories and ensure that they met the principles of narrative.

In summary, four types of stories (medical timing, medical collaboration, NeoCITIES timing, and NeoCITIES collaboration) and a control group (no story) were constructed. The four stories allowed us to test the effect of story content (timing versus collaboration) as well as story context (medical versus emergency crisis management) on team cognition. In the medical context, the deep level structure was similar to NeoCITIES, but the surface level structure was distinct (metaphorical storytelling). In contrast, in the NeoCITIES context, the surface and deep level structures were similar (analog storytelling) (Mancuso, Parr, et al., 2011). Participants viewed a five minute video on their computer. A narrator's voice was heard, whereas various pictures of key points were displayed.

Reflexivity

We sought to bring clarity to the literature by developing and testing guided reflexivity interventions that involved the inducement of team members' engagement in a systematic reflection exercise (Gurtner et al., 2007). We manipulated both team and individual reflexivity.

Guided Team Reflexivity

Participants were given six minutes to collectively reflect on their Scenario 1 performance via instant computer messaging. Several questions were used to guide the discussion (e.g., How well do you think you and your team just performed in performance Scenario 1? What went right? What went wrong?), and the group was asked to brainstorm strategies for how to improve their performance in Scenario 2. In the control group, the participants discussed an unrelated topic on technology and social relationships for an identical amount of time (Tesler et al., 2011).

Guided Individual Reflexivity

Participants were given a structured opportunity to make connections between the story and how it could help improve performance in NeoCITIES. Specifically, they were given an exercise to match important elements from the storytelling/control video to their correct counterparts representing the meanings of those elements as they related to successful performance in NeoCITIES (Tesler, Mohammed, Mancuso, Hamilton, & McNeese, 2012).

SUMMARY OF KEY FINDINGS AND ACROSS THE FOUR EXPERIMENTS

Below, we present key findings and conclusions across experiments as organized by our three research objectives. Readers are referred to other publications (noted below) for specific details regarding each study's findings.

Research Objective 1: To Understand the Differential Effects of Multiple Types of Team Cognition

Despite various calls for studies to simultaneously assess multiple forms of team cognition (e.g., DeChurch & Mesmer-Magnus, 2010; Mohammed et al., 2010), few previous studies have done so. In response, we measured cognitive process (information sharing) as well as dynamic (shared SA) and more enduring (TMMs) emergent cognitive structures. Controlling for Scenario 1 performance, there was a significant positive relationship between team performance and team mental model similarity when measured as concept maps (Tesler et al., 2011) and both concept maps and similarity ratings (Mohammed et al., 2015; Tesler et al., 2012). Another study showed that shared SA (using SAGAT) and TMMs (using concept maps) evidenced a reciprocal relationship, such that shared SA in Scenario 1 contributed to more accurate TMMs for Scenario 1, which contributed to more accurate shared SA in Scenario 2, which then contributed to more accurate TMMs in Scenario 2. TMMs mediated the relationship between shared SA and team performance. Other results revealed that information sharing played a moderating role in that the relationship between analog storytelling and team mental model similarity was strengthened under higher information sharing (Mohammed, Tesler, Hamilton, Mancuso, & McNeese, 2014).

RESEARCH OBJECTIVE 2: TO INFUSE A TEMPORAL FOCUS INTO THE STUDY OF TEAM COGNITION

Infusing a temporal focus into the study of team cognition and team outcomes proved fruitful in better reflecting the time-based context of many organizational and military teams. Temporal TMMs were distinct from taskwork and teamwork categories and positively and uniquely contributed to team performance beyond these traditionally measured domains. In addition, temporal TMMs assessed later in a team's development exerted stronger effects on team performance than those assessed earlier (Mohammed et al., 2015). Given the salience of time-based demands in work teams with these promising results, future research should expand the conceptualization of team cognition to represent who is going to do what, how, and *when* (Mohammed et al., 2015).

RESEARCH OBJECTIVE 3: TO INVESTIGATE THE EFFECTIVENESS OF INTERVENTIONS DESIGNED TO IMPROVE TEAM COGNITION

Results provided encouraging evidence that storytelling and reflexivity interventions may help overcome the collaborative obstacles faced by team members in distributed environments, particularly when administered at the group level. Storytelling had a positive influence on team mental model similarity, but only for the analog story (Mohammed et al., 2014). The effectiveness of the analog story over the metaphorical/medical story in helping team members *get on the same page* highlights the importance of maintaining a close connection between the story and the target domain. In addition, the relationship between receiving the analog story and team mental model similarity was strengthened under higher information sharing (Mohammed et al., 2014).

In another study, analog storytelling also increased team mental model similarity over a nonstory format, which in turn, increased team performance (Tesler et al., 2011). However, storytelling was most useful when teams were additionally given an opportunity to discuss and come to a consensus on the story's meaning while developing strategies to improve future performance (Tesler et al., 2011). In enhancing the positive effects of storytelling, guided team reflexivity provided the mechanism by which the story content could be fully processed.

In another experiment, guided team reflexivity had a significantly larger positive effect on team mental model similarity than guided individual reflexivity (Tesler et al., 2012). That is, allowing team members to communally reflect on their performance and strategies was more effective in getting members on the same page than individual reflection. Storytelling plus guided team reflexivity yielded the highest level of team mental model similarity out of all of the tested conditions (Tesler et al., 2012).

Individual reflexivity did not provide incremental benefits when combined with group reflexivity, and actually hurt outcomes when storytelling was also present (Tesler et al., 2012). Team mental model similarity mediated the relationship of guided reflexivity type (team versus individual) and performance. Based on these results, group reflexivity was preferable to individual reflexivity.

LESSONS LEARNED

THE VALUE OF AN INTEGRATIVE PERSPECTIVE

This overall research initiative combined research on team cognition, temporal dynamics, and storytelling toward the goal of improving team coordination and performance in distributed decision-making teams. In addition, we integrated across multiple forms (information sharing, shared SA, and TMMs) and operationalization of team cognition. For example, TMMs were measured through a more sophisticated lens than previous studies in that three content domains (taskwork, teamwork, and temporal) were assessed. Furthermore, both concept mapping and similarity ratings were measured for temporal TMMs. Similarly, we operationalized SA in three ways (SAGAT, MARS, and SART). Adding to the complexity, concept maps and the three SA metrics were measured twice in each study. Information sharing was captured continuously through chat logs. Assessing multiple forms and operationalization was no small task in that many of our team cognition measures were context-dependent and therefore had to be developed specifically for the NeoCITIES simulation (e.g., information sharing coding, SAGAT, similarity ratings, and concept maps). However, emphasizing the importance of triangulation, the type of team cognition and operationalization differentially affected team outcomes. Across studies, results revealed that various types of team cognition positively influenced team performance indices, but results differed depending on the type of team cognition (e.g., information sharing, shared SA, and TMMs), the operationalization of team cognition (e.g., concept maps, similarity ratings), and the team outcome assessed (overall team performance and timing outcomes).

THE VALUE OF AN INTERDISCIPLINARY PERSPECTIVE

From the outset, an interdisciplinary perspective was interwoven into all aspects of the research process. The integration of psychology and information sciences and technology fields was essential to obtaining grant funding from the Office of Naval Research. Crossing disciplinary boundaries facilitated the freedom to venture into nonexplored territory on multiple fronts, including empirically examining the effect of planned storytelling as a team-training intervention and pioneering the operationalization of a temporal TMM. In addition, an interdisciplinary focus empowered us to combine constructs that are not commonly measured together. Whereas TMMs derive from the Industrial/Organizational Psychology literature, SA derives from aviation, human factors, and cognitive psychology (Mohammed et al., 2010). Thus, assessing in both our experiments enabled us to contribute uniquely to multiple literatures.

Building a research team consisting of graduate students from both psychology and information sciences and technology was critical to carrying out our experimental objectives. Conducting experiments with complex designs necessitating extensive piloting to check manipulations and large team sample sizes to run advanced statistical analyses is the strength of doctoral training in psychology. The expertise needed to design, upgrade, and modify NeoCITIES to meet the needs of each experiment came from graduate students in information sciences and technology. Indeed, NeoCITIES 3.0 and the setup of the client server architecture was under construction for more than nine months, requiring more than 10,000 lines of code and more than a month of pilot testing the new interface. In addition, we continued to upgrade and modify NeoCITIES throughout the four studies to accommodate new scenario development and a new adaptive interface. The interdisciplinary nature of this research is also evidenced through our publication outlets (e.g., *International Journal of Gaming and Computer-Mediated Simulations, European Journal of Work and Organizational Psychology,* chapters in *Theories of Team Cognition: Cross Disciplinary Perspectives*) and conference presentations (e.g., *Society of Industrial/Organizational Psychology, Academy of Management, Human Factors and Ergonomics Society, Interdisciplinary Network of Group Researchers, and the International Science of Team Science*).

CONCLUSION

Using the living laboratory framework, this program of research expanded team cognition research into previously unexplored avenues identified as theoretically fertile, but not investigated empirically. As storytelling has been underresearched in the team literature, we investigated how the powerful benefits of ad hoc stories could be leveraged as a team-training tool to allow team members to get on the same page more efficiently. As failure to understand *when* can seriously jeopardize final team outcomes, we expanded the definition of team cognition to include a temporal dimension to better reflect the temporal context of many organizational and military teams. Across four programmatic experimental studies, research questions were tested using NeoCITIES, an ideal, adaptable interface with which to measure multiple aspects of team cognition. Results provided encouraging evidence that storytelling and guided reflexivity interventions may help overcome the collaborative obstacles faced by team members in distributed environments, particularly when administered at the group level. Infusing a temporal focus into the study of TMMs proved fruitful in predicting unique variance beyond traditional teamwork- and taskwork-content domains.

REVIEW QUESTIONS

1. What were the three research objectives of this program of research?
2. Describe the differences between information sharing, shared SA, and TMMs.
3. Describe the differences between metaphorical and analog storytelling.
4. What were the key findings across the four experiments conducted?
5. What did you find most interesting?

ACKNOWLEDGMENTS

The research presented in this paper was supported by Grant Number N000140810887 from the Office of Naval Research. We are appreciative of this funding that enabled the completion of this research. The opinions expressed in this paper are those of the authors only and do not necessarily represent the official position of the Office of Naval Research, the U.S. Navy, the U.S. Department of Defense, or The Pennsylvania State University.

REFERENCES

Ancona, D. G., Goodman, P. S., Lawrence, B. S., & Tushman, M. L. (2001). Time: A new research lens. *Academy of Management Review, 26*(4), 645–663.

Ang, C. S., & Rao, G. S. V. R. K. (2008). Computer game theories for designing motivating educational software: A survey study. *International Journal on E-Learning, 7*(2), 181–199.

Badke-Schaub, P., Neumann, A., Lauche, K., & Mohammed, S. (2007). Mental models in design teams: A valid approach to performance in design collaboration? *CoDesign, 3*, 5–20.

Bartel, C. A., & Garud, R. (2009). The role of narratives in sustaining organizational innovation. *Organization Science, 20*(1), 107–117.

Bransford, J. D., & Stein, B. S. (1984). *The ideal problem solver. A guide for improving thinking, learning, and creativity.* New York, NY: Freeman.

Cannon-Bowers, J. A., & Salas, E. (2001). Reflections on shared cognition. *Journal of Organizational Behavior, 22*, 195–202.

DeChurch, L. A., & Mesmer-Magnus, J. (2010). The cognitive underpinnings of effective teamwork: A meta-analysis. *Journal of Applied Psychology, 95*, 32–53.

Denning, S. (2001). *The springboard: How storytelling ignites action in knowledge-era organizations.* Boston, MA: Butterworth Heinemann.

Endsley, M. R. (1995). Measurement of situation awareness in dynamic systems. *Human Factors, 37*(1), 65–84.

Espinosa, J. A., & Clark, M. A. (2012). Team knowledge: Dimensional structure and network representation. In E. Salas, S. Fiore, & M. Letsky (Eds.), *Theories of team cognition: Cross disciplinary perspectives* (pp. 289–312). New York, NY: Taylor and Francis Group, LLC.

Fiore, S. M., McDaniel, R., & Jentsch, F. (2009). Narrative-based collaboration systems for distributed teams: Nine research questions for information managers. *Information Systems Management, 26*(1), 28–38.

Fiore, S. M., & Salas, E. (2006). Team cognition and expert teams: Developing insights from cross-disciplinary analysis of exceptional teams. *International Journal of Sport and Exercise Psychology, 4*(4), 369–375.

Gilson, L. L., Maynard, M. T., Young, N. C., Vartiainen, M., & Hakonen, M. (2015). Virtual teams research: 10 years, 10 themes, and 10 opportunities. *Journal of Management, 41*(5), 1313–1337.

Glantz, E. J. (2017). *Cognitive systems engineering: An integrative living laboratory framework.* Boca Raton, FL: CRC Press.

Gurtner, A., Tschan, F., Semmer, N. K., & Nägele, C. (2007). Getting groups to develop good strategies: Effects of reflexivity interventions on team process, team performance, and shared mental models. *Organizational Behavior and Human Decision Processes, 102*, 127–142.

Hamilton, K., Mancuso, V., Minotra, D., Hoult, R., Mohammed, S., Parr, A., Dubey, G., MacMillan, E., & McNeese, M. (2010). Using the NeoCITIES 3.1 simulation to study and measure team cognition. In *Proceedings of the 54th annual meeting of the human factors and ergonomics society* (pp. 433–437). San Francisco, CA: Human Factors and Ergonomics Society.

Hellar, D. B., & McNeese, M. (2010). NeoCITIES: A simulated command and control task environment for experimental research. In *Proceedings of the 54th meeting of the human factors and ergonomics society* (pp. 1027–1031). San Francisco, CA: Human Factors and Ergonomics Society.

Klein, G. (1998). *Sources of power.* Cambridge, MA: MIT Press.

Kozlowski, S. W. J., & Bell, B. S. (2003). Work groups and teams in organizations. In W. C. Borman, D. R. Ilgen, & R. J. Klimoski (Eds.), *Handbook of psychology (Vol. 12): Industrial and organizational psychology* (pp. 333–375). New York, NY: Wiley.

Lee, J. D., & Kirlik, A. (Eds.) (2013). *The Oxford handbook of cognitive engineering (Oxford library of psychology).* New York, NY: Oxford University Press.

Mancuso, V., Hamilton, K., McMillan, E., Tesler, R., Mohammed, S., & McNeese, M. (2011). What's on "their" mind: Evaluating collaborative systems using team mental models. In *Proceedings of the 55th annual meeting of the human factors and ergonomics society* (pp. 1284–1288). Las Vegas, NV: Human Factors and Ergonomics Society.

Mancuso, V., Parr, A., McMillan, E., Tesler, R., McNeese, M., Hamilton, K., & Mohammed, S. (2011). Once upon a time: Behavioral, affective and cognitive effects of metaphorical storytelling as a training intervention. In *Proceedings of the 55th annual meeting of the human factors and ergonomics society* (pp. 2113–2117). Las Vegas, NV: Human Factors and Ergonomics Society.

Matthew, M. D., & Beal, S. A. (2002). *Assessing situation awareness in field training exercises.* U.S. Army Research Institute for the Behavioural and Social Sciences. Ft Belvoir, VA.

McComb, S. A. (2008). Shared mental models and their convergence. In M. P. Letsky, N. W. Warner, S. M Fiore, & C. A. P. Smith (Eds.), *Macrocognition in teams: Theories and methodologies* (pp. 35–50). Aldershot: Ashgate Publishing.

McNeese, M. D. (1996). An ecological perspective applied to multi-operator systems. In O. Brown & H. L. Hendrick (Eds.), *Human factors in organizational design and management - VI* (pp. 365–370). Amsterdam: Elsevier.

McNeese, M. D., Bains, P., Brewer, I., Brown, C., Connors, E. S., Jefferson, T., Jones, R. E. T., & Terrell, I. (2005). The NeoCITIES simulation: Understanding the design and experimental methodology used to develop a team emergency management simulation. In *Proceedings of the human factors and ergonomics society 49th annual meeting.* Santa Monica, CA: HFES.

McNeese, M. D., Mancuso, V., McNeese, N., Endsley, T., & Forster, P. (2013, May). Using the living laboratory framework as a basis for understanding the next generation analyst work. *Proceedings SPIE 8758, Next-Generation Analyst,* 87580F.

McNeese, M. D., Perusich, K., & Rentsch, J. R. (2000). Advancing socio-technical systems design via the living laboratory. In *Proceedings of the industrial ergonomics association/human factors and ergonomics society (IEA/HFES) 2000 congress* (pp. 2-610–2-613). Santa Monica, CA: Human Factors and Ergonomics Society.

Mohammed, S., & Dumville, B. (2001). Team mental models in a team knowledge framework: Expanding theory and measurement across disciplinary boundaries. *Journal of Organizational Behavior, 22,* 89–106.

Mohammed, S., Ferzandi, L., & Hamilton, K. (2010). Metaphor no more: A 15-year review of the team mental model construct. *Journal of Management, 36*(4), 876–910.

Mohammed, S., Hamilton, K., & Lim, A. (2009). The incorporation of time in team research: Past, current, and future. In E. Salas, G. F. Goodwin, & C. S. Burke (Eds.), *Team effectiveness in complex organizations: Cross-disciplinary perspectives and approaches* (pp. 321–348). New York, NY: Routledge Taylor and Francis Group. (Organizational Frontiers Series sponsored by the Society for Industrial and Organizational Psychology).

Mohammed, S., Hamilton, K., Tesler, R., Mancuso, V., & McNeese, M. (2015). Time for temporal team mental models: Expanding beyond "what" and "how" to incorporate "when." *European Journal of Work and Organizational Psychology* (special issue, dynamics of Team Adaptation and Cognition), *24*(5), 693–709.

Mohammed, S., & Harrison, D. (2013). The clocks that time us are not the same: A theory of temporal diversity, task characteristics, and performance in teams. *Organizational Behavior and Human Decision Processes, 122*(2), 244–256.

Mohammed, S., & Nadkarni, S. (2014). Are we all on the same temporal page? The moderating effects of temporal team cognition on the polychronicity diversity-team performance relationship. *Journal of Applied Psychology, 99*(3), 404–422.

Mohammed, S., Tesler, R., & Hamilton, K. (2012). Time and shared cognition: Towards greater integration of temporal dynamics. In E. Salas, S. Fiore, & M. Letsky (Eds.), *Theories of team cognition: Cross disciplinary perspectives* (pp. 87–116). New York, NY: Taylor and Francis Group, LLC.

Mohammed, S., Tesler, R., Hamilton, K., Mancuso, V., & McNeese, M. (2014, May). Improving temporal mental model similarity in distributed decision making teams. In M. S. Prewett (Chair), *Strategies for improving virtual team processes and emergent states.* A symposium presented to the Twenty-Ninth Annual Conference of the Society for Industrial/Organizational Psychology, Honolulu, HI.

Rouse, W. B., Cannon-Bowers, J. A., & Salas, E. (1992). The role of mental models in team performance in complex systems. *IEEE Transactions on Systems, Man, and Cybernetics, 22*(6), 1296–1308.

Salas, E., Rosen, M. A., Burke, C. S., Goodwin, G. F., & Fiore, S. M. (2006). The making of a dream team: When expert teams do best. In K. A. Ericsson, N. Charness, P. J. Feltovich, & R. R. Hoffman (Eds.), *The Cambridge handbook of expertise and expert performance* (pp. 439–453). New York, NY: Cambridge University Press.

Salas, E., & Wildman, J. L. (2009). Ten critical research questions: The need for new and deeper explorations. In E. Salas, G. F. Goodwin, & C. S. Burke (Eds.), *Team effectiveness in complex organizations* (pp. 525–547). New York, NY: Routledge, Taylor & Francis Group.

Schank, R. C. (1995). *Tell me a story: Narrative and intelligence.* Evanston, Illinois: Northwestern University Press.

Tannenbaum, S. I., Mathieu, J. E., Salas, E., & Cohen, D. (2012). Teams are changing: Are research and practice evolving fast enough? *Industrial and Organizational Psychology: Perspectives on Science and Practice, 5*(1), 2–24.

Taylor, R. M. (1990). Situational awareness rating technique (SART): The development of a tool for aircrew systems design. Situational awareness in aerospace operations. AGARD-CP-478. Neuilly Sur Seine, France: NATO-AGARD, 3/1–3/17.

Tesler, R., Mohammed, S., Hamilton, K., Mancuso, V., Parr, A., MacMillan, E., & McNeese, M. (2011, April). *The effects of storytelling and reflexivity on team mental models.* A poster presented to the Twenty-Sixth Annual Conference of the Society for Industrial/ Organizational Psychology, Chicago, IL.

Tesler, R. M., Mohammed, S., Mancuso, V., Hamilton, K., & McNeese, M. D. (2012, April). *Improving team mental models: Individual versus team reflexivity and storytelling.* A poster presented to the Twenty-Seventh Annual Conference of the Society for Industrial/Organizational Psychology, San Diego, CA.

Undre, S., Sevdalis, N., Healey, A. N., Darzi, S. A., & Vincent, C. (2006). Teamwork in the operating theatre: Cohesion or confusion? *Journal of Evaluation in Clinical Practice, 12*, 182–189.

Wellens, A. R. (1993). Group situation awareness and distributed decision making: From military to civilian applications. In N. J. Castellan Jr. (Ed.), *Individual and group decision making: Current issues* (pp. 267–291). Hillsdale, NJ: Lawrence Erlbaum Associates.

10 Modeling Cognition and Collaborative Work

Rashaad Jones

CONTENTS

Introduction 200
   Overview of Cognition Modeling 200
      Fuzzy Cognitive Maps 201
Living Lab-Generated FCMs in Practice 204
   Application #1: FCM to Study Emergency Crisis-Management Teams 204
   Application #2: The SA–FCM Modeling Approach 207
   SA–FCM for Decision-Aiding Support 208
   SA–FCM for Predictive Assessments 212
   SA–FCM for Measuring Cognitive Readiness 213
   A Conceptual Model of the SA–FCM for Data Fusion 214
Conclusion 215
Review Questions 216
References 216

ADVANCED ORGANIZER

This chapter describes a cognitive-modeling approach that was used in conjunction with the living lab framework to create models of decision making. The fuzzy cognitive mapping technique was utilized to represent the relationship between cognitive requirements and decisions. This modeling technique was used in a variety of decision-making research that is explained below. This chapter begins with an introductory of the fuzzy cognitive maps (FCM) cognitive-modeling technique and its relationship within the living lab framework. Next, the chapter presents cognitive modeling and specifically details the FCM-modeling approach. Several use cases are then presented to illustrate how this cognitive-modeling approach has been used in research. Finally, the chapter concludes with benefits, limitations, and future studies.

INTRODUCTION

The scaled-world component of the living lab framework provides a setting for researchers to study a variety of constructs related to cognition, decision making, and teamwork. Scaled worlds are a type of virtual environment that are useful in simulating real-world complex tasks derived from situated activities (Angles et al., 2012; Gray, 2002). A chief advantage of scaled worlds is their ability to *scale-up* to meet the demands of the work domain. McNeese and associates (McNeese, Perusich, & Rentsch, 1999; McNeese, Zaff, Citera, Brown, & Whitaker, 1995) have cited several examples of using scaled worlds to examine socio-organizational constraints, modeling teamwork processes, and measuring team schema development. Researchers have used scaled worlds to study the mutual interplay of understanding, modeling, and measuring teamwork within complex systems (McNeese, Perusich, & Rentsch, 2000; Jones, 2006). Furthermore, scaled worlds that employ cognitive models can simulate situations necessary to inform the cognitive system engineering design and to make sense of human centered requirements.

Particularly useful for research in decision making and team performance, researchers can use scaled worlds to control complexity to experimental tasks performed by human subjects. Often explicitly expressed in scenarios or *dialed-in* during research tasks, scaled worlds offer a way for researchers to set and control complexity during their examination. One such complexity is decision-aids that are designed to facilitate decision-making tasks and collaborative activities.

Decision-aids can be described as intervention tools to support decision making for human-related tasks. There are numerous techniques to develop decision-aids (see Jones, 2006). Initially, decision-aids were developed using *technocentric approaches* where requirements were developed to meet the needs of the system. This oftentimes has resulted in aids being nonadaptive to human user needs and produce *automation surprises* that leads to errors (McNeese, Rentsch, & Perusich, 2000; Woods, Johannesen, Cooke, & Sarter, 1994).

The view that this chapter takes is a human-centered, or user-centric, approach to designing decision-aids. These types of decision-aids are capable to utilize contextualized information that can enact human intentions. One such technique is FCM that is an artificial-intelligence (AI) approach that involves modeling human decision making to support teamwork. The remainder of this chapter is organized as follows: the next section will provide an overview of cognitive-modeling approaches as related to team decision making, then the chapter will review specific FCM-modeling approaches that represent team decision making, and finally, the chapter will conclude with closing remarks that highlight the benefits and limitations FCMs offer to cognitive modeling.

OVERVIEW OF COGNITION MODELING

Cognitive modeling is a discipline of computer science that involves creating computational representations of human cognition to simulate human problem solving and mental task processes (Badler, Phillips, & Zeltzer, 1993; Funge, Tu, & Terzopoulos, 1999). The products of cognitive modeling are cognitive models, which can then be used in a variety of ways, including decision-aids to support decision-making tasks (Dhukaram & Baber, 2016; Todd & Benbasat, 1994; Vitoriano, Montero de Juan, & Ruan, 2013).

There is a broad range of cognitive-modeling approaches (see Busemeyer & Diederich, 2010) and a complete overview is outside the scope of this chapter. Instead, this chapter will briefly discuss similar characteristics shared by decision-making models. (It is important to note that this overview is not a prescription, but rather a description of certain core characteristics of decision-making models). Each model represents human decision making by dividing the decision-making process into the following five primary subprocesses defined below:

1. *Recognition* involves receiving new information from the input space that can come in various forms, including structured reports, events, or unstructured information.
2. *Situation assessment* involves an understanding of an identification/diagnosis of the current situation and how it fits within the context.
3. *Situational analysis* is an analysis routine that involves a reasoning mechanism that compares the current situation to what is already known. It also contains a formulation of a set of goals that are needed to address the current situation.
4. *Planning/evaluation* is a cognitive process that transforms goals into a plan by decomposing the goals into tasks.
5. *Decision/act* that performs a specific activity that impacts the output space.

Fuzzy Cognitive Maps

FCM are a kind of cognitive model that uses fuzzy casual graphs to model the relationship between inputs and outputs of a given system. Introduced by Bart Kosko in 1986, FCMs combine fuzzy logic (introduced by Lotfi A. Zadeh in 1965) and cognitive mapping (introduced by Robert Axelrod in 1976). Fuzzy logic is based on fuzzy set theory that extends classical (or Boolean) logic, where values may only be 0 (false) or 1 (true), to represent the notion of partial values, where values may be any real number between −1 and 1 (defined as partial truths).[*]

Cognitive mapping is based on graph theory and was designed to represent social scientific knowledge, where a cognitive map represents a mental landscape of a system (Jacobs & Schenk, 2003; Kitchin, 1994; O'Keefe & Nadel, 1978). In practice, cognitive maps are graphical tools that allow people to organize and structure knowledge (Goldstein, 2010; Gray, Gray, Cox, & Henly-Shepard, 2013; Knight, 2002; Kosko, 1986; Manns & Eichernbaum, 2009; to name a few). Cognitive maps represent concepts, such as ideas and information, as geometric shapes, typically boxes or circles that connect to other concepts with labeled arrows. The labeled arrows denote the relationship between concepts and can be articulated in linking phrases, such as *causes*, *requires*, or *contributes to*.

---

[*] A fuzzy value of −1 means an inverse relationship, 0 means no relationship, and 1 means a direct relationship.

A FCM is a signed fuzzy-directed graph that consists of nodes that represent elements (e.g., concepts and events.) of a system. Each node is connected to other nodes by a directed line (i.e., line with an arrow) that indicates the relationship of the nodes. The direction of the arrow denotes which node influences or impacts other nodes, where the concept and where the arrow is flowing from, impacts or influences the node where the arrow is flowing toward. Each directed line is labeled with a fuzzy value (weight value) to indicate the strength of the causal conditions between nodes.

Computationally, a node has a value of 0 when the node is inactive (or the state is *false* if the node is an input node) and has a value of 1 when the node is active (or the state is *true* if the node is an input node). A threshold (or squashing) function is used to activate all noninput nodes. In practice, many different squashing functions can be used and the FCMs presented in this chapter used a binary function that was suggested by Kosko (1986), Dickerson & Kosko (1993), and Taber (1991). With the binary function, all noninput nodes become inactive and can only become active when the sum of all connected nodes multiplied by their corresponding weight value is greater than or equal to 1. Formally, the equation is represented in the Equation 10.1. Formula to compute activation value for a single FCM concept is given in Equation 10.1 as

$$A(c) = \sum_{k=1}^{n} a_k * w_k \tag{10.1}$$

where:

c is a noninput concept

a_k represents the activation value for node k (that is connected to c)

w_k represents the weight value between c and k

Consider the FCM in Figure 10.1 below that models how a student can receive an A in a class. For this FCM, there are five input concepts: *do homework*, *attend class*, *meet with instructor*, *read textbook*, and *study for tests*. These input nodes are factors

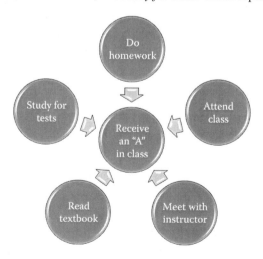

FIGURE 10.1   Sample FCM.

to the output concept, *receive A in class*. Each input concept has a weight that defines the strength of influences on the output concept. For example, concepts *do homework*, *attend class*, and *study for tests* each influence the output by 0.30.

For *receive A in class* to be true, the sum of all input nodes multiplied by their corresponding weight must be greater than or equal to 1. For this to occur, A, B, D, and E must be active. In other words, for *receive A in class* to be true, *do homework*, *attend class*, *read textbook*, and *study for tests* must all be true. No one concept has a strong enough value to influence the output without the help of others and the concept, *meet with instructor* does not have a weight value to influence the output at all.[*]

FCMs represent a soft AI approach to modeling systems. Hard AI approaches, such as neural networks, Bayesian networks, and so on, use symbolic processing to represent knowledge (Jones, 2006; Kosko, 1993). Kosko (1986) has argued that symbolic processing is limited during the *recognition* process when uncertainty exists (due to ambiguous, unknown, or uncertain information). The reason for this is that symbolic processing requires knowledge to be exact and has a hard time representing knowledge with inexact states. Conversely, the fuzzy-graphical structures of FCMs offer the capability to represent uncertainty by representing knowledge with hazy degrees of states.

As a cognitive model, FCMs model the cause-and-effect relationships that define a complex system. Unique to FCMs is their ability to represent attributes of the decision-making process as qualitative states, rather than numerical values, which is more aligned with how humans make decisions. For example, a person may turn on an air conditioner when the temperature is hot and not know the exact numerical temperature value and for team decision making, humans typically rely on qualitative states rather than numerical values. Furthermore, by using qualitative states, FCMs can model emergent and nonlinear qualities of a complex system. Thus, this transformational grammar that translates numerical values into qualitative states is a significant characteristic used by FCMs and has been highlighted by several researchers (Perusich & McNeese, 1998).

It has been a popular misconception to confuse fuzzy logic with probability. The distinction between fuzzy logic and probability involves how each refers to uncertainty. Fuzzy logic is specifically designed to deal with imprecision of facts (fuzzy logic statements), whereas probability deals with chances of an event occurring (but still considering the result to be precise). In other words, fuzzy logical statements represent membership in vaguely defined sets, whereas probabilities represent the likelihood of some event or condition. To illustrate the difference, consider the following scenario: Bob is in a house with two adjacent rooms: the kitchen and the dining room. In probabilistic terms, Bob is either in the kitchen or not in the kitchen (for this specific scenario, there is a 50% chance that he is in the kitchen). However, fuzzy logic statements will not refer to his chances of being in the kitchen; instead, it will convey the degree of Bob being in the kitchen. This becomes especially important when Bob is in the doorway that is between the two rooms where he would be considered both *partially in the dining room* and *partially in the kitchen*.

---

[*] In practice, to reduce complexity, concept C could be removed from the FCM because it does not impact the output.

MIDWAY BREATHER

- The FCM-modeling approach was applied to the scaled-world component of the living lab framework to study decision making and to support user decision making.
- A FCM combines concept mapping and fuzzy logic to model how cognitive requirements influence decision making. Cognitive requirements are represented as concepts and the strength of influence for each concept is represented as a weight value between −1 and 1.

LIVING LAB-GENERATED FCMs IN PRACTICE

Since their introduction 30 years ago, FCMs have been used in a wide range of applications, and due to their ability to represent uncertainty and causal relationships that are uncertain, ambiguous, or nondeterministic, FCMs have been particularly applicable to modeling decision making in soft knowledge domains, including political science, military science, history, international relations, and organization theory (Kosko, 1993; Mkrtchyan & Ruan, 2011; Niskanen, 2005; Papaioannou, Neocleous, & Schizas, 2013; Rodriguez-Repiso, Setchi, & Salmeron, 2007; Rodriguez, Vitoriano, Montero, & Keeman, 2011; Taber, 1991; Taylor, 2006; Vasantha-Kandasamy & Smarandache, 2003; to name a few).

The living lab framework provides a seamless methodology to generate FCM models that can be used as decision-aids for decision-making research. Even though FCMs are used in the *scaled-worlds* component of the living lab, the *knowledge-elicitation* component provides researchers with the opportunity to identify parameters in the decision space, including requirements, decision points, and decisions. During this phase of the research project, researchers can interview with SMEs to learn information requirements (that can be represented as concepts) and the qualitative states of these concepts (that can be represented as logical states). For example, when driving, one may say they determine how fast to drive their car based on weather conditions, which may have the following states: *rainy, foggy or dark*, or *good*. Drivers generally do not know numerical values for these states, but can easily express how these states impact their driving. In addition, researchers can draw their map during interview sessions, which allows the SMEs to provide feedback.

The section mentioned below list examples where FCMs have been developed to study decision making. The first describes a general FCM using for emergency crisis teams and the remaining describe the situation awareness–FCM (SA–FCM), which is a special type of cognitive models that support decision making by focusing on developing SA.

APPLICATION #1: FCM TO STUDY EMERGENCY CRISIS-MANAGEMENT TEAMS

This section describes a research effort that involved the development of a FCM to study teams engaged in an emergency crisis-management (ECM)-simulated task. The ECM work domain is an interesting area to examine teamwork because it covers a multitude of situations that can arise from a wide range of circumstances including natural disasters (e.g., tornados, wildfires, landslides, thunderstorms, blizzards, and volcanic eruptions) to human-caused mishaps such as mine subsidence, toxic spills, nuclear meltdowns, and even terrorism. Each type of emergency requires decision makers to take immediate steps to aid response and recovery. Similarly, emergencies call for preventative measures, such as planning and preparedness, as well as sustained efforts to reduce long-term risk to hazards or mitigation. In addition, decision makers must make appropriate decisions because errors can *snowball* into larger disasters and even become catastrophic (e.g., the disaster at Three Mile Island; see, Woods, Johannesen, Cooke, & Sarter, 1994).

Gredler (1994) conceptualizes the ECM domain as having three major characteristics: (1) the high-priority goals of a decision-making unit are threatened, (2) the time available to decision makers for reacting to situations before they transform is restricted, and (3) high situation awareness from the decision makers is required. Unlike nonemergency situations where decision makers may include long-range plans, the ECM work domain typically excludes extended strategy planning as a means of resolution. Instead, decision makers are required to accurately assess the situation as quickly as possible while they consider ways to apply available resources, and the allocation of these resources is also accompanied by periodic reassessments to determine whether changes in response are needed.

Cannon-Bowers and Salas (2001) contend that in a crisis-management domain, the nature of work is so complex that it would be impossible for any single team member to hold all the knowledge required to succeed. Rouse, Cannon-Bowers, and Salas (1992) contend that this kind of environment can be turbulent, with time-varying goals and requirements. Further, such environments often require the coordinated efforts of a team operating within an organizational or hierarchical context. According to Rasmussen and associates (1994), effective organizations must be flexible and adaptive because it is doubtful that coordination *by plan* can satisfy all of the contingencies likely to arise. Moreover, the ECM domain is highly emergent and complex, and requires people to work in teams interacting through various levels of collaborative and communicative technologies. Several research efforts have identified the major behaviors of teams working within this type of domain: (1) teams operate effectively when individual members share the workload, (2) team members continually monitor the work behavior of other members, and (3) team members develop and contribute expertise on subtasks (Mathieu, Heffner, Goodwin, Salas, & Cannon-Bowers, 2000; Cooke, Salas, Cannon-Bowers, & Stout, 2000).

ECM offers a real-world domain that served as an excellent area for examining teamwork and team cognition. Jones (2006) noted that information in this domain was found to be highly dynamic with incomplete and ambiguous data elements and had high degrees of uncertainty. Cognitive-modeling research in this area that utilized hard AI techniques generally resulted in decision-aids that would make inaccurate and erroneous recommendations. Consequently, Jones (2006) utilized a FCM-modeling approach that produced a computational cognitive model that conceptually is not limited by exact values and measurements, and is capable to represent uncertainty and relatively unstructured knowledge and causalities expressed in imprecise forms.

In his research, Jones created the NeoCITIES task, which is a scaled-world emergency crisis-management simulation (see Jones [2006] for a description of the NeoCITIES task). Jones then performed research to examine FCM decision-aids that supported teamwork in this task. The FCM model was developed from previous research that performed knowledge elicitation on counter terrorism, crisis-management operations (e.g., preventing the spread of the Bubonic plague), and intelligence analysis (see Jones, Jefferson, Connors, McNeese, & Perusich, 2005; McNeese et al., 2005). From the knowledge elicitation sessions, Jones developed a FCM model to support one role for each team. As this team member received events, the model assessed the likelihood of the event involving terrorism. All events that the model determined to have a terrorism status (i.e., a severe event requiring numerous resources) were highlighted for the team member. As the FCM only provided a prediction, team members had to decide to trust the FCM or ignore the model output.

Utilizing a technique described by Perusich, McNesse, and Rentsch (1999), Jones developed a FCM via the knowledge elicitation component of the living lab. A follow-up validation task was completed to verify the accuracy of the weight values. The FCM model included the following information requirements:

- Low resources are available—a low status (<10%) of resources are available.
- Advisory information is present—recent transmission of local, state, or federal warnings occurred.
- Current event has high severity—the most recent event has characteristics suggest that the event requires multiple teams and resource types (e.g., a riot).
- History of attacks suggests terrorism—emerging trends with severe events have recently occurred.
- Feedback from prior events indicate terrorism—information obtained by dispatched resources to older events suggest terrorism ties (e.g., a Police event that included apprehending a perpetrator with ties to a terrorist organization).

The objective of the research examined how the presence of the FCM impacted team cognition. The examination included 34 sessions consisting of approximately 800 undergraduate students. Due to limitations with the enrollment of participants, all of the sessions consisted of two teams instead of the full three team operation. After assignments were made, the participants received training based on their team and role. The experimental sessions commenced with the completion of an IRB-mandated inform consent document. Next, participants were randomly assigned to a role (i.e., information officer or resource manager; see Jones [2015] for a brief overview of NeoCITIES) and held the same role throughout the entire session. Dyads (i.e., a team consisting of two members) were randomly assigned to one of the three NeoCITIES response team types (i.e., police, fire/EMS, or hazards). The participants were then given a briefing session that introduced and described the experiment and the game. Following the briefing, participants were then asked to participate in a short training session. Similar to a walkthrough in a video game, these sessions served to acclimatize the participants' orientation of the interface, how to use the features/components provided by the interface, and how to utilize the decision-aid functionality (a training segment of the FCM was only provided to the teams who were assigned to a session that included full AI support). The training concluded with an activity that involved three single-team events per team.

The study included one independent variable that represented the level of AI support (full or none). Half of the sessions included a FCM model for each team (i.e., full-support). Metrics were automatically recorded throughout the game, including *team performance, number of decision-making errors, and the number of failed events*. At the conclusion of the scenario, team members were provided with a teamwork similarity questionnaire to measure their team mental model (see Rentsch & Hall, 1994). This survey asked each player to rate 15 items on a seven-point Likert scale from 1 (extremely unimportant) to 7 (extremely important). Each individual completed the measure with respect to his or her own views on teamwork.

Although certain hypotheses were not able to be rejected, results did indicate that the FCM impacted team performance. Consequently, the significance of the research effort can be measured in various ways. First, the NeoCITIES task paved the way for future research projects (some are discussed in this book). Second, the capabilities of the FCM cognitive model were leveraged in subsequent research efforts to create the situation awareness–FCM, a special type of FCM model that builds SA. These efforts are explained below.

APPLICATION #2: THE SA–FCM MODELING APPROACH

The situation awareness–fuzzy concept map (SA–FCM) cognitive model is a unique and innovative modeling technique that creates models of SA to support user decision making. The SA–FCM integrates SA theory, which is the ability to identify, comprehend, and project critical elements in the environment, into the original FCM-modeling technique to create a computational cognitive model of SA. Formally, *situation awareness* is defined as "the perception of environmental elements with respect to time or space, the comprehension of their meaning, and the projection of their status" (Endsley, 1988). SA begins with user goals/subgoals and includes SA Requirements on three levels for each basic goal: (1) Perception: the most basic level of SA that involves simple monitoring and recognition of elements in the environment, (2) Comprehension: involves combining Level 1 SA elements to understand how these impact human goals, and (3) Projection: involves projecting the future actions of elements in the environment. As a cognitive construct, SA has been noted in decision-making research as an essential driving mechanism to decision making (Endsley, 1993).

The SA–FCM model defines the relationship that SA Requirements have on decisions. This cognitive model was developed to provide decision-aiding support that utilizes fuzzy logic and concept mapping to represent the partial or fuzzy relationships between significant entities and elements of various systems. FCMs are a suitable approach for modeling SA because they provide a mechanism to express relationships between relevant pieces of information that are needed to form higher cognitive levels required for effective decision making. Jones et al. (2010) posits that the SA–FCM is a significant cognitive-modeling approach because traditional modeling approaches represent only Level 1 SA, whereas the SA–FCM represents

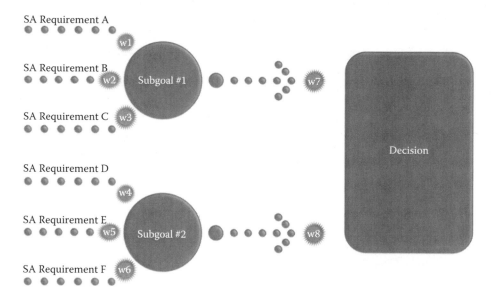

FIGURE 10.2   The SA–FCM is an actionable model of SA and includes (1) a model of SA, which is represented by a GDTA that defines information requirements needed to make decisions and (2) a FCM, which represents the relationships between the requirements. (Jones, R. E. T. et al., Incorporating the human analyst into the data fusion process by modeling situation awareness using fuzzy cognitive maps. *Proceedings of the Twelfth International Conference on Information Fusion*, Seattle, WA, 2009.)

all three levels of SA. Consequently, the SA–FCM assists the human decision maker to recognize *the what* (Level 1 SA), understand the *so what* (Level 2 SA), and—project *the now what* (Level 3 SA). Integrating SA theory and soft cognitive modeling, the SA–FCM represents a useful technique for modeling the dynamic and multifaceted nature of decisions for complex systems and thus, is particularly well suited to model dynamic work domains with a few being highlighted below (Figure 10.2).

SA–FCM for Decision-Aiding Support

A critical component to decision making understands how the availability of data at any given time impacts higher-level cognitive processing and decision making because not all data is accessible or known at the same time. To address the challenges found in dynamic work domains, such as decision making for operations that involve Army infantry platoons, SA–FCMs provide an adaptive modeling framework that affords qualitative reasoning as assessed from the current levels or states of a complex system along with quantitative elements.

A SA–FCM cognitive model was developed to provide decision-aiding support to infantry platoon leaders. Conducting knowledge elicitation tasks with SMEs (i.e., Army soldiers), the SA–FCM models the relationship between user goals, decisions, and SA Requirements. By defining the relationship between the input space (i.e., current situation) and the output space (i.e., correct action) that includes the goals of the decision maker, the decisions that must be carried out to accomplish the goals, and cognitive requirements (i.e., SA Requirements) that are needed to make correct decisions. Unlike other cognitive models and cognitive architectures, the SA–FCM cognitive model was designed to conform to the decision-making process defined by Army doctrine and thus, it provides a layer of transparency for the decision maker to understand how the cognitive model arrived at a decision. The methodology that was utilized to develop the SA–FCM model for this effort is as follows:

1. *Translate SA Requirements into FCM nodes*: The SA Requirements were identified from interviews with SMEs, and each SA Requirement was translated into a node. The model utilizes both top-down (i.e., goal driven) and bottom-up (i.e., data-driven) approaches. Specifically, the top-down approach begins at the goal node, which influences what the leader perceives from the available data in the world (i.e., the Level 1 SA node). Similarly, the leader's goal also influences the Level 2 SA node through (1) how much is comprehended (quantity) and (2) which data items are comprehended (quality), thereby impacting the nature of the comprehensions. Furthermore, the leader's goal also has the same influence on projections (i.e., the Level 3 SA node).

   The SA–FCM map consists of several submaps. From the top, the goals submap defines the relationships of the main goal, its subgoals, and how each goal influences the other goals (i.e., the activation of one goal can cause the activation of other related goals). For example, a decision may have several subgoals under their main goal. The overall SA–FCM details the causal relationships between these main goal and subgoals, with each goal (and subgoal) representing a node in the map. For this effort, a total of 15 causal relationships with preliminary weight placeholders were mapped between the nodes. For each of the subgoals, 2nd level subgoals were created and causal relationships were defined for them.

   At the bottom of the map contained the Level 1 SA Requirements that identify the relevant elements of the environment or background. Army doctrine categorizes these requirements as mission, environment, terrain and weather, troops, time available, and civil considerations (METT–TC factors). Data elements within these categories constituted the essential information requirements needed for good decision making. In addition, the SA–FCM model incorporated the METT–TC factors to define relationships that linked requirements to decisions.

2. *Determine relationships between nodes*: Although there are several methods that can be utilized to determine the polarity of the relationship between nodes (i.e., direct or inverse), SME interviews were used to determine these relationships (i.e., which concepts were influenced by other concepts and which concepts were impacted by other concepts).

3. *Identify weights*: Similar to the previous step, SME interviews were utilized to quantify the strength of relationship between nodes. Specifically, SMEs were asked to determine binary values for each relationship where

strong relationships were given a *1* value and weak (or no) relationships were given a *0* value. SMEs were given a FCM map that contained a graphical illustration of the concepts and their relationships. Working through the map in a systematic manner that involved beginning with the input nodes (i.e., Level 1 SA Requirements) and working up to higher-level SA Requirements, SMEs were asked to assign a binary value to each relationship until all relationships had a value. Although this step may appear to be a time-consuming activity, in practice the SMEs were able to do this very quickly because the map allowed them to visually see the set of lower-level concepts that influenced higher-level concepts and thereby they were able to quickly identify the concepts that had an effect. All relationships with a 0 value were removed and all empty nodes (nodes that did not have a *1* weight value to other concepts) were removed from the map.

4. *Convert weight values to fractional values*: The binary values obtained from the previous step were then converted to a fractional value. These fractional values indicate the strength of the relationship between nodes. The weights are relative values, which are determined in conjunction with our SME, who prioritized the terrain-related factors.

   An example is provided to demonstrate how the weights were determined using the methodology defined by Kosko (1987). Our procedure parallels the methodology employed in the development of a FCM that modeled the behaviors of dolphins, fish, and sharks in an undersea virtual world (Dickerson & Kosko, 1994). For terrain considerations, specifically understanding areas of concealment, an Army infantry platoon leader may want to know the following factors: humidity, type of road, and dew point. The infantry platoon leader interprets this information to understand if the road is traversable for covert and stealth operations. A lower dew point combined with a high humidity generally means that a dirt road would more than likely be wet, and therefore quieter, which is preferable for stealth operations.

   For this particular example, the critical factor to stealth movement is identifying the type of the road. Once it is established that a road is a dirt road, the platoon leader can then consider the dew point and humidity as factors, and the impact of those on stealth movement. As explained by the SME, even though the dew point and humidity are related, the platoon leader is more interested in the dew point, and only cares about the humidity in extreme situations. Thus, the condition for conducting stealth movements is primarily dependent on the road type being dirt and the dew point being low. Consequently, the weight values for those factors are set such that if the nodes for road type is dirt and dew point is low are true, the road permits for quiet movement node will be activated.

   It is important to note that this process of prioritizing factors parallels the cognitive processes that humans naturally employ. It is easier to characterize an event by prioritizing the conditions that must be present for an event to occur. Conversely, the use of traditional probabilistic modeling approaches requires quantifying events in terms of probabilities by associating an event to a set of conditions. This is a major difference in fuzzy logic compared to probabilistic reasoning. For example, using probabilistic reasoning, a SME would be required to provide the likelihood that the road permits quiet movement given the conditions that the humidity is high, the dew point is low, and the road type is dirt. In order to do this, there would have to be historical data, or theoretical information, that would help quantify the probabilities, which for this particular case is unavailable and difficult to determine. In other words, for this particular scenario, humans would not try to determine if a road permits quiet movement using a probabilistic approach that involves assigning probabilities to known factors (i.e., humidity, dew point, and road type). Instead, humans would reason that the road will permit quiet movement based on the presence of several known factors. Our SA–FCM represents this cognitive processing by prioritizing these known factors and relates them to an outcome.

5. *Extend nodes*: In a follow-up discussion with SMEs, they were then asked to review and evaluated the new map to determine if new nodes or relationships must be added to the model.
6. *Validate map*: The SA–FCM was successfully validated using a Turing test. A set of scenarios were created and the SA–FCM model was used to determine a decision for each scenario. The same scenarios were given to half of the SMEs and they were instructed to anonymously record their decision. The remaining SMEs were then given each scenario, the decision from the SA–FCM and the decisions from the SMEs. They were then asked to identify which decision belonged to the SA–FCM. As the SMEs were unable to distinguish the decisions made by the SA–FCM and the decisions made by their human counterparts, the model passed the Turing test.

The SA–FCM comprised of several submaps. For example, the top map defines the relationships of the main goal, its subgoals, and how each goal influences the other goals (i.e., the activation of one goal can cause the activation of other related goals). A total of 15 causal relationships (represented as arcs) with preliminary weight placeholders (e.g., w_{16}) were mapped between the nodes. For each of the subgoals, additional *sub-FCMs* using the subgoals as nodes and defined the causal relationships between subgoal nodes with an average of approximately 12 causal relationships/placeholder weights per submap.

A second submap maps the SA Requirements at each SA level. The model accomplishes this by maintaining the hierarchical relationship of each SA Requirement in accordance with its corresponding SA Level. The nodes *data element A, B, and C* are Level 1 SA Requirements tied to the Level 2 SA element *comprehension ABC*. The specific weights for this map were obtained from discussions with SMEs, where they were asked to rank the importance of each concept.

From this example, it is clear that to have good SA, *projection ABCDE* must be active. *Projection ABCDE* is only active if *comprehension ABC* and *comprehension DE* are both active. As this is a simple case, it should be clear to determine that *data element A, D, and E* are the most significant concepts, which means that it is impossible to have good SA without those data elements being presented to the user in a meaningful way.

SA–FCM for Predictive Assessments

In another effort described by Jones, Strater, Riley, Connors, and Endsley (2009), a SA–FCM was developed to make predictive assessments for Next Generation (NextGen) Air Transportation System tools. The model received as input workload parameters for a given aviation system tool and would produce as the output quantitative and qualitative assessments of function allocation assignments on critical operational parameters, including operator workload and SA.

The SA–FCM for this effort is a computational cognitive model of SA that represents SA-oriented design principles and experimental results from more than 40 years of research in aviation and 20 years of research in SA. The design of the SA–FCM began with *knowledge elicitation* sessions with SMEs who has extensive experience in SA theory and aviation. They were asked to identify the high-level categories that are involved with SA and the relationship between those categories and SA. They also helped to define the model's output value, termed the SA index value, which indicates how well NextGen tools support SA. The values were normalized to fit within a range of 0 (lowest score) to 100 (highest score). In addition, based on input parameters and SA index, the SA–FCM would produce recommendations, from a predefined list, on how to improve the SA index value.

Each category was discussed in great detail and deconstructed into additional core factors. This process continued until all of the core factors were identified. Each core factor was further deconstructed into states that represent possible levels of measurement. For example, the core factor *mode consistency* was deconstructed into three possible measurement levels (states): *poor, medium*, and *good*. The final step assigned a numerical value that each state has on the output SA index value. The discussions led to illustrations, such as charts, tables, and graphs that were translated into the SA–FCM.

The final design of the SA–FCM resulted in a feed forward model with 66 concepts. The FCM is a complex map organized into eight high level categories. These categories are represented as submaps that pertain to specific aspects of work relevant to SA for pilots and ATC personnel: *SA Requirements, automation aids, task workload, overall workload, alarms, general design principles, uncertainty*, and *complexity*.

The SA–FCM cognitive model was used to produce an output that provides a quantitative and qualitative assessment of a tool's impact on operator SA. The user, either a human factors engineer or a systems engineer, entered model inputs via a front-end user interface (UI), including the following parameters: (1) operator (e.g., pilot, ATC, and TRACON), (2) specific goal or work activities, (3) information requirements that are supported by the tool and how it supports those requirements, and (4) tool characteristics, such as alarms and automation implementation. The front-end UI saved the user selections to an XML file and was translated into the SA–FCM to compute the model output.

The significance of this effort is measured in several ways. First, traditional approaches to function allocation focus on the workload impact of allocating tasks to humans or automation. However, research has established that automation can have a significant impact on operator SA, which can leave the operator out-of-the-loop, subject to performance degradation and high consequence failures, even when workload itself is manageable (Jones, Strater, et al., 2009). In order to be successful, function allocation must consider both the workload and SA implications of allocation options to adequately predict performance. Consequently, the SA–FCM model is capable to direct the development of tools in achieving increase in capacity and throughput in the terminal area, while maintaining the highest possible standards of safety. Second, in addition to the aviation work domain, this model can be applied to other areas of aerospace to evaluate similar human factors related issues.

SA–FCM for Measuring Cognitive Readiness

Morrison and Fletcher (2002) define cognitive readiness as the "mental preparation (including skills, knowledge, abilities, motivations, and personal disposition) an individual needs to establish and sustain competent performance in the complex and unpredictable environment of modern military operations" (p. I-3). This definition implies that there is a threshold for cognitive readiness that is not only unknown, but also varies for different situations and events. In addition, to assess the impact of cognitive readiness on modern military operations, where the complexity of both the technologies and the information requires team coordination and decision making, cognitive readiness must be viewed from a team perspective. Adapting the definition above, team cognitive readiness can be viewed as the degree of mental preparation (including the skills, knowledge, abilities, and attitudes) that a team possesses collectively to support team collaboration and performance in complex and unpredictable environments. Thus, effective cognitive models of both individual and team cognitive readiness is needed to support prediction, monitoring, and improvement of team performance in complex military operations.

To effectively assess team cognitive readiness, team mental models (TMM) provide a framework for understanding the interplay among team members of their respective knowledge, skills, abilities (KSAs), and attitudes. TMM can be considered the team members' shared and organized assessment of team knowledge about key elements of the current situation (Cannon-Bowers, Salas, & Converse, 1993; Mohammed & Dumville, 2001). Current methods of assessing TMM, however, tend to focus primarily on knowledge structures, or another single aspect of the TMM, rather than the composite of the variables. The need for a direct measure of TMM that incorporates all these variables into a single assessment is significant. Another issue for existing measures of TMM is that most TMM analysis is performed post hoc, and has not been used for either predictive or prescriptive analysis. For this effort, a SA–FCM model was developed as part of a cognitive readiness assessment tool. This tool provides a comprehensive assessment of TMM that can be used to predict team performance in a variety of operational environments, and identify areas of team convergence and divergence that allows prescriptive action. The tool uses the SA–FCM to integrate measures of both explicit and tacit knowledge structures, team member skills and abilities, and team member attitudes, to produce a cognitive readiness assessment value.

A Conceptual Model of the SA–FCM for Data Fusion

A conceptual SA–FCM model was developed to provide a theoretical illustration on how to address challenges to effectively keep the human involved in the data-fusion process. Previous efforts within the domain did not typically include the human user (i.e., analysts and decision makers) as being an integral part of the system. For example, the original JDL model developed in the mid-1980s was the first model for data fusion, but the design was approached from an automation perspective with the research goal to identify how automation can integrate data bits for the user. However, this led to an absence of a true understanding of the requirements for the human user. Subsequent models have been developed to address the gaps identified within the JDL, including the SAW model, the Salerno Reference model, the λJDL model, Stotz Decision Fusion Frame model, Steinberg model, and Kokar's Fusion Process Reference model (Jones, Connors, & Endsley, 2009); however, these models do not provide adequate decision-making support, particularly decisions involving Level 2 and Level 3 SA Requirements.

The SA–FCM maps information requirements into SA Requirements and are represented as nodes in the map. Specifically, perceived inputs that are Level 0 and Level 1 from the JDL model are mapped to Level 1 SA Requirements. Similarly, how the user makes comprehensions, Level 2 from the JDL model is mapped to Level 2 SA Requirements, and projections, Level 3 from the JDL model are mapped to Level 3 SA Requirements. The conceptual model describes the relationship between what the user sees, comprehends, and projects to support their decisions and goals.

The model utilizes both top-down (i.e., goal-driven) and bottom-up (i.e., data-driven) approaches. Specifically, the top-down approach begins at the goal node, which influences what the operator perceives from the available data in the world. Similarly, the operator's goal also influences (1) how much is comprehended (quantity) and (2) what data items are comprehended (quality), which describes the nature of the comprehensions (i.e., the *so what* of the data). Furthermore, the operator's goal also has the same influence on the projection node. The aggregate SA from these nodes affects the action of the operator, which then influences the next goal of the operator. The operator's expertise and the amount of SA Demons, which are distractors to developing high SA (described by Endsley, 1993), are nodes that can degrade or enhance the operator's SA. For example, a novice operator may have trouble achieving the same level of high SA as an experienced operator given the same conditions. In addition, an increased amount of SA Demons will limit the SA of the operator, whereas a low amount of SA Demons will not have significant impact on the operator's SA.

The bottom-up approach begins at the data node that includes the SA Requirements needed to make a decision. The available data determines the goal, which then influences each level of SA. Similar to the top-down approach, the operator's SA is affected by the operator's expertise and the amount of SA Demons. The resulting action is impacted by the operator's SA, which then influences the goal. Moreover, each top-level node represents a submap that contains concepts and relationships that determines the output of its map. For brevity, only a brief description of the goals submap, and the SA Requirements submap are provided.

CONCLUSION

This chapter discusses the FCM cognitive-modeling approach and its relationship to the living lab framework by highlighting research efforts that utilize FCM approaches to improve collaborative decision making. Particularly, FCMs are put into practice for use, and for further testing/validation to complete the cycle of the living lab framework. The focus of the living lab would involve testing these models in the context of practice and continue to improve them, which represents the full complement of application of the living lab framework. Furthermore, observing how a FCM performs in an intended field of practice also relates back to the central problem set with the model was designed to assuage, and determines whether that came to pass in reality. Hence, this provides the theory–problems–practice coupling in the living lab framework.

The values added by these efforts demonstrate the efficacy of FCM models to decision making. FCM models provide a method for analyzing and depicting human perception of a given system, and these models are suitable for representing the human decision-making process. In addition, FCMs are dynamic tools that include cause–effects relations and feedback mechanisms. As demonstrated with the applications, FCMs can be produced by one individual map with many nodes or by merging multiple submaps into a larger map. Furthermore, the research efforts discussed the relative ease to build and understand FCM models, which is an important attribute when involving SME participation. As FCMs are a soft-modeling approach, there is a time-intensive activity that is required to capture expert knowledge. In addition, some decision-making scenarios may involve inter and intrapersonal characteristics, which is subjective and difficult to capture. For example, an individual's aversion (or lack thereof) to risk may impact certain decisions and to capture this accurately would require capturing expert knowledge from individuals who are risk-takers and those who are risk-averse. Furthermore, FCMs model only monotonic causal relations between concepts. A FCM model can represent the influence of one concept (i.e., the cause node) has on another node (i.e., the effect node); however, there are real-world scenarios where relationships are nonmonotonic. For example, how far one has to travel has a nonmonotonic relationship to the speed, where the speed for both very short and very long distances may be slower, but for short, medium, and long distances, the speed may be faster. Thus, future efforts should seek to extend the capabilities of the FCM approach to overcome these limitations. For example, mechanisms that can rapidly capture expert knowledge via crowd sourcing have been developed, which facilitates the creation of computational FCM models from a variety of users (with unique perspectives) in an efficient way (Pfaff, Drury, & Klein, 2015). In addition, advanced knowledge-mapping approaches have been examined to augment FCM's capabilities to model nonmonotonic relationships. For example, research on the knowledge map (KM) technique (Allen & Marczyk, 2012) has been used to model the financial health of companies (Shou, Fan, Liu, & Lai, 2013). These efforts have showed great promise to improving the capabilities of FCMs and new research should be done that (1) uses these approaches in other decision-making domains and (2) merges both approaches to examine the combined payoff.

REVIEW QUESTIONS

1. Describe the difference between technocentric and user-centric approaches.
2. Describe the difference between how FCMs represent uncertainty compared to traditional probabilistic approaches.
3. Define the process involved for decision making.
4. Even though the FCM are developed for scaled worlds, describe how the knowledge-elicitation component has been used for the design and development of FCMs.
5. Define the three levels of SA.

REFERENCES

Allen, G., & Marczyk, J. L. (2012). *Tutorial on complexity management for decision-making.* Great Falls, VA: Complex Systems Engineering.

Angles, J., Trochez, G., Nakata, A., Smith-Jackson, T., Hindman, D., & Harold, V. (2012). *A scaled-world prototype for structural and usability testing in residential construction.* Proceedings of the 56th Human Factors and Ergonomics Society Annual Meeting, Santa Monica, CA.

Badler, N. I., Phillips, C., & Zelter, D. (1993). *Simulating humans.* New York, NY: Oxford University Press.

Busemeyer, J. R., & Diederich, A. (2010). *Cognitive modeling.* New York, NY: Sage.

Cannon-Bowers, J. A., & Salas, E. (2001). Reflections on shared cognition. *Journal of Organizational Behavior, 22*(2), 195–202.

Cannon-Bowers, J. A., Salas, E., & Converse, S. (1993). Shared mental models in expert team decision-making. In N. J. Castellan (Ed.), *Individual and group decision making: Current issues* (pp. 221–246). Hillsdale, NJ: Lawrence Erlbaum.

Cooke, N. J., Salas, E., Cannon-Bowers, J. A., & Stout, R. (2000). Measuring team knowledge. *Human Factors, 42*, 151–173.

Dhukaram, A. V., & Baber, C. (2016). A systematic approach for developing decision aids: From cognitive work analysis to prototype design and development. *System Engineering, 19*(2), 79–100.

Dickerson, J. A., & Kosko, B. (1994). Virtual worlds as fuzzy cognitive maps. *Presence: Teleoperators & Virtual Environments, 3*(2), 173–189.

Endsley, M. R. (1988). Situation awareness global assessment techniques (SAGAT). *Proceedings of the IEEE Aerospace and Electronics Conference, 3*, 789–795.

Endsley, M. R. (1993). Toward a theory of situation awareness requirements in air-to-air combat fighters. *The International Journal of Aviation Psychology, 3*, 157–168.

Funge, J., Tu, X., & Terzopoulos, D. (1999). *Cognitive modeling: Knowledge, reasoning and planning for intelligent characters.* Proceedings of the 26th Annual Conference on Computer Graphics and Interactive Techniques, SIGGRAPH, Los Angeles, CA.

Goldstein, E. (2010). *Cognitive psychology: Connecting mind, research and everyday experience.* Scarborough: Nelson Education.

Gray, S. A., Gray, S., Cox, L. J., & Henly-Shepard, S. (2013, January). *Mental modeler: A fuzzy-logic cognitive mapping modeling tool for adaptive environmental management.* Proceedings of the 46th Hawaii International Conference on System Sciences, IEEE, Honolulu, HI.

Gray, W. D. (2002). Simulated task environments: The role of high-fidelity simulations, scaled worlds, synthetic environments, and laboratory tasks in basic and applied cognitive research. *Cognitive Science Quarterly, 2*, 205–227.

Gredler, M. (1994). *Designing and evaluating games and simulations: A process approach.* Houston, TX: Gulf Publishing Company.

Jacobs, L. F., & Schenk, F. (2003). Unpacking the cognitive map: The parallel map theory of hippocampal function. *Psychological Review, 110*(2), 285–315.

Jones, R. E. (2006). The development of an emergency crisis management simulation to assess the impact a fuzzy cognitive map decision-aid has on team cognition and team decision making. (Unpublished doctoral dissertation). The Pennsylvania State University. doi:https://etda.libraries.psu.edu/catalog/7192.

Jones, R. E., Strater, L. D., Riley, J. M., Connors, E. S., & Endsley, M. R. (2009). Assessing automation for aviation personnel using a predictive model of SA. In *AIAA SPACE 2009 Conference and Exposition.* Reston, VA: AIAA.

Jones, R. E. T., Connors, E. S., and Endsley, M. R. (2009). *Incorporating the human analyst into the data fusion process by modeling situation awareness using fuzzy cognitive maps.* Proceedings of the Twelfth International Conference on Information Fusion, Seattle, WA.

Jones, R. E. T., Connors, E. S., Mossey, M. E., Hyatt, J. R., Hansen, N. J., & Endsley, M. R. (2010). *Modeling situation awareness for Army Infantry Platoon Leaders using fuzzy cognitive mapping techniques.* Proceedings of the 19th Conference on Behavior Representation in Modeling and Simulation (BRIMS), Sundance, UT. (Awarded Best Conference Paper).

Jones, R. E. T., Jefferson, Jr., T., Connors, E. S., McNeese, M. D., & Obieta, J. F. (2005, May 16–19). *Exploring fuzzy cognitive maps for use in a crisis-management simulation.* Conference on Behavior Representation in Modeling and Simulation, Seattle, WA.

Kitchin, R. M. (1994). Cognitive maps: What are they and why study them? *Journal of Environmental Psychology, 14*(1), 1–19.

Knight, P. (2002). *Conspiracy nation: The politics of Paranoia in Postwar America.* New York: New York University Press.

Kosko, B. (1986). Fuzzy cognitive maps. *International Journal of Man-Machine Studies, 24,* 65–75.

Kosko, B. (1987). Adaptive inference in fuzzy knowledge networks. In *IEEE conference on neural networks* (Vol. 2, pp. 261–268). San Diego, CA: SOS Printing.

Manns, J., & Eichenbaum, H. (2009). A cognitive map for object memory in the hippocampus. *Learning & Memory, 16,* 616–624.

Mathieu, J., Heffner, T. S., Goodwin, G. F., Salas, E., & Cannon-Bowers, J. A. (2000). The influence of shared mental models on team process and performance. *Journal of Applied Psychology, 85*(2), 273–283.

McNeese, M., Bains, P., Brewer, I., Brown, C., Connors, E., Jefferson, T., Jones, R., & Terrell, I. (2005). The NeoCities simulation: Understanding the design and experimental methodology used to develop a team emergency management simulation. In *Proceedings of the human factors and ergonomics society 49th annual meeting.* Orlando, FL: Human Factors and Ergonomics Society.

McNeese, M. D., Connors, E. S., Jones, R. E. T., Terrell, I., Jefferson, T., Brewer, I., & Bains, P. (2005, June 22–27). *Encountering computer-supported cooperative work via the living lab: Application to emergency crisis management.* Paper read at 11th International Conference on Human-Computer Interaction, Las Vegas, NV.

McNeese, M. D., Perusich, K., & Rentsch, J. R. (1999). What is command and control coming to? Examining socio-cognitive mediators that expand the common ground of teamwork. In *Proceedings of the 43rd annual meeting of the human factors and ergonomic society* (pp. 209–212). Santa Monica, CA: Human Factors and Ergonomics Society.

McNeese, M. D., Perusich, K., & Rentsch, J. R. (2000). Advancing socio-technical systems design via the living laboratory. In *Proceedings of the Industrial Ergonomics Association / Human Factors and Ergonomics Society (IEA/HFES) 2000 congress* (pp. 2-610–2-613). Santa Monica, CA: Human Factors and Ergonomics Society.

McNeese, M. D., Perusich, K., & Rentsch, J. R. (2000). Advancing socio-technical systems design via the living laboratory. *Proceedings of the Industrial Ergonomics Association/Human Factors and Ergonomics Society(IEA/HFES) 2000 Congress* (pp. 2-610–2-613), Santa Monica, CA: Human Factors and Ergonomics Society.

McNeese, M. D., Zaff, B. S., Citera, M., Brown, C. E., & Whitaker, R. (1995). AKADAM: Eliciting user knowledge to support participatory ergonomics. *The International Journal of Industrial Ergonomics, 15*(5), 345–363.

Mkrtchyan, L., & Ruan, D. (2010). *Belief degree distributed fuzzy cognitive maps.* In Proceedings of 2010 International Conference on Intelligent Systems and Knowledge Engineering (ISKE2010), 159–165, Hangzhou, China.

Mohammed, S., & Dumville, B. C. (2001). Team mental models in a team knowledge framework: Expanding theory and measurement across disciplinary boundaries. *Journal of Organizational Behavior, 22*, 89–106.

Morrison, J. E., & Fletcher, J. D. (2002). *Cognitive readiness*, p. 3735. Alexandria, VA: Institute for Defense Analyses.

Niskanen, V. A. (2005). Application of fuzzy linguistic cognitive maps to prisoner's dilemma. In *Proceedings of the international conference on intelligent computing* (pp. 725–730). IEEE.

O'Keefe, J., & Nadel, L. (1978). *The hippocampus as a cognitive map.* Oxford: Oxford University Press.

Papaioannou, M., Neocleous, C., & Schizas, C. (2013). *A fuzzy cognitive map model for estimating the repercussions of Greek PSI on Cypriot bank branches in Greece.* IFIP International Conference on Artificial Intelligence Applications and Innovations, 597–604.

Perusich, K., & McNeese, M. D. (1998). *Understanding and modeling information dominance in battle management: Applications of fuzzy cognitive maps* (AFRL-TR98–0040). Wright-Patterson Air Force Base, OH: Air Force Research Laboratory.

Perusich, K., McNeese, M. D., & Rentsch, J. (1999). Qualitative modeling of complex systems for cognitive engineering. *In Proceedings of the SPIE* (Alex F. Sisti, Ed.), Vol. 3996, 240–249.

Pfaff, M. S., Drury, J. L., & Klein, G. L. (2015). *Crowdsourcing mental models using DESIM (Descriptive to Executable Simulation Modeling).* International Conference on Naturalistic Decision Making, McLean, VA.

Rasmussen, J., Pejtersen, A. M., & Goodstein, L. P. (1994). *Cognitive systems engineering.* New York, NY: J. Wiley.

Rentsch, J. R., & Hall, R. J. (1994). Members of great teams think alike: A model of the effectiveness and schema similarity among team members. *Advances in Interdisciplinary Studies of Work Teams, 1*, 223–261.

Rodriguez, J. T., Vitoriano, B., Montero, J., & Keeman, V. (2011). A disaster-severity assessment DSS comparative analysis. *OR Spectrum, 33*, 451–479.

Rodriguez-Repiso, L., Setchi, R., & Salmeron, J. L. (2007). Modeling IT projects success with fuzzy cognitive maps. *Expert Systems with Applications, 32*(2), 543–559.

Rouse, W. B., Cannon-Bowers, J. A., & Salas, E. (1992). The role of mental models in team performance in complex systems. *IEEE Transactions on Systems, Man, and Cybernetics, 22*, 1296–1308.

Shou, W., Fan, W., Liu, B., & Lai, Y. (2013). Knowledge map mining of financial data. *Tsinghua Science and Technology, 18*(1), 68–75.

Taber, R. (1991). Knowledge processing with fuzzy cognitive maps. *Expert Systems with Applications, 2*(1), 83–87.

Taylor, W. (2006). *Lethal American confusion: How bush and the pacifists each failed in the war on terrorism.* iUniversve, Inc. (FCM application is discussed in Chapter 14).

Todd, P., & Benbasat, I. (1994). The influence of decision aids on choice strategies: An experimental analysis of the role of cognitive effort. *Organizational Behavior and Human Decision Processes, 60*(1), 36–74.

Vasantha-Kandasamy, W. B., & Smarandache, F. (2003). *Fuzzy cognitive maps and neutrosophic cognitive maps.* Ann Arbor, MI: Xiquan.

Vitoriano, B., Montero de Juan, J., & Ruan, D. (2013). *Decision aid models for disaster management and emergencies.* Amsterdam, the Netherlands: Atlantis Press.

Woods, D. D., Johannesen, L. J., Cook, R. I., & Sarter, N. B. (1994). *Behind human error: Cognitive systems, computers, and hindsight. Wright-Patterson Air Force Base,* Crew Systems Ergonomics Information Analysis Center, OH.

11 Cognitive Architectures and Psycho-Physiological Measures

Christopher Dancy

CONTENTS

Introduction 222
Cognitive Architectures 223
    ACT-R 224
    Soar 227
    EPIC 228
    Cognitive Architectures and the Driving Story 229
Psycho-Physiological Measures and Cognitive Architectures 230
    Electroencephalography 230
    Functional Magnetic Resonance Imaging 231
    Electrodermal Activity 233
An Example of a Stressed Serial Subtraction Model 234
Discussion 235
Summary 236
Review Questions 237
References 237

ADVANCED ORGANIZER

Imagine you have just purchased a new Tesla Model S to take advantage of all of that new technology Tesla has packed into the latest iteration of their flagship car model and to achieve great fuel economy on your upcoming long trip. Excited, yet inundated with the new systems and technology in your Tesla, you embark on a cross-country trip. One system that had not previously taken much of your attention, but has now peaked your attention and interest, is the autopilot feature. The recent interest has arisen from current external and internal contexts: your *appreciation* that you are in heavy traffic and running behind schedule, and you are beginning to get drowsy. You remember reading somewhere that you *should* be attentive and to have your hands on the wheel while in

(Continued)

ADVANCED ORGANIZER (Continued)

autopilot mode, but instead you decide to allow the autopilot system to navigate you through this crawling traffic while you take a quick nap. Minutes later you wake up to a sudden jolt. You have just been involved in an auto accident that has ruined the right side of your new car and the guard rail that got in its way.

INTRODUCTION

The above story (which is fictional, but see Guarino, 2016, for a similar news report from mid-2016) illustrates just how important it is to consider the human-side of system design and implementation. A car may have an autopilot mode that is meant to be used only with human supervision, but how do you ensure that the system's human users read both the manual and follow directions? How can you understand the effects of various changes to system components as they interact with user behavior (which may be moderated by states, like being sleepy)? Could we design a safer system without having to run through a plethora of very expensive human tests?

Of course this story and these follow-up questions can be generalized, to look at the problem from a more generic system point of view. How can one understand the effects of technological system design and implementation changes on human performance in a realistic and tractable manner? In addition, how can one accomplish the understanding of these effects in a way that allows one to revisit and test the effects of system changes, as design and implementation morphs across-time before, during, and after human-in-the-loop simulations and testing (e.g., using a spiral model of design; Pew & Mavor, 2007). One way to study the effects of system design on user behavior and performance before completing costly user studies is to develop computational process models to simulate user interaction with a system. In this chapter, we provide an overview of a particular modeling and simulation approach to human behavior, cognitive architectures, some psycho-physiological measures, and discuss examples of ways the two have been used together.

Cognitive architectures are unified theories of cognition instantiated in a computational system. They define the knowledge and action space for computational-process models of human behavior. These architectures are typically implemented in software that can be run on a standard personal computer, giving one the opportunity to better understand the assumptions hidden in a designed system. Cognitive models that run within an architecture will use the provided action and knowledge space to execute tasks within a simulated system environment, thereby allowing one to study both the predicted performance of a user performing a given set of tasks within that system, and the changes in cognitive processing that lead to those output behaviors that result in the performance.

Computational cognitive process models that run within these architectures can also force the designer to consider moderators (Ritter, Reifers, Klein, & Schoelles, 2007) of behavior, whether exogenous (e.g., caffeine; Ritter, Kase, Klein, Bennett, & Schoelles, 2009) or endogenous (e.g., sleep homeostasis; Gunzelmann,

Moore, Salvucci, & Gluck, 2011*). However, one issue that often plagues cognitive architectures is a lack of theory and representation for emotional, affective, or physiological modulation of cognition; this is despite much evidence that these modulating systems are integral to human behavior (e.g., see Pfaff in this book for a discussion of affect and mood or Panksepp & Biven, 2012, for a more general discussion on affect).

In the next sections, we give an overview of some of the widely used cognitive architectures. Although these systems do not present the complete picture of cognitive architectures (indeed, this would be very difficult given the many extensions, modifications, and all together new architectures that appear, and disappear, periodically), those listed generally have been used in various contexts and should be a good starting point for any scientist, engineer, or designer who may wish to develop computational process models of human behavior. After our discussion of various architectures, we discuss how psycho-physiological measures and cognitive architectures may fit together, and how they have been used in concert with existing work.

Though the connection may not always be clear, it may be useful to lay out the ways discussed systems relate to the living laboratory framework. Beyond the fairly straightforward connection to models, cognitive models and simulation of moderators of cognition capture some impacts of external and internal changes on *use*. These models also represent the *problem* in the form on the knowledge and behavior simulated, allowing them to become a means to couple our theory with the problem and current practices. Cognitive models also improve our *design* whether it be explicitly (e.g., testing out a change to a system component through simulation with a cognitive model), or implicitly (e.g., the very development of the model, changing the way we think about the problem the system design is addressing.)

COGNITIVE ARCHITECTURES

Cognitive architectures, and the computational process models developed to run within these architectures, are a useful tool in the living lab approach. They allow one to study and predict many behavioral and performance effects of the design of systems before, during, and after the involvement of specific human-in-the-loop experimentation and testing. Not only does one get the predicted behavior, but also traces of the processes that interacted to result in such a behavior, giving any designer or engineer the opportunity to systematically change the interactions in sociotechnical systems. Developing these models often leads to a better understanding of the specific cognitive processes involved, or at least a more robust understanding of our premises, which can propagate to other portions of the living laboratory framework.

ACT-R (Anderson, 2007), Soar (Laird, 2012), and Epic (Kieras & Meyer, 1997) are discussed in the following sections. These architectures all fill particular niches in the cognitive architecture space (particularly the *functional-level* architecture space) and have slightly different perspectives that have colored their development. However, all architectures have notable features that can make them useful, depending on one's need.

---

* If the reader was intrigued by the fictional story at the beginning of this chapter, this paper may also be of interest. The researchers studied the effects of sleep loss on driving performance using computational modeling and simulation.

ACT-R

ACT-R is a hybrid modular cognitive architecture that uses symbolic and subsymbolic representations to store and process information. There are two main memory systems within the architecture: a procedural memory system (composed of the procedural module, utility module, and production-compilation module) and a declarative memory system (declarative module). In many ways, the architecture has been built around these memory systems, for example, the perceptual-motor modules are more recent additions to the system (Byrne, 2001). The architecture also contains a goal module, which modulates a representation of current goals, and an imaginal module that is responsible for imagined representations, that are also central to information processing. Figure 11.1 gives a high-level picture of the architecture. We focus on the memory systems in this overview of ACT-R; see Byrne (2001) and Anderson (2007) for a discussion of other systems in ACT-R.

The procedural memory system provides memory in the form of state–action pairs, implemented (in the software) as if–then style rules. Procedural memories, or production rules, have subsymbolic properties, with each rule having a certain *utility* that can be used with reinforcement-learning (Fu & Anderson, 2006) and also procedural compilation (Taatgen & Lee, 2003); procedural compilation is a way for a model to learn a skill more quickly by compiling sequences of rules into a single rule. All rules that completely match any given state (as determined by information in *buffers* during a stage called *conflict resolution*) can be fired, but only one rule (the rule with the highest at-time utility) will actually fire on a given cycle. If partial

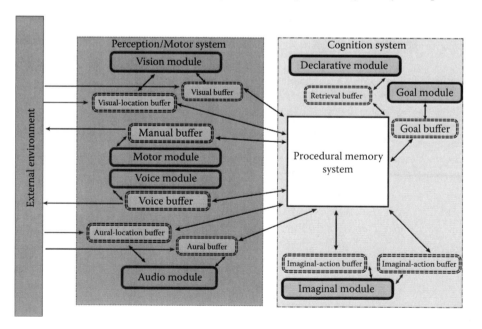

FIGURE 11.1 A high level picture of the ACT-R architecture, the architecture contains several functional modules that use buffers for communication with the external environment and with the procedural memory system.

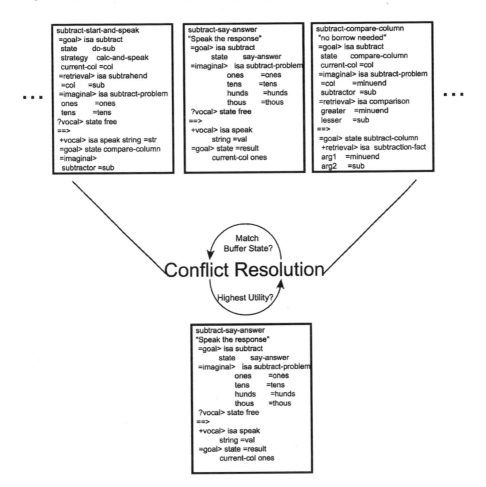

FIGURE 11.2   The ACT-R conflict resolution matches existing production rules against the current buffer states and picks the rules that have the highest utility.

matching is enabled within the architecture (using a parameter, *:ppm*) rules that only partially match the current state can also be selected during conflict resolution. Figure 11.2 gives an idea of how the current state (as specified by perceptual-motor and central-system buffers) may cause certain rules to be selected, and fired, during conflict resolution.

In Figure 11.2, the conflict resolution cycle results in a few production rules to be selected (in this case, we assume partial matching is turned off). Also during this conflict-resolution cycle, one rule is selected and fired. After symbolic pattern matching (via the current state of the buffers), subsymbolic memory comes into play through utility values attached to each rule, the rule with the highest utility is selected and fired. A form of stochasticity can be modeled and simulated using the procedural memory *noise* parameter (represented as *egs* in the canonical, LISP version of ACT-R). Explicit in the theory of the architecture is that only one rule can fire at a time, that is, procedural memory represents a serial bottleneck on central cognition. All other modules (and corresponding buffers) can operate in parallel during a given decision cycle. Thus, the architecture is mostly parallel in its operation before and after the conflict resolution.

Procedural memory selection is dependent on activity in buffers of all of the modules (central and perceptual motor). In current ACT-R models, more focus is typically placed on central modules (i.e., the goal, imaginal, and especially the declarative modules.) Declarative memory is important in this mechanistic process as these *chunks of knowledge* not only hold facts about the world, but also internal representations, such as goals.

Declarative memory is used by the architecture to retrieve knowledge about the world, and indeed the model itself. The module is parallel in nature both in terms of its use of symbolic memory and subsymbolic memory. Declarative knowledge is represented on a functional level as certain *chunks* that can have their subsymbolic value strengthened by continuous retrieval over time, and also retrieval by *associated* chunks. The subsymbolic activation value of each chunk in declarative memory is represented using Equation 11.1. This means that the probability of each declarative memory is potentially affected by several factors: time since last retrieval of the initial base-level activation of that memory (B_i), contextual activation of that memory either from both a deliberate memory probe, that is, *trying to remember something* (P_i), or from state information contained in buffers, that is, being *primed* to remember something (S_i), and declarative memory noise, which decreases the probability of retrieving the *correct* memory (ε_i).

$$A_i = B_i + S_i + P_i + \varepsilon_i \qquad (11.1)$$

As a way to illustrate the declarative-memory process in ACT-R, consider the following set of events. You are taking an exam, perhaps a stressful exam that is adding *noise* to your retrieval process, making it difficult to remember the correct facts. The instructor has set up the question so that it provides certain hints; the words used provide a context for the correct answer. In an ACT-R theory, these words are perceived and then processed using central modules. This processing, which includes the holding of such information in buffers, causes an increased activation of the correct declarative facts in memory that allow you to answer the question. Both similar contexts and recency will increase the probability of your retrieving the facts to solve your problem (assuming, of course, that you have correctly learned such facts).

The procedural and declarative memory systems are typically used in concert for more complex cognitive-process models. For example, the models described by Trafton et al. (2013) use multiple modules (including both memory systems), because the tasks employed tend to be more complex and have a longer temporal scale. Those models simulate learning and behavior across weeks, months, and years. Gunzelmann et al. (2011) also use a more complex model to simulate driving, but this model has less dependency on declarative learning than the procedural memory system. Dancy (2014) used both procedural and declarative memory systems and both symbolic and subsymbolic representations in those specific systems to simulate learning and decision making during a modified version of the Iowa gambling task. Despite the power mechanisms of learning within these two memory systems, process models and intelligent agents run within the ACT-R architecture tend to primarily focus (and thus use) either declarative memory or procedural memory.

ACT-R is a hybrid modular cognitive architecture that represents cognitive processes on a functional level. ACT-R evolved as psychological theory has also evolved, making it useful for those looking to have a cognitively realistic simulation with computational-process models. Though ACT-R has two memory systems, each with symbolic and subsymbolic learning, ACT-R models/agents typically focus on one of the systems and some of the learning mechanisms. Despite this, sometimes frustrating, limitation in generality of some models, the architecture as it stands, remains a reliable and useful choice for cognitive system simulations. ACT-R's open-source code and resources like the ACT-R website (currently http://act-r.psy.cmu. edu/) that hold a database of existing sorted related publications and models, make the architecture more approachable for non-experts.

SOAR

Soar (Laird, 2012) is a cognitive architecture that is centered on the goal of providing agent functionality with cognitive plausibility; matching human data is not a primary goal of Soar. Thus, although the Soar cognitive architecture does have an underlying unified theory of cognition (Newell, 1990), it also has significant influence from goals in the AI domain (e.g., models of general intelligence). In the past Soar has relied on production rules as its sole form of long-term memory, however, more recently it has been extended to provide explicit symbolic representations for long-term declarative memory (Laird, 2008).

Traditional long-term memory structures in Soar (i.e., pre Soar 9.0) consisted solely of procedural knowledge implemented as if–then style rules. Soar models select an operator based on a condition that (usually) results in an action that retrieves information to be placed into a short-term memory buffer. The amount of information that can go into this buffer is not inherently limited by the architecture and thus must be limited explicitly by the modeler.

Short-term memory is represented in a symbolic graph structure that allows the representation of symbol properties and relations. The short-term memory structures are used by the model for checking conditions of the current state of the model's representations within a given decision cycle. Within Soar, rules are used to propose, evaluate, and apply operators. Thus a rule is used to propose an operator, which creates a symbolic structure in the short-term memory that has an associated condition for selection.

Figure 11.3 displays the structure of typical Soar agents; these computational models are composed of problem spaces. Problem spaces are composed of operators that are condition–action pairs, these operators change the state of working memory and consequently allow the agent to move between problem spaces. Agents will typically move between subproblem spaces to accomplish a subtask to accomplish an overall goal (that will be specified with the current overarching problem space). This representation can be very useful for understanding how to solve tasks, and the problem–space operator representation allows an agent to reuse many of the

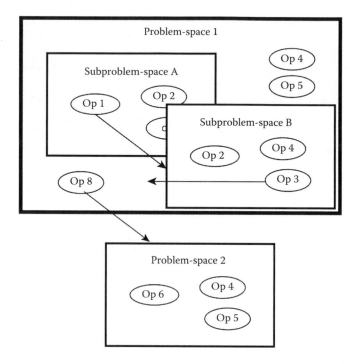

FIGURE 11.3 An high-level picture of an example Soar agent. Agents are composed of problem spaces and operators that change working memory structures and transition of the agent between problem spaces.

rules (operators) within different problem–spaces (i.e., to solve different, but perhaps related, goals).

Soar provides an architecture that straddles the line between a focus on human cognition and behavior and a focus on general agent intelligence. Though it has a foundation in human cognition, Soar is built with less of a concern for simulating and replicating human experimental data than other architectures like ACT-R. Nonetheless, Soar continues to present a cognitive architecture that is attractive for developing organized intelligent agents based on a representation of human cognition. Of particular interest may be the prospect of developing intelligent agents that process some information in an environment to enhance information processing performance by human users of a system.

EPIC

EPIC is a computational cognitive architecture with theories of perceptual, motor, and cognitive processing that constrain process models (Kieras & Meyer, 1997). The EPIC architecture is composed of several information processors that are responsible for handling different types of information during any given process resulting in behavior: a cognitive processor, an auditory processor, a visual processor, an ocular motor processor, a vocal motor processor, a manual motor processor, and a tactile processor.

The cognitive processor is composed of two main components: a production rule interpreter and a working memory system. Similar to the procedural-memory system in ACT-R architecture, long-term memories in EPIC are represented as production rules, that is, if–then style rules that prescribe an action given in a particular state. A model's state is specified by the working memory system that contains representations to give an account of all perceptual and motor processors. The central production rule system in EPIC has one important distinction from ACT-R (and indeed, Soar): the production system does not have a serial constraint, production rules can be selected and fired in parallel.

The EPIC theory constrains the bottleneck to the various working memory systems. One way to think about this in different perspective is that when completing a task on a web interface, say navigating to your desired destination page, there is no limit on how many rules may fire, as determined by the state of your processor-based working memory. Instead, there is a limit to the working memory itself, which must be used to accomplish actions throughout the environment; one only has the mental (and physical) resources to move the mouse to select one particular link on a web interface. Thus, two rules may compete for the same resources, or one of the fired rules may use the needed working-memory resources, causing a bottleneck of action.

The perceptual and motor processors are very similar to those in ACT-R (in fact, ACT-R/PM, the original extension used before the new modules were absorbed into the architecture and distributed with canonical ACT-R, was based, in part, on theory from the EPIC architecture). Perceptual processors in the architecture include an auditory processor (used to perceive sounds, both as determined due to external stimuli or internal representations placed into auditory working memory) and a visual processor that controls perceptions of visual information (as well as internal visual representations placed in visual working memory). EPIC contains three motor processors: a manual motor processor (hands), a vocal motor processor (voice), and an oculomotor processor (eyes). All processors have a general preparation phase in which resources are prepared to complete the action and an actual action phase where the desired command (as specified by some action of a production is carried out).

EPIC provides a fairly powerful perception and motor system, and takes an interesting stance on production-rule firing. The architecture itself has been used in several studies to understand and predict human visual processing (e.g., Kieras & Hornof, 2014). Nonetheless, EPIC does have drawbacks that limit its usefulness in all contexts (e.g., a little stance on how learning should occur within the architecture and no learning mechanism). This and a much smaller user community than some other architectures, has led to less architectural development. Nonetheless, the architecture can be very useful for predicting the interactions between perceptual and motor processes, and a simulated system.

Cognitive Architectures and the Driving Story

It is important to be able to put these architectures into context and to consider the places where they may be useful. Thus it may be useful to reflect on a few questions related to the story introduced at the beginning of this chapter. How might we use an architecture to model and simulate the effects of changes of the autopilot system on driver behavior and performance? How could we study the way sleep moderate this interaction? Which features of the interaction should we focus on (e.g., should we focus on the ways the autopilot alerts the user that it is enabled, or should we focus on the way we deliver the directions for use of the autopilot feature.) Perhaps the most pressing question involves the choice in our architecture for simulation of these interactions. Indeed, our answer to the previous questions would affect which architecture and simulation method we choose to employ. Consider how much money we can save by not paying multiple users to test out multiple instantiations of system designs through the design cycle!

Though modeling and simulation using cognitive architectures has clear advantages, collection of user data in the design of the system is still important wherever possible. One aspect of data collection that will increasingly complement using cognitive architectures in simulation is the collection of psycho-physiological measures. Reflecting on the story at the beginning of this chapter, an understanding of the ways that changes physiological measures related to sleepiness interact with the decision to run the autopilot in a risky manner may have helped us design a system that did not allow such an easy path to the risky behavior.

PSYCHO-PHYSIOLOGICAL MEASURES AND COGNITIVE ARCHITECTURES

As sensors and analysis have become cheaper (both monetarily and computationally), there has been an increased use of psycho-physiological measurements in behavioral experiments and human-in-the-loop simulations. These measurements provide extra context for the cognitive and behavioral process; psycho-physiological measurements give one an additional window into the processes, both unconscious and conscious, that would be otherwise difficult to measure through self-report or introspection. They also allow computational modelers to connect functional processing to physiological systems, to ground these processes in activity within systems, and communication between networks of physiological systems.

Below, I give a brief overview of three particular psycho-physiological measures that are useful for cognitive systems engineering in general, and can be used with cognitive architectures in particular.

ELECTROENCEPHALOGRAPHY

Electroencephalography (EEG) is a functional technique that involves placing sensors over the surface of a study participant's head to continuously record (i.e., with a temporal resolution in the millisecond range) electrical brain activity. This activity is associated with firing of synapses, causing an electrical signal that can be detected by the sensors. Though the temporal resolution is very useful for understanding responses to specific events (called event-related potentials, or ERPs), the temporal resolution can make it difficult to localize activity to some neural structures (and for others, it is virtually impossible.)

General EEG patterns are categorized to represent (on average) certain behavior in people; this activity is categorized as alpha (8–12 Hz), beta (18–30 Hz), gamma (30–70 Hz), theta (5–7 Hz), and delta (0.5–4 Hz) activity. Relaxed individuals typically show alpha activity or rhythmic waves of 8–12 Hz. Beta activity signifies alertness in a participant, whereas gamma rhythms mark an ability to integrate multiple stimuli into a coherent whole stimulus. It has been suggested that theta activity may signify either global inactivity (e.g., falling asleep) or an overlearned behavior process (resulting in a lower frequency EEG). Consequently, delta waves are associated with healthy sleep in humans.

Event-related potentials, or ERPs, are used to measure a particular response to stimuli. Areas that particularly respond to the stimuli (near the scalp surface) will have higher average signal amplitude after a stimulus response. ERP analysis is segmented at deflections (i.e., when the sign of a wave slope changes), and is named by whether the deflection is positive or negative, and roughly how long after the stimulus the deflection occurred; for example, a P300 ERP component is on a positive deflection and is roughly 300 ms after the stimulus.

It has been suggested that EEG can be useful for studying differences in time course of activation for specific systems in cognitive architectures; in the case of ACT-R, activity in specific modules and buffers may be correlated with specific EEG oscillation profiles (e.g., beta or theta) and also ERP data. van Vugt (2012) used an ACT-R model that completes an attentional blink task to correlate architecture behavior with activity in modules and buffers. They found location-specific correlations between theta activity, and both the imaginal and procedural modules. Correlations between delta waves and the declarative module, and between gamma activity and the visual module, were also found. Though these correlations can be useful for understanding some of the possible areas of the brain used across time and for generalizing ACT-R activity to other EEG data, a method with a higher temporal resolution is needed to more clearly localize the neural activity. The results found here and elsewhere (e.g., Cassenti, Kerick, & McDowell, 2011; van Vugt, 2013) would be even more useful with a complementarily increased understanding of potentially localized neural structures that are active that interact during predicted cognitive architecture system activity; that is, an increase in spatial resolution is also required.

FUNCTIONAL MAGNETIC RESONANCE IMAGING

Though the use of blood oxygen level dependent (BOLD) functional magnetic resonance imaging (fMRI) achieves a temporal resolution that is lower than EEG (seconds vs. milliseconds, respectively), the spatial resolution one can achieve from its use is highly valuable for observing subcortical activity (as shown in Figure 11.4). The temporal resolution limitation of fMRI is due to a reliance on the hemodynamic response function (hrf).

The hrf reflects a change in oxygenated blood in area of the brain due to activity in that area. Due to an initial dropout of signal in the first few seconds, temporal resolution is often limited to 3–4 seconds. It may be possible to achieve a lower temporal resolution depending on the type of experimental design one chooses to employ, that is, event-related or block design.

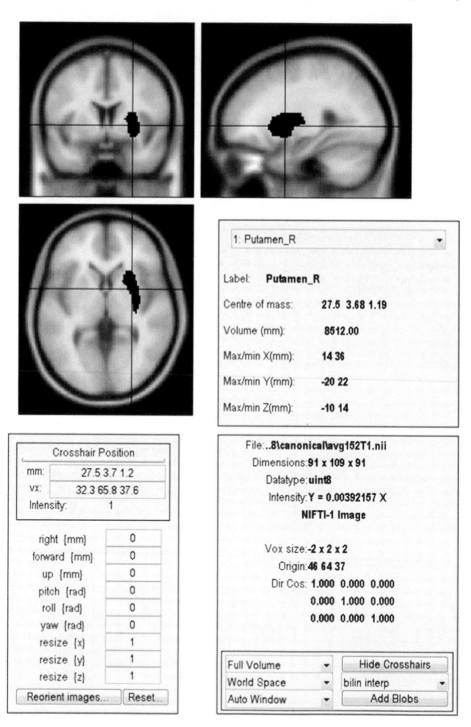

FIGURE 11.4 An example image output using the SPM fMRI analysis package (Wellcome Trust Centre for Neuroimaging). This image reflects activity in the Putamen.

An event-related design allows one to record activation in the brain in response to specific stimuli. Using a block design has an advantage of a decreased study time and an increase in signal-to-noise ratio. However, a huge disadvantage to using a block design is the use of cognitive subtraction, which results in a high loss of temporal resolution.

Despite the increased temporal resolution of event-related designs, concerns have been raised over the possibility of having too high of a granularity for some behavior. It has been posited that it may take up to 10s for activation in the brain to begin due to an emotional response (Liotti & Panksepp, 2004), possibly making an event-related paradigm inappropriate (and making a block design the appropriate choice). One may also raise an opposite problem, processing could be too rapid (e.g., LeDoux, 1996) to correctly capture in even an event-related design. Though these concerns and others (e.g., Kosslyn, 1999) about fMRI and its use in experiments are rightfully raised, fMRI still seems the most attractive option for observing activity in deeper structures during affective and cognitive processing given its relative availability, low invasiveness, and adequate spatial resolution (as compared to EEG).

The higher temporal resolution has been used in ACT-R to connect neural structures with functional systems specified in the ACT-R theory (Anderson, Fincham, Qin, & Stocco, 2008). This connection to actual neural structures makes ACT-R a very attractive architecture for any human-in-the loop simulations that include a measure from some physiological sensor. Though one may not be measuring the brain directly, understanding the *process* can lead one to a principled theory and model of behavior while using a system that provides a realistic and tractable account of both the performance and physiological data.

ELECTRODERMAL ACTIVITY

Electrodermal activity (EDA) is used to measure peripheral physiological activity in reaction to stimuli. This method has been used in many experiments and is useful due to its relatively low effect on the participant. EDA primarily records electrical activity in the eccrine sweat glands that are concentrated in the hands and feet, and are innervated by the ANS (solely the sympathetic branch). EDA measurement is relatively cheap, unobtrusive, and easy to achieve in an experimental setting.

The usual placement of instrumentation (hands or wrist) makes it a viable measurement tool to combine with other behavioral measurement tools like an eyetracker and/or instrument to conduct pupillometry. These peripheral tools give an experimenter the opportunity to run more experiments on participants due to the cost (usually) being independent of time and likely related to the price of the actual equipment. Thus a researcher may find it useful to combine (neural-activity and ANS-activity) experimental methodologies to see the different effects a given stimuli or task has on a participant.

Though EDA is a useful measure and has been used in many behavioral studies, less work has been completed on using EDA in concert with computational cognitive architectures. This is likely due to the lack of physiological representation that would allow a tractable use of those data in nearly all cognitive architectures. One of the few architectures that possess the computational physiological processes to make the integration straightforward is an extended version of the ACT-R architecture (ACT-R/Φ, Dancy, 2013; Dancy, Ritter, Berry, & Klein, 2015).

The recent increase in pervasiveness of fitness devices that combine EDA with other measures (e.g., heart-rate and skin temperature) make EDA an especially attractive low-cost solution to understanding changes in physiological processes that correlate with conscious and nonconscious behavior. As the sensor data becomes more pervasive, these data can be more realistically used both in concert with cognitive models of the user to develop and assess a system design for the user, and also for integrating within the system itself.

AN EXAMPLE OF A STRESSED SERIAL SUBTRACTION MODEL

The importance of understanding the interactions between physiological processes and psychological processes led to the development of the ACT-R/Φ architecture (Dancy, 2013; Dancy et al., 2015) that integrates the ACT-R cognitive architecture and the HumMod physiological model and simulation system (Hester et al., 2011). We have used the architecture to computationally study several *moderators* of behavior (e.g., thirst, stress, and sleep deprivation) and this has proved useful as we have been able to take lessons learned from each model developed.

In a previous study, researchers examined interactions between stress, caffeine, and cognitive performance using series of tasks that included the Trier Social Stressor Task (TSST; Kirschbaum, Pirke, & Hellhammer, 1993) and a mental arithmetic task, collecting both behavioral and psycho-physiological measures (Klein, Bennett, Whetzel, Granger, & Ritter, 2010; Ritter et al., 2009). Below, we briefly describe a model that simulates this task and provide some important lessons from that work (see Dancy et al., 2015, for a more complete picture of the task, model functionality, results, and insights.)

In using the system to examine certain effects of stress on cognition (more particularly effects on declarative memory and learning processes), we were able to study relations between physiological changes (e.g., changes to levels of epinephrine in the blood) and performance during a cognitive task. Figure 11.5 shows an expected range in declarative memory noise* *over-time* for two different versions of a *stressed* physio-cognitive model. The changes across time are based on the average state of physiological variables at each point in time of the task (e.g., at 175s into the task, the model had a physiological state that resulted in an average declarative memory noise of 0.53.) Figure 11.5 also shows points in the task in which variability in performance are likely to be high (that functionally occurs when the declarative memory noise tends to be higher.)

With this type of model and simulation, one can examine not only individual performances aggregated at the end of the task, but one can also quantify how well a person is doing well at that particular time given a physiological state. Thus, no one can identify the points across time in a task when one may be predisposed to an unacceptable performance due to a suboptimal physiological state, even in tasks

---

* As a reminder, declarative memory noise controls the stochasticity that affects a model's ability to always retrieve the *correct* memory given a context (see the section on ACT-R and particularly Equation 11.1).

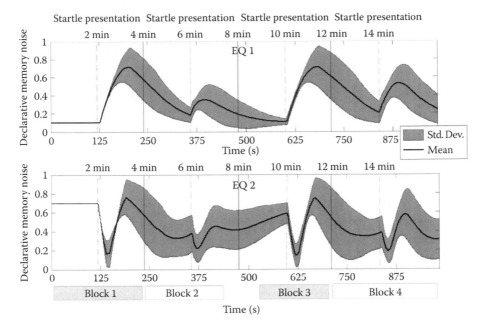

FIGURE 11.5 Resulting declarative memory noise values from running two different versions of a hybrid physio–cognitive model. (From Dancy, C. L., et al., *Computational and Mathematical Organization Theory*, 21, 90–114, 2015. With Permission.)

where the important measureable performance is available at the end of the task. In many ways, this approach differs radically from the majority of other cognitive models that assume static parameters across time.

As we continue to move toward the modeling and simulation of interactions between physiological and cognitive processes, points of integration for these dynamic processes need to be identified. Another decision that has to be made during the process is the determination of what temporal, spatial, and functional levels are needed for the particular system analysis. As an example, the ACT-R/Φ architecture could provide a fairly straightforward simulation for heart rate, but may have a more difficult time simulating EDA data as the physiological system does not provide a representation for activation of eccrine sweat glands, thereby making connection to EDA data more difficult. Nonetheless, as we continue to collect psycho-physiological data using increasingly pervasive and powerful sensors, it will be important to have a theory to guide and, at times, constrain the understanding of those data; hybrid cognitive architectures used in simulation will be an important tool to develop theory and make our premises clearer.

DISCUSSION

Though there are limitations with all cognitive architectures and psycho-physiological measures, these methods offer an experimenter or system designer a unique opportunity to go beyond strictly behavioral analysis to parse out how physiological states may react to and interact with environmental changes (e.g., different systems designs). One can understand both the conscious and unconscious processes that may be affecting behavior, independent of a participant's feedback. Before deciding which, if any, psycho-physiological measure will be sufficient for one's needs, a system designer must decide how important fidelity is for the particular human-in-the-loop simulation; some measures, like fMRI, make it very difficult to achieve a high fidelity environment, whereas others, like EDA, can be measured using relatively cheap devices.

Computational cognitive architectures offer an opportunity to get a good handle on the process that leads to and results from endogenous and exogenous behavior, relative to the person, or people, interacting with the system. Process models of people interacting with the systems can give precise, quantitative predictions about user–system performance and behavior. These models can be a key in the system design process, especially as the design matures and becomes more concrete. Nonetheless, these process models require a certain skill and knowledge set that can sometimes off-set their advantages (though some work has been done to lower this development cost, see Cohen, Ritter, & Haynes, 2010). Furthermore, current cognitive architectures can be difficult to combine with many psycho-physiological measurements. Although some headway has been made for some of the common neural measurements (e.g., Anderson et al., 2008; van Vugt, 2013), it can be argued that there is a lack of physiological representation to realistically model and simulate the effects of many physiological systems, especially as these systems interact (e.g., the psycho-physiological change that results from a combination of stress and sleep deprivation). Extensions, like that presented with ACT-R/Φ may provide a way to realistically and tractably simulate more of the physiological changes, making the psycho-physiological measures more useful.

In developing a technological system, it is important to understand as many of the assumptions inherent in the system as possible before releasing the system. Computational cognitive models that run within cognitive architectures make these assumptions more clear and can give quantitative predictions related to behavior and performance during human–system interactions. Psycho-physiological measures add dimensions to collected data, giving a system designer a deeper understanding of the effects of the human–system interaction and integration. Though cognitive architectures and psycho-physiological measures are not always suitable to be used together in system design and simulation, and they both can present large costs and learning curves. If they are used correctly, these two approaches can combine to make a powerful tool for prediction and understanding in system design. These two general methods can reveal the hidden implicit assumptions about a system or user, and provide powerful quantitative predictions for principled adjustment of technological system design.

SUMMARY

In this chapter, we presented computational modeling and simulation, and relations to psycho-physiological measures. Both of these methods can be used to gain useful knowledge and understanding in the living lab approach, and the two approaches become more powerful when combined in useful and realistic ways. Below, the reader will find a few questions that should help them review the material presented and to hopefully think beyond the discussion here toward ways that they may apply the information to their own interests and problems.

REVIEW QUESTIONS

1. What are the main memory systems in ACT-R and why might they be important to the living lab approach?
2. In what ways are EPIC and ACT-R similar? How do they differ?
3. In the story at the beginning of the chapter, where might the Soar cognitive architecture be useful? In what ways might EPIC be of use?
4. What is fMRI and why might we use it instead of EEG? In what cases would EEG be a better fit?
5. What part of the bodily function and/or system does electrodermal activity measure?
6. Why should we care about computational process models?

REFERENCES

Anderson, J. R. (2007). *How can the human mind occur in the physical universe?* New York, NY: OUP.

Anderson, J. R., Fincham, J. M., Qin, Y., & Stocco, A. (2008). A central circuit of the mind. *Trends in Cognitive Sciences, 12*(4), 136–143.

Byrne, M. D. (2001). ACT-R/PM and menu selection: Applying a cognitive architecture to HCI. *International Journal of Human-Computer Studies, 55*(1), 41–84.

Cassenti, D. N., Kerick, S. E., & McDowell, K. (2011). Observing and modeling cognitive events through event-related potentials and ACT-R. *Cognitive Systems Research, 12*(1), 56–65.

Cohen, M. A., Ritter, F. E., & Haynes, S. R. (2010). Applying software engineering to agent development. *AI Magazine, 31*, 25–44.

Dancy, C. L. (2013). ACT-RΦ: A cognitive architecture with physiology and affect. *Biologically Inspired Cognitive Architectures, 6*(1), 40–45.

Dancy, C. L. (2014). *Why the change of heart? Understanding the interactions between physiology, affect, and cognition and their effects on decision-making* (PhD). Penn State, University Park, PA.

Dancy, C. L., Ritter, F. E., Berry, K. A., & Klein, L. C. (2015). Using a cognitive architecture with a physiological substrate to represent effects of a psychological stressor on cognition. *Computational and Mathematical Organization Theory, 21*(1), 90–114.

Fu, W., & Anderson, J. R. (2006). From recurrent choice to skill learning: A reinforcement-learning model. *Journal of Experimental Psychology: General, 135*(2), 184–206.

Guarino, B. (2016). Man appears to snooze at the wheel of his Tesla while the car drives itself on L.A. highway. *The Morning Mix.* Retrieved from https://www.washingtonpost.com/news/morning-mix/wp/2016/05/26/man-appears-to-snooze-at-the-wheel-of-his-tesla-while-the-car-drives-itself/

Gunzelmann, G., Moore Jr., L. R., Salvucci, D. D., & Gluck, K. A. (2011). Sleep loss and driver performance: Quantitative predictions with zero free parameters. *Cognitive Systems Research, 12*(2), 154–163.

Hester, R. L., Brown, A. J., Husband, L., Iliescu, R., Pruett, D., Summers, R., & Coleman, T. G. (2011). HumMod: A modeling environment for the simulation of integrative human physiology. *Frontiers in Physiology, 2*(12).

Kieras, D. E., & Hornof, A. J. (2014). *Towards accurate and practical predictive models of active-vision-based visual search.* Paper presented at the Proceedings of the 32nd Annual ACM Conference on Human Factors in Computing Systems, Toronto, ON.

Kieras, D. E., & Meyer, D. E. (1997). An overview of the EPIC architecture for cognition and performance with application to human-computer interaction. *Human-Computer Interaction, 12*(4), 391–438.

Kirschbaum, C., Pirke, K.-M., & Hellhammer, D. H. (1993). The "Trier Social Stress Test": A tool for investigating psychobiological stress responses in a laboratory setting. *Neuropsychobiology, 28*(1–2), 76–81.

Klein, L. C., Bennett, J. M., Whetzel, C. A., Granger, D. A., & Ritter, F. E. (2010). Caffeine and stress alter salivary α-amylase activity in young men. *Human Psychopharmacology: Clinical and Experimental, 25*(5), 359–367.

Kosslyn, S. M. (1999). If neuroimaging is the answer, what is the question? Philosophical Transactions of the Royal Society of London. Series B: Biological Sciences, *354*(1387), 1283–1294.

Laird, J. E. (2008). *Extending the soar cognitive architecture.* Proceedings of the First Conference on Artificial General Intelligence, Memphis, TN.

Laird, J. E. (2012). *The soar cognitive architecture.* Cambridge, MA: MIT Press.

LeDoux, J. E. (1996). *The emotional brain: The mysterious underpinnings of emotional life.* New York, NY: Simon & Schuster.

Liotti, M., & Panksepp, J. (2004). Imaging human emotions and affective feelings: Implications for biological psychiatry. In M. Liotti & J. Panksepp (Eds.), *Textbook of biological psychiatry* (pp. 33–75). Hoboken, NJ: Wiley.

Newell, A. (1990). *Unified theories of cognition.* Cambridge, MA: Harvard University Press.

Panksepp, J., & Biven, L. (2012). *The archeology of mind: Neuroevoloutionary origins of human emotions.* New York, NY: W.W. Norton & Company.

Pew, R. W., & Mavor, A. S. (2007). *Human-system integration in the system development process: A new look.* Washington, DC: National Academies Press.

Ritter, F. E., Kase, S. E., Klein, L. C., Bennett, J., & Schoelles, M. (2009). *Fitting a model to behavior tells us what changes cognitively when under stress and with caffeine.* Proceedings of the Biologically Inspired Cognitive Architectures Symposium at the AAAI Fall Symposium Series. Keynote presentation, Washington, DC.

Ritter, F. E., Reifers, A. L., Klein, L. C., & Schoelles, M. J. (2007). Lessons from defining theories of stress for cognitive architectures. In W. D. Gray (Ed.), *Integrated models of cognitive systems* (pp. 254–262). New York, NY: OUP.

Taatgen, N. A., & Lee, F. J. (2003). Production compilation: A simple mechanism to model complex skill acquisition. *Human Factors: The Journal of the Human Factors and Ergonomics Society, 45*(1), 61–76.

Trafton, G. J., Hiatt, L., Harrison, A., Tamborello, F., Khemlani, S., & Schultz, A. (2013). ACT-R/E: An embodied cognitive architecture for human-robot interaction. *Journal of Human-Robot Interaction, 2*(1), 30–55.

van Vugt, M. K. (2012). *Relating ACT-R buffer activation to EEG activity during an attentional blink task.* Proceedings of the 11th International Conference on Cognitive Modeling, Berlin, Germany.

van Vugt, M. K. (2013). *Towards a dynamical view of ACT-R's electrophysiological correlates.* Proceedings of the 12th International Conference on Cognitive Modeling, Ottawa, ON.

Section V

Scaled-World Simulations

12 Evaluating Team Cognition

How the NeoCITIES Simulation Can Address Key Research Needs

Katherine Hamilton, Susan Mohammed,
Rachel Tesler, Vincent Mancuso,
and Michael D. McNeese

CONTENTS

Introduction 242
    The NeoCITIES Simulation 243
        Overview 243
        Evolution 243
        Simulation Structure 244
        Interface 245
        Gameplay 246
    NeoCITIES as a Computer Simulation for Team Research 246
    NeoCITIES as a Test Bed for Team Cognition Research 248
Team Mental Models 248
Team Situation Awareness 250
Team Information Sharing 253
Conclusion 254
Review Questions 255
Acknowledgments 255
References 255

ADVANCED ORGANIZER

This chapter uses the living lab framework to demonstrate how the NeoCITIES simulation can be used as an effective test bed for conducting team cognition research. A brief description of the simulation is provided, which includes the evolution of the software, simulation structure, interface, and gameplay. This chapter also highlights the efficacy of the simulation in studying teams through its use of flexible scenario scripting, ability to explore concepts that would be difficult or unethical to examine in the real world, and enhanced data collection abilities. It concludes with future research directions on how the simulation can be used to study multiple team cognition constructs, namely team mental models, team situation awareness, and team information sharing.

INTRODUCTION

Team cognition refers to the collective knowledge shared within a team (Cannon-Bowers, & Salas, 2001; Klimoski & Mohammed, 1994). Meta-analytic evidence has shown that team cognition has a meaningful positive impact on team effectiveness beyond team's motivational and behavioral processes (DeChurch & Mesmer-Magnus, 2010a). Research on the importance of team cognition has spanned a broad array of industries, including the military (e.g., Rafferty, Stanton, & Walker, 2010), medicine (e.g., Santos et al., 2012), aviation (e.g., Bell & Kozlowski, 2011), sports (e.g., Gershgoren, Filho, Tenenbaum, & Schinke, 2013), computer science (e.g., Schreiber & Engelmann, 2010), engineering (e.g., Patterson & Stephens, 2012), and management (e.g., Maynard & Gilson, 2014).

Despite the prevalence of research on team cognition, multiple difficulties exist in effectively evaluating the construct (Mohammed, Ferzandi, & Hamilton, 2010). Measures of team cognition tend to be both labor and time intensive in their design and administration. However, in the tradition of the living lab framework (McNeese, Perusich, & Rentsch, 2000), the investigation of such research questions are greatly facilitated through the use of simulated task environments, such as NeoCITIES (Hamilton et al., 2010). Well-designed experimental simulations have the advantage of providing rich contextual information that mimics the key decision-making processes that take place in natural settings (McGrath, 1982). They can therefore aid researchers in being able to better generalize findings to a given context more so than lab experiments (McGrath, 1982). Experimental simulations are also associated with higher levels of precision than field experiments (McGrath, 1982).

The popularity of simulation-based studies in the team cognition literature is evident by the vast array of simulations that have been used. These have ranged from commercially available software such as, Space Fortress (e.g., Edwards, Day, Arthur, & Bell, 2006), SimCity (e.g., Resick, Murase, Randall, & DeChurch, 2014), Gunship (e.g., Stout, Cannon-Bowers, Salas, & Milanovish, 1999), and Freelancer (e.g., Resick, Dickson, Mitchelson, Allison, & Clark, 2010) to simulate task environments that have been created for the purpose of research or training. According to McGrath (1982), the creation of experimental simulations can be both costly and time intensive but can be worth the investment if done well. The purpose of this chapter is therefore to introduce the NeoCITIES simulation as a test bed for conducting team cognition research. This chapter will highlight how unique aspects of the simulation allow it to address multiple research needs in the team cognition literature. Through addressing this purpose, the current chapter will rely on the tenets of the living lab framework to explore and understand key theoretical and practical concerns in the team cognition literature. The following sections will provide a brief description of the NeoCITIES simulation, review the efficacy of the simulation in studying teams, and concludes with ways the simulation can be used to study multiple types of team cognition constructs.

THE NeoCITIES SIMULATION

Overview

NeoCITIES is a simulated task environment that emulates the decision-making processes of an emergency response team. The simulation is unique in that it requires both individual work (vis-à-vis specific team member roles) and teamwork to enable successful performance response. The simulation is based on Wellen's (1993) CITIES game (command, control, and communications interactive task for identifying emerging situations) that served as a platform for studying team decision making in crisis management teams. Typically NeoCITIES is designed to facilitate interdependent team cognition that is distributed and is synchronous in play. Updates to the CITIES game conformed to the living lab framework. These updates were heavily informed by data collection from field observations and interviews with government intelligence analysts, police, and fire authorities (Jones, McNeese, Connors, Jefferson, & Hall, 2004). This qualitative data served to help replicate the cognitive processes in which members of emergency response teams are engaged. There have been several instantiations of the simulation since its inception (e.g., Balakrishnan, Pfaff, McNeese, & Adibhatla, 2009; Hamilton et al., 2010; Hellar, 2009; Mancuso, 2012; Pfaff, 2008). The current section focuses on the basic simulation structure, interface, and typical gameplay found in NeoCITIES. These features are common across multiple versions of the software but were most evident in NeoCITIES 3.1.

Evolution

There have been several instantiations of the simulation since its inception (e.g., Balakrishnan et al., 2009; Hamilton et al., 2010; Hellar, 2009; Mancuso, 2012; Pfaff, 2008). The original NeoCITIES simulation (McNeese et al., 2005) was developed based on data collected from ethnographic research and knowledge elicitations from emergency dispatch centers (Terrell, McNeese, & Jefferson, 2004). This iteration was developed as a platform for studying intelligent group interfaces, information overload, and team communications. Based on the findings of the initial round of research, experiences with the platform, and new research questions, NeoCITIES 2.0 was developed with a focus on geocollaborative tools to study perceptual anchoring and spatial communications (Balakrishnan et al., 2009). The most current major iteration, NeoCITIES 3.0 was further refined, and was built to be a more flexible research tool than previous versions, designed specifically to support multiple research questions (Hellar & McNeese, 2010).

In addition to emergency dispatch, NeoCITIES was also used as a basis for studying decision making, and team cognition within cyber space. Using the NeoCITIES 3.0 infrastructure, the NeoCITIES Experimental Task Simulator (NETS) was developed to support research in the cyber domain (Mancuso, Minotra, Giacobe, McNeese, & Tyworth, 2012). Similar to the NeoCITIES simulations, NETS was developed based on ethnographic research with cybersecurity professionals and cyber operations (Tyworth, Giacobe, Mancuso, McNeese, & Hall, 2013). NETS was built on the same principles of situation assessment and resource allocation as previous NeoCITIES simulations, but was expanded to account for the complexities and unique nature of the cyber environment. In this iteration, the NeoCITIES platform, was used to study cyber situation awareness (Giacobe, 2013), team knowledge profiles (Mancuso & McNeese, 2012), and workload balancing (Minotra, 2012).

Simulation Structure

NeoCITIES is designed using a hierarchical structure. The two main modifiable components of this structure are the scenarios and roles. Scenarios are made up of events, which are composed of several different attributes. Modifiable event attributes include its name, brief description of what happened, time when it occurred, severity of the event, number and type of resources needed to resolve it, sequence in which resources should arrive, and deadline by which resources should arrive. These attributes enable researchers to vary the length, difficulty, and level of interdependence required in each scenario. The flexible scripting of events also allows researchers to create scenarios that are tailored to the organizational affiliations of the participants thereby increasing their level of engagement during gameplay. NeoCITIES has a host of pretested scenarios with varying levels of interdependence. Even though pretested scenarios exist, NeoCITIES offers the opportunity to adapt events in those scenarios unique ways, or to even design new scenarios that satisfy the requirements needed by a researcher. This provides an element of flexibility in the overall scaled-world model on which NeoCITIES is based. New scenario design often requires iterations of pilot testing to ensure that no one participant is overloaded, underwhelmed, or receives events in a predictable fashion.

The traditional roles in NeoCITIES include police, fire, and hazardous materials. Each role is associated with its own unique resources. For example, individuals in the role of police have investigators, squad cars, and bomb squads at their disposal. Roles are typically balanced in the number and types of resources they have at their disposal. This is done to ensure that all roles are weighted equally within the simulation. However, modifiable role attributes include the number and types of roles and the number, types, and functions of resources available to each role.

A core element of the simulation structure is the scoring model. Gameplay in NeoCITIES is considered to be an intellective task (McGrath, 1984) in which there is only one correct answer on how to respond to an event. Success in the simulation is dependent on how quickly participants are able to identify which resources and of what type are required to resolve each event. All participants begin the simulation with a perfect score, which decreases based on the length of time taken to resolve each event. Once dispatched, the severity of events grows over time until it reaches its maximum penalty or the correct resources are applied. The scoring model for the simulation is also based on the original CITIES task (Wellens, 1993). Subcomponents of the overall scoring model have also been used to analyze different elements of performance (e.g., identify certain aspects of team performance or errors that indicate certain states within the overall simulation).

Interface

The traditional NeoCITIES interface is made up of five key components that each member uses for their individual work associated with the role they are playing. At the same time, the interface also represents a group interface (Connors, 2006), in that, it provides status and updates of information at the team level of work. These interface components include the chat panel (see Figure 12.1, section A), dispatch panel (see Figure 12.1, section B), team monitor (see Figure 12.1, section C), unit monitor (see Figure 12.1, section D), and event tracker (see Figure 12.1, section E). Each component of the interface has a unique function that ties into the gameplay of the simulation.

The purpose of the chat panel is to enable team members to communicate with each other. This is the main tool that teams rely on to coordinate their responses to interdependent events. The chat panel allows for distributed, synchronous text-based communication among team members.

The dispatch panel enables participants to see the type and number of resources that they have at their disposal. It is also used to allocate resources to events and recall resources from events. Resources are updated in this section once an event is complete or the resource is recalled.

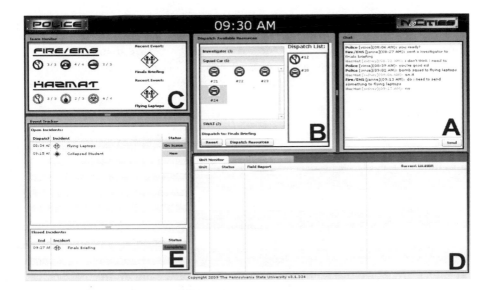

FIGURE 12.1   The NeoCITIES 3.1 interface.

The team monitor is used to keep track of the most recent event to which team members have responded. This monitor is particularly useful when addressing events that require team resources in a particular order (e.g., fire, hazardous material, and then police) or by a specific simulation time (e.g., 9:40 am deadline). The team monitor also enables participants to keep track of the number and type of resources that teammates have at their disposal.

The unit monitor displays information on the status and location of each resource dispatched. The feedback provided in this section enables participants to keep track of the accuracy of their decisions. Reports include whether or not the resource sent was able to help resolve the event, or if more resources are needed. Participants are able to use this feature to recall unneeded resources from an event.

The event tracker is one of the most dynamic elements of the interface. It provides a list of both active and resolved events. Event attributes that are displayed to participants include the name, icon, description, status, and time stamp of events. Event status is displayed as new, en route, on scene, or complete. Both text and colored cues are used to highlight the progression of events. Events that are closed appear in the lower section of the tracker. Failed events are highlighted in black, whereas completed events are highlighted in green. During simulation gameplay, events pop-up sequentially and are balanced according to the number of players so that no one individual's resources are over or underutilized during a given run. Events continue to appear across most of the simulation at specified time intervals.

Gameplay

Typical gameplay begins with individuals being randomly assigned to either the role of the police, fire, or hazardous material personnel in the emergency management team. These individuals receive video-based training on the functions of the various components of the simulation's interface and detailed descriptions of his/her unique resources and its functions. Participants then have the opportunity to engage in practice sessions to familiarize themselves with the simulation. These sessions often include fewer events that are delivered at a slower pace than full scenarios. No prior expertise in emergency response is needed in order to perform effectively in the NeoCITIES simulation. Proficiency in using the simulation is typically demonstrated at the end of the training and practice sessions. Once gameplay begins, participants need to review events, determine whether or not they are relevant to their or their teammates' resources, allocate resources to relevant events, determine if resources are effectively resolving events and if not, recall ineffective resources, and coordinate with team members to respond to interdependent events.

NeoCITIES as a Computer Simulation for Team Research

NeoCITIES can be classified as a computer simulation for team research (CSTR). CSTRs are "team task environments simulated with the use of computer software and hardware that serve as controlled environments for the study of teams" (Marks, 2000, p. 656). A key difference between CSTRs and simulations that develop cognitive architectures for computational modeling, such as ACT-R (Anderson et al., 2004; Smart, Sycara, & Tang, 2014), is that the data collection process includes responses from actual human participants as opposed to being simulated based on prespecified parameters. CSTRs also differ from simulations of reality (e.g., Emergo Train System; Nilsson et al., 2013), which focus more on capturing the fidelity of the actual team setting as opposed to the deeper set of interactions and decision processes that take place in these contexts (Marks, 2000). CSTRs are therefore meant to be a reference system of specific work contexts that enables researchers to explore a broader set of relationships in the nomological net without having to rely on populations with a specific skill set (Marks, 2000). This places a greater emphasis on theory evaluation (that is one of the main sources of knowledge elaborated on as part of the living laboratory framework) as opposed to training transfer.

As a CSTR, the validity of NeoCITIES for studying team research is not dependent on its physical fidelity but on its psychological reality, structural validity, process validity, and predictive validity (Marks, 2000; Raser, 1969). These components of the simulation were enhanced through an ethnographic study of emergency dispatch workers, which involved conducting field observations and interviews (Jones et al., 2004). During this process, the focus was placed on understanding the decision-making processes, cognitive demands, and core responsibilities of each role in the emergency response team (Jones et al., 2004). This helped to enhance the validity of the simulation in several aspects. The psychological reality of NeoCITIES is evidenced by the way the simulation represents the core demands of the team environment and task. Without much prompting, participants in the simulation are able to see the interdependencies needed in order to be successful in the simulation. They are able to view the simulation as a team performance environment and are therefore more likely to display the processes that are considered a part of the team setting. Ensuring that the roles and functions map back to the actual roles and functions of an emergency management team enhances the structural validity of the situation. The process validity of the simulation was enhanced by making sure the decision-making processes, the levels and types of information weighed during the decision-making process was consistent with that of emergency response teams. Finally, predictive validity was enhanced by ensuring that the system predicted outcomes consistent with the referent system (e.g., speed, accuracy, and performance).

Features of CSTRs that enhance team research include flexibility in scenario scripting, opportunities to study concepts that may be difficult to examine in the real world, and enhanced data collection abilities (Marks, 2000). NeoCITIES enables the scripting of scenarios that would be difficult or unethical to manipulate in the real world. These include being able to manipulate workload, feedback ambiguity, event ambiguity, environment complexity, level of task, goal, or reward interdependence, and team size. This information can be valuable in understanding how context affects teams and how teams adapt. Enhanced data collection in NeoCITIES is composed of rich action histories, which includes information such as which resources were dispatched, how quickly they were dispatched, whether or not they were correct, whether or not a resource was recalled, length of time it took to recall ineffective resources, and whether or not it was recalled in error. The communication data are available for a variety of communication analyses (e.g., Pfaff, 2008). The communication history is part of a rich sequential data stream that is archived and available for its own analysis to complement the embedded performance measures. For interdependent events, it is also possible to determine the order in which resources were sent. This rich source of data enables researchers to explore many questions, most notably how team emergent states and processes vary longitudinally (within and across performance episodes) and across levels of analysis (within and across individual and team responses). Such questions represent some of the most pressing research concerns within the team effectiveness literature (Mathieu, Maynard, Rapp, & Gilson, 2008).

MIDWAY BREATHER

- The use of simulated task environments is an integral part of the living lab framework. NeoCITIES is presented as an example of how simulation studies can be used to identify and address key research needs in the team cognition literature.
- Key components of the simulation, such as the ability to script dynamic scenarios, study concepts that are difficult/controversial to examine in the real world, and enhanced data collection abilities, enable NeoCITIES to be a valuable test bed for conducting team research.

NEoCITIES AS A TEST BED FOR TEAM COGNITION RESEARCH

Multiple forms of team cognition have been studied in the literature on team effectiveness (DeChurch & Mesmer-Magnus, 2010a). Three of the most researched forms of team cognition are team mental models (TMM), team situation awareness, and team information sharing (Mesmer-Magnus, DeChurch, Jimenez-Rodriguez, Wildman, & Shuffler, 2011; Mohammed et al., 2010; Wildman, Salas, & Scott, 2013). The following sections will provide a brief overview of each form of cognition, highlight key areas for future research, and describe how the NeoCITIES simulation can be used to address them.

TEAM MENTAL MODELS

TMM refer to "team members' shared, organized understanding and mental representation of knowledge about key elements of the team's relevant environment" (Mohammed & Dumville, 2001, p. 90). This form of team cognition varies based on both its content and properties (Mohammed et al., 2010). The content relates to the type of information that is organized. The two key types of content of TMMs evaluated in the literature are taskwork (i.e., shared knowledge on *what* the team should accomplish) and teamwork (i.e., shared knowledge on *how* the team should go about accomplishing their goal). More recent but less researched TMM types include temporal TMMs (i.e., shared knowledge on *when* things should be accomplished; Mohammed, Tesler, & Hamilton, 2012) and situational TMMs (i.e., shared knowledge of the patterns of events taking place in the team's external environment; Hamilton, 2009; Waller, Gupta, & Giambatista, 2004).

TMMs have been evaluated in a variety of ways. These include card-sorting (e.g., Smith-Jentsch, Campbell, Milanovich, & Reynolds, 2001), qualitative measures (e.g., McComb, Kennedy, Perryman, Warner, & Letsky, 2010), Likert-based surveys (e.g., Ellwart, Konradt, & Rack, 2014), concept mapping (e.g., Burtscher, Kolbe, Wacker, & Manser, 2011), and pair-wise comparison ratings (e.g., Randall, Resick, & DeChurch, 2011). Given that the conceptualization of TMMs includes both the content and structure of knowledge, measures that capture the true nature of the construct are those that evaluate both the content and structure of the construct (DeChurch & Mesmer-Magnus, 2010b; Mohammed et al., 2010). The most popular measure of TMMs is pair-wise comparison ratings, which capture both the content and structure of knowledge (DeChurch & Mesmer-Magnus, 2010b). The development of pair-wise comparison ratings involves first conducting a thorough team task analysis to identify such information as the key task requirements, role functions, and goal interdependencies among team members (Mohammed & Hamilton, 2012). The resultant dimensions and their descriptions are then organized into a grid. Individual team members then evaluate the extent to which each of the dimensions is related. These data are aggregated to the team level using structural measures that represent the strength of sharedness among team members' ratings, such as QAP correlations or Euclidean Distances (Mohammed & Hamilton, 2012).

With regards to the evaluation of TMMs, multiple opportunities for future research exist. First, more research is needed that evaluates multiple types of TMMs. The most traditional forms of TMMs examined are taskwork and teamwork. Research has shown that taskwork TMMs have a positive effect on team performance (Cooke, Kiekel, & Helm, 2001), team viability (Resick et al., 2010), and team adaptability (Uitdewilligen, Waller, & Zijlstra, 2010). Teamwork TMMs have been shown to positively predict team performance (Rentsch & Klimoski, 2001), team viability (Rentsch & Klimoski, 2001), and team innovativeness (Reuveni & Vashdi, 2015). Studies that have evaluated both types have found that taskwork TMMs have a stronger effect on performance than teamwork TMMs (e.g., Cooke, Kiekel, & Helm, 2001). However, little is known on how these types of TMMs interact to predict the effectiveness of teams. The limited research in this area has limited our ability to determine the contextual boundaries surrounding the question, what knowledge needs to be shared in order for teams to be effective? Future research is needed to more deeply explore this research question.

Second, more research is needed that compares the effectiveness of different measures of TMMs. Given the complexity in the administration of TMM measures, both in their design and administration, oftentimes only one type of measure of TMM is used in any single study (Mohammed et al., 2010). However, a meta-analysis on the construct showed that different types of measures produce different results (DeChurch & Mesmer-Magnus, 2010b). Measures that capture only the content (e.g., surveys) have a weaker effect on team performance than measures that capture both the content and structure of TMMs (e.g., pair-wise comparison ratings). Future research is needed that evaluates how different forms of measurement of TMMs impact various team outcomes (i.e., emergent states, team processes, and team performance) while keeping the knowledge content of the TMM and sample characteristics constant.

Third, more research is needed that broadens the types of knowledge examined to those outside of taskwork and teamwork TMMs (Mohammed et al., 2010). TMM researchers have begun to develop a better understanding of how more specific types of TMMs, particularly those related to contextual elements, such as time and the external environment impact team performance. Research has shown that temporal TMMs, that is, shared knowledge on when to do something, explain a unique variance in team performance beyond taskwork and teamwork TMMs (Mohammed, Hamilton, Tesler, Mancuso, & McNeese, 2015). Research has also shown that situational TMMs, that is, shared knowledge on elements in the external environment, has a unique and positive effect on team performance (Hamilton, 2009). Future research is needed that continues to broaden the conceptual domain of TMMs explored in order to better address what needs to be shared in order for teams to be effective.

The NeoCITIES simulation helps to address many of these research needs. First, the simulation has clear role specifications and descriptions of task interdependencies that facilitate the evaluation of both taskwork and teamwork TMMs. The short but intensive duration of scenarios create cognitively rich and engaging experiences that are repeatable within a short duration. This enables the measurement of multiple types of measures of TMM over a shorter time span than would be possible in a field setting. The flexibility in the scripting of scenarios helps to facilitate the development of storylines within the simulation. Scenarios can be temporally scripted to include information related to deadlines, pacing, and order that are easily tied to key performance outcomes. Scenarios can also be scripted to include themes in events that reward team members who identify patterns with higher levels of team performance. This flexibility in the scripting of scenarios helps researchers to evaluate both temporal and situational TMMs, respectively.

TEAM SITUATION AWARENESS

Situation awareness (SA) can be defined as "the perception of the elements in the environment within a volume of time and space, the comprehension of their meaning, and the projection of their status in the near future" (Endsley, 1995a, p. 36). SA is therefore composed of three levels. Level 1 relates to the perception of information in the external environment. An example of this would be noticing that multiple events relate to robberies. Level 2 SA involves a deeper level of cognitive engagement that focuses on moving past detection to understanding what the information means. For example, once individuals are aware that many events are robberies and the next step would be deciphering that the robberies seem to target recreational facilities. Level 3 SA goes beyond perception and comprehension to projection. Individuals at this stage engage in determining what would happen in the future based on previous patterns that they detected. An example of Level 3 SA is predicting where the next robbery will take place based on the patterns in previous events. Even though much of the empirical work on SA has been conducted at the individual level of analysis, SA has also been conceptualized to exist at the team level of analysis. The original CITIES and in turn NeoCITIES was predicated on the study of group situation awareness. SA has been theorized to impact team decision making (Wellens, 1993). In these studies, teams were considered to have high SA if they were able to detect patterns in events and predict what events would occur next (Wellens, 1993). The use of simulation studies provides a basis for discovery of higher order elements of situations and can be defined using flexible scripting of scenarios. The coupling of certain events together can represent awareness of the bigger picture (common operation picture) that informs strategies on resource allocation across the team. The area of hidden knowledge as relatable to the script itself represents another form of SA that can spread from one individual to the team and inform performance improvements.

Despite these early developments, team SA is less understood in both its definition and measurement. Definitions vary based on whether SA is viewed as overlapping among team members (e.g., Wellens, 1993), distributed among team members (e.g., Artman & Garbis, 1998), or shared between team members and technological artifacts in the team's environment (Stanton, Salmon, Walker, & Jenkins, 2010). In this chapter, we define team SA as the extent to which members of a team develop a common perspective on all three levels of SA (Wellens, 1993). Measures of team SA also vary based on the evaluation of either the product (i.e., what the team knows) or process (i.e., how the team attains the knowledge) of team SA. Between these two forms of measurement, product measures are the most popular, within which the Situation Awareness Global Assessment Technique (SAGAT; Endsley, 1995b) is the most utilized (Salmon, Stanton, Walker, & Green, 2006).

A key limitation to the study of team SA relates to the difficulties involved in its measurement, particularly as it relates to the administration of the product measure SAGAT. SAGAT involves freezing the team's task at random intervals and asking each team member questions that relate to their perception, comprehension, and projection of elements in the team's environment (Endsley, 1995b). These are knowledge-based questions and therefore only have one correct answer. The use of SAGAT is associated with strict administration guidelines. The questions should only be asked during a freeze-in performance, during the freeze no visual cues should be available that would enable a participant to correctly guess an answer, there should be no communication among participants during this time, and the freeze should take place at a random time to prevent participants from predicting and thus preparing for when it will occur (Endsley, 1995b). The types of questions asked should be based on information in the team's environment that will be relevant to their performance. This information needs to be determined ahead of time using a hierarchical goal-directed task analysis (Endsley, 1995b). Given the complexity in the design and administration of the SAGAT, it is best utilized in simulated task environments (Salmon, Stanton, Walker, & Green, 2006).

Opportunities for future research in the study of team SA include the need to better understand the relationship among subjective and objective measures of the construct. Product measures of team SA that are often used in field settings include the Crew Awareness Rating Scale (CARS; McGuinness & Foy, 2000) and the Mission Awareness Rating Scale (MARS; Matthews & Beal, 2002). Both of these rating scales are more subjective than SAGAT. They involve having individual team members evaluate their own ability to perceive, comprehend, and project the status of elements in their external environment. These scores are then aggregated to the team level using the mean. Given the fact that SAGAT is often administered when teams are working in simulated environments and subjective rating scales are more often used in field settings, it is not often that data are available on the comparative effects of these types of measures on team effectiveness. Studies that have been conducted at the individual level of analysis have found little to no correlation between SAGAT and subjective ratings of SA (e.g., Jones & Endsley, 2004; Lichacz, 2008; Salmon et al., 2009). These two sets of measures have been found to explain unique variability in team performance (Rousseau, Tremblay, Banbury, Breton, & Guitouni, 2010). These findings underscore the possibility that objective measures of SA, such as SAGAT, capture situational knowledge, whereas subjective measures of SA capture situational metacognition (Rousseau et al., 2010). Additional research at the team level of analysis is needed to better understand the differences among these measures.

An additional area for future research on team SA is to better understand its temporal nature. Time is implicit in the conceptualization of SA in the fact that at the highest level of the construct, individuals should be able to predict what will happen in the future (Endsley, 1995a). However, the temporal nature of the construct (i.e., Level 3 SA) is not often featured. Exceptions to this include Hauland (2008) who found that participants working in an air traffic control simulation were more likely to develop Level 3 SA when they focused on future-oriented information in the simulation than information focused on the present environment. While studying participants engaged in an emergency management simulation, Wellens (1993) also found that participants, who were able to detect patterns in the events, thus demonstrating Level 3 SA, were able to predict and respond more quickly to upcoming events. Research is needed that more deeply explores the temporal nature of team SA (Mohammed et al., 2012). Little is known about the awareness of teams with regards to temporal information, such as deadlines, sequencing, or pacing. Members of diverse teams may vary in their interpretations of such temporal elements in the external environment, which would impact the SA and performance of the team.

The NeoCITIES simulation helps to address several of these limitations and research needs. First, both the design and administration of the SAGAT is facilitated by the use of simulated task environments such as NeoCITIES. The ability to script scenarios with patterned events coupled with the rich contextual information provided within the interface enable the development of a broad array of situational-based questions. The simulation also has the capability to freeze the scenarios at random intervals and present a list of SAGAT questions based on participant's recent performance. During this time, participants are not able to view the interface or communicate with team members. The scoring of the SAGAT is facilitated through the presence of rich data collection, such as a detailed action history of participant responses to events. Second, the short but intense nature of scenarios enables the administration of multiple measures of team SA at multiple time points. Third, events are easily infused with temporal information related to responding to events by a specific time (i.e., deadline), arriving in a particular order (i.e., sequencing), or completing an event within a specific period of time (i.e., pacing), to facilitate the evaluation of temporally-based team SA. Fourth, team cognition can be assessed through multiple exposures over a given timeframe to determine the extent of learning that evolves with team members. For example, university participants with a given course may be exposed to the simulation, 4 to 8 times over the course of an academic semester (e.g., see Mohammed, Hamilton, Mancuso, Tesler, & McNeese, this volume, Chapter 9). This enables understanding many of the team cognition constructs at an even deeper level.

TEAM INFORMATION SHARING

Team information sharing refers to how teams collectively manage information (Wittenbaum, Hollingshead, & Botero, 2004). The focus of the construct is on the distribution of task-related information or expertise within the team (Mohammed & Dumville, 2001). A key finding in the information sharing literature is that individuals working in teams have a tendency to discuss information that they have in common and to avoid sharing information that is unique to them (Stasser & Stewart, 1992). However, meta-analytic evidence has shown that higher decision making quality in teams occurs when team members openly discuss information that is unique to them (Mesmer-Magnus & DeChurch, 2009). The extent to which teams engage in sharing unique or common information has been predominantly evaluated within the context of a hidden-profile task (Sohrab, Waller, & Kaplan, 2015). Hidden-profile tasks have taken different forms, the most common of which has been a murder mystery task in which team members are given information packets on a list of suspects that vary in its degree of uniqueness and commonalities with other team members and asked to decide who committed a murder (Stasser & Stewart, 1992). These tasks are designed so that the optimal choice can only be reached if team members share the information that is unique to them.

There are multiple opportunities for future research on team information sharing. First, research is needed that moves beyond the traditional hidden-profile task to evaluate team information sharing across broader tasks and contexts (Mohammed & Dumville, 2001; Sohrab et al., 2015; Wittenbaum et al., 2004). The hidden-profile task has been criticized for not accurately capturing naturalistic decision-making processes (Sohrab et al., 2015). Participants are often provided with packets containing the necessary information they need to make a decision and given a clear definition of the problem that needs to be solved (Stasser & Stewart, 1992). This decision-making process is not representative of situations in which individuals need to first identify that there is a problem, define the problem, then gather/seek the necessary information needed before making a decision. The added complexity of problem detection, problem definition, and information gathering can be difficult to replicate in a lab setting. These processes are more likely to occur in field settings but such environments may pose another set of challenges in the precision of measurement of information sharing. A potential solution to this problem is to study these decision-making processes within the context of a rich simulated task environment (Sohrab et al., 2015), which helps to replicate some of the unstructured and ambiguous elements of the decision-making process. In such environments, team members have to first define what the problem is, identify resources to help resolve the problem, and sort through information in order to find the optimal solution.

Second, research is needed that evaluates information sharing within distributed contexts. Many explanations have been provided on why team members tend to focus on information that they have in common and have less of a discussion of unique information. These explanations have focused on the existence of mutual enhancement (Henningsen & Henningsen, 2004), social validation (Boos, Schauenburg, Strack, & Belz, 2013), and the likelihood of the information being sampled from memory (Stasser, 1992). Many of these explanations relate to the influence of group norms and information permanence. However, these are factors that tend to have less of an impact on distributed team performance. In fact, meta-analytic evidence has shown that teams high in virtuality are more likely to share unique information than face-to-face teams (Mesmer et al., 2011). Future research is needed to help strengthen this claim.

Third, research is needed that more deeply explores the role of time in information sharing. Previous research has shown that both time pressure and length of discussion impacts the amount of unique information shared (Wittenbaum et al., 2004). Team members that had low time pressure were more likely to solve a hidden-profile task and share more unique information than teams that worked under high time pressure (Bowman & Wittenbaum, 2012). Team members are also more likely to mention information that they have in common early in the team's discussion and to communicate unique information later in the discussion (Larson, 1997). Information sharing therefore seems to be strongly impacted by time. Future research is needed that embeds time within the construct (Mohammed et al., 2012). This would involve determining how teams share information that is temporally related and to examine how this information sharing varies across the lifecycle of the team.

The NeoCITIES simulation helps to address many of these key future research needs. First, NeoCITIES serves as a rich test bed in which to study decision making that moves beyond the highly structured hidden-profile task. Within the simulation, the degree of information sharing among participants can be manipulated through the delivery of information briefings. These briefings can coincide with particular events that are dispatched at different types in the scenarios. The nature of the briefings can be manipulated so that information varies on the level of shared and unique elements distributed among team members. The recording and evaluation of team information sharing is enhanced in NeoCITIES through the presence of the chat feature. The frequency and type of information shared can be qualitatively coded using the chat logs that are recorded during each performance session. The type of information team members share could easily be coded from each session, which may include such details as the resources required for an event, the temporal ordering of resources needed, the deadline in which to respond to an event, or the pacing needed for the team response. Second, distributed collaboration within NeoCITIES primarily takes place virtually through computer-mediated chat logs. Studying team information sharing within this context, therefore, helps to address future research needs on evaluating the team-information processes within distributed environments. Third, through the scripting of events coupled with the use of temporally-related information briefings on deadlines, pacing, and sequencing, NeoCITIES can be easily used to study how temporal information is shared in teams across multiple performance sessions.

CONCLUSION

An integral element of the living lab framework is the role of simulated task environments to better explore and understand key theoretical and practical concerns in research. The current chapter demonstrates how the NeoCITIES simulation provides the foundational baseline for exploring collaboration, utilizing group interface modalities, and assessing models/aid that may help teams with decision-making process. This chapter highlights how NeoCITIES lends itself to the study of multiple forms of team cognition, including three of the most researched types (i.e., TMMs, team situation awareness, and team information sharing). The basic NeoCITIES infrastructure and information architecture has been the basis for evolving additional individual and team work simulations involving cognition within the complex domain of cybersecurity/cyber situation awareness (e.g., see Mancuso, this volume, Chapter 3; Minotra, Burns, Dikmen, & McNeese, this volume, Chapter 13), studying cultural differences (Endsley, Forster, & Reep, this volume, Chapter 7). Through the use of NeoCITIES, future researchers will be better able to answer many key research needs, including those related to how individual and team cognition constructs operate longitudinally, how different types of measures interrelate, and how temporally-infused elements of team cognition impact team effectiveness (e.g., see Mohammed, Hamilton, Mancuso, Tesler, & McNeese, this volume, Chapter 9).

REVIEW QUESTIONS

1. Differentiate among the roles of the five main components of the NeoCITIES interface.
2. Describe the sequence of the typical gameplay in NeoCIITES.
3. Compare and contrast the three most researched forms of cognition studied in teams.
4. Describe the key research needs in the teamwork literature.
5. Describe the key research needs in the team cognition literature.
6. Explain how simulation-based studies can be used to address the research needs in each of these areas.

ACKNOWLEDGMENTS

The research presented in this paper was supported by Grant Number N000140810887 from the Office of Naval Research. The opinions expressed in this paper are those of the authors only and do not necessarily represent the official position of the Office of Naval Research, the U.S. Navy, the U.S. Department of Defense, or The Pennsylvania State University.

REFERENCES

Anderson, J. R., Bothell, D., Bryne, M. D., Douglass, S., Lebiere, C., & Qin, Y. (2004). An integrated theory of the mind. *Psychological Review, 111*(4), 1036–1060.

Artman, H., & Garbis, C. (1998). Situation awareness as distributed cognition. In T. Green, L. Bannon, C. Warren, & J. Buckley (Eds.), *Cognition and cooperation. Proceedings of 9th conference of cognitive ergonomics* (pp. 151–156). Le Chesnay, France: European Association of Cognitive Ergonomics (EACE).

Balakrishnan, B., Pfaff, M., McNeese, M. D., & Adibhatla, V. (2009). NeoCITIES geo-tools: Assessing impact of perceptual anchoring and spatially annotated chat on geo collaboration. In *Proceedings of the human factors and ergonomics society 53rd annual meeting* (pp. 294–298). Thousand Oaks, CA: SAGE Publications.

Bell, B., & Kozlowski, S. (2011). Collective failure: The emergence, consequences, and management of errors in teams. In D. A. Hoffman & M. Frese (Eds.), *Errors in organizations* (pp. 113–141). Hoboken, NJ: Routledge Academic.

Boos, M., Schauenburg, B., Strack, M., & Belz, M. (2013). Social validation of shared and nonvalidation of unshared information in group discussions. *Small Group Research, 44*, 257–271.

Bowman, J. M., & Wittenbaum, G. M. (2012). Time pressure affects process and performance in hidden-profile groups. *Small Group Research, 43*, 295–314.

Burtscher, M. J., Kolbe, M., Wacker, J., & Manser, T. (2011). Interactions of team mental models and monitoring behaviors predict team performance in simulated anesthesia inductions. *Journal of Experimental Psychology: Applied, 17*(3), 257–269.

Cannon-Bowers, J. A., & Salas, E. (2001). Reflections on shared cognition. *Journal of Organizational Behavior, 22*, 195–202.

Cooke, N. J., Kiekel, P. A., & Helm, E. E. (2001). Measuring team knowledge during skill acquisition of a complex task. *International Journal of Cognitive Ergonomics, 5*(3), 297–315.

Connors, E. S. (2006). *Intelligent group interfaces: Envisioned designs for exploring team cognition in emergency crisis management* (Unpublished doctoral dissertation). University Park, PA: The Pennsylvania State University.

DeChurch, L. A., & Mesmer-Magnus, J. R. (2010a). The cognitive underpinnings of effective teamwork: a meta-analysis. *The Journal of Applied Psychology, 95*(1), 32–53.

DeChurch, L. A., & Mesmer-Magnus, J. R. (2010b). Measuring shared team mental models: A meta-analysis. *Group Dynamics: Theory, Research, and Practice, 14*(1), 1–14.

Edwards, B. D., Day, E. A., Arthur, W., & Bell, S. T. (2006). Relationships among team ability composition, team mental models, and team performance. *Journal of Applied Psychology, 91*, 727–736.

Ellwart, T., Konradt, U., & Rack, O. (2014). Team mental models of expertise location: Validation of a field survey measure. *Small Group Research, 45*(2), 119–153.

Endsley, M. (1995a). Toward a theory of situation awareness in dynamic systems. *Human Factors, 37*(1), 32–64.

Endsley, M. (1995b). Measurement of situation awareness in dynamic systems. *Human Factors, 37*(1), 65–84.

Gershgoren, L., Filho, E. M., Tenenbaum, G., & Schinke, R. J. (2013). Coaching shared mental models in soccer: A longitudinal case study. *Journal of Clinical Sport Psychology, 7*, 293–312.

Giacobe, N. A. (2013). A picture is worth a thousand alerts. In *Proceedings of the human factors and ergonomics society 57th annual meeting* (Vol. 57, No. 1, pp. 172–176). Thousand Oaks, CA: SAGE Publications.

Hamilton, K. (2009). *The effect of team training on team mental model formation and team performance under routine and non-routine environmental conditions* (Unpublished doctoral dissertation). University Park, PA: The Pennsylvania State University.

Hamilton, K., Mancuso, V., Minotra, D., Hoult, R., Mohammed, S., Parr, A., … McNeese, M. (2010). Using the NeoCITIES 3.0 simulation to study and measure team cognition. In *Human factors and ergonomics society annual meeting proceedings* (Vol. 54, pp. 433–437). Thousand Oaks, CA: SAGE Publications.

Hauland, G. (2008). Measuring individual and team situation awareness during planning tasks in training of en route air traffic control. *The International Journal of Aviation Psychology, 18*(3), 290–304.

Hellar, B. (2009). *An investigation of data overload in team-based distributed cognition systems* (Unpublished doctoral dissertation). University Park, PA: The Pennsylvania State University.

Hellar, D. B., & McNeese, M. (2010). NeoCITIES: A simulated command and control task environment for experimental research. In *Proceedings of the human factors and ergonomics 54th society annual meeting* (Vol. 54, No. 13, pp. 1027–1031). Thousand Oaks, CA: SAGE Publications.

Henningsen, D. D., & Henningsen, M. L. M. (2004). The effect of individual difference variables on information sharing in decision-making groups. *Human Communication Research, 30,* 540–555.

Jones, D. G., & Endsley, M. R. (2004). Use of real-time probes for measuring situation awareness. *The International Journal of Aviation Psychology, 14,* 343–367.

Jones, R. E. T., McNeese, M. D., Connors, E. S., Jefferson, T., & Hall, D. L. (2004). A distributed cognition simulation involving homeland security and defense: The development of NeoCITIES. In *Proceedings of the human factors and ergonomics society 48th annual meeting* (pp. 631–634). Thousand Oaks, CA: SAGE Publications.

Klimoski, R., & Mohammed, S. (1994). Team mental model: Construct or metaphor? *Journal of Management, 20,* 403–437.

Larson, J. R. (1997). Modeling the entry of shared and unshared information into group discussion: A review and basic language computer program. *Small Group Research, 28,* 454–479.

Lichacz, F. M. J. (2008). Augmenting understanding of the relationship between situation awareness and confidence using calibration analysis. *Ergonomics, 51,* 1489–1502.

Mancuso, V. (2012). *An interdisciplinary evaluation of transactive memory in distributed cyber teams* (Unpublished doctoral dissertation). University Park, PA: The Pennsylvania State University.

Mancuso, V. F., & McNeese, M. D. (2012). Effects of integrated and differentiated team knowledge structures on distributed team cognition. In *Proceedings of the human factors and ergonomics society 56th annual meeting* (Vol. 56, No. 1, pp. 388–392). Thousand Oaks, CA: SAGE Publications.

Mancuso, V. F., Minotra, D., Giacobe, N., McNeese, M., & Tyworth, M. (2012). idsNETS: An experimental platform to study situation awareness for intrusion detection analysts. In *2012 IEEE international multi-disciplinary conference on cognitive methods in situation awareness and decision support* (pp. 73–79). Piscataway, NJ: IEEE.

Marks, M. A. (2000). A critical analysis of computer simulations for conducting team research. *Small Group Research, 31*(6), 653–675.

Mathieu, J., Maynard, M. T., Rapp, T., & Gilson, L. (2008). Team effectiveness 1997–2007: A review of recent advancements and a glimpse into the future. *Journal of Management, 34,* 410–476.

Matthews, M. D., & Beal, S. A. (2002). *Assessing situation awareness in field training exercises.* U.S. Army Research Institute for the Behavioural and Social Sciences. Research Report 1795.

Maynard, M. T., & Gilson, L. L. (2014). The role of shared mental model development in understanding virtual team effectiveness. *Group and Organization Management, 39*(1), 3–32.

McComb, S., Kennedy, D., Perryman, R., Warner, N., & Letsky, M. (2010). Temporal patterns of mental model convergence: Implications for distributed teams interacting in electronic collaboration spaces. *Human Factors: The Journal of the Human Factors and Ergonomics Society, 52*(2), 264–281.

McGrath, J. E. (1982). Dilemmatics: The study of research choices and dilemmas. In J. E. McGrath, J. Martin, & R. A. Kulka (Eds.), *Judgment calls in research* (pp. 69–102). Beverly Hills, CA: Sage Publications.

McGrath, J. E. (1984). *Groups: Interaction and performance.* Englewood Cliffs, NJ: Prentice Hall.

McGuinness, B., & Foy, L. (2000). A subjective measure of SA: The crew awareness rating scale. In D. B. Kaber & M. R. Endsley (Eds.), *Human performance, situation awareness and automation: User centered design for the new millennium*. Atlanta, GA: SA Technologies.

McNeese, M. D., Bains, P., Brewer, I., Brown, C., Connors, E. S., Jefferson, T., ... Terrell, I. (2005). The NeoCITIES simulation: Understanding the design and experimental methodology used to develop a team emergency management simulation. In *Proceedings of the human factors and ergonomics society 49th annual meeting* (Vol. 49, No. 3, pp. 591–594). Thousand Oaks, CA: SAGE Publications.

McNeese, M. D., Perusich, K., & Rentsch, J. R. (2000). Advancing socio-technical systems design via the living laboratory. In *Proceedings of the industrial ergonomics association/human factors and ergonomics society (IEA/HFES) 2000 congress* (pp. 2-610–2-613), Santa Monica, CA: Human Factors and Ergonomics Society.

Mesmer-Magnus, J. R., & DeChurch, L. A. (2009). Information sharing and team performance: A meta-analysis. *Journal of Applied Psychology, 94*, 535–546.

Mesmer-Magnus, J. R., DeChurch, L. A., Jimenez-Rodriguez, M., Wildman, J., & Shuffler, M. (2011). A meta-analytic investigation of virtuality and information sharing in teams. *Organizational Behavior and Human Decision Processes, 115*, 214–225.

Minotra, D. (2012). *The effect of a workload-preview on task-prioritization and task-performance* (Unpublished doctoral dissertation). University Park, PA: The Pennsylvania State University.

Mohammed, S., & Dumville, B. C. (2001). Team mental models in a team knowledge framework: Expanding theory and measurement across disciplinary boundaries. *Journal of Organizational Behavior, 22*, 89–106.

Mohammed, S., Ferzandi, L., & Hamilton, K. (2010). Metaphor no more: A 15-year review of the team mental model construct. *Journal of Management, 36*(4), 876–910.

Mohammed, S., & Hamilton, K. (2012). Studying team cognition: The good, the bad, and the practical. In A. B. Hollingshead & M. S. Poole (Eds.), *Research methods for studying groups and teams: A guide to approaches, tools, and technologies* (pp. 132–153). New York, NY: Routledge.

Mohammed, S., Hamilton, K., Tesler, R., Mancuso, V., & McNeese, M. (2015). Time for temporal team mental models: Expanding beyond "what" and "how" to incorporate "when." *European Journal of Work and Organizational Psychology. Special Issue: Dynamics of Team Adaptation and Team Cognition, 24*, 693–709.

Mohammed, S., Tesler, R., & Hamilton, K. (2012). Time and shared cognition: Towards greater integration of temporal dynamics. In E. Salas, S. Fiore, & M. Letsky (Eds.), *Theories of team cognition: Cross disciplinary perspectives* (pp. 87–116). New York, NY: Taylor and Francis Group, LLC.

Nilsson, H., Jonson, C-O., Vikström, T., Bengtsson, E., Thorfinn, J., Huss, F., ... Sjöberg, F. (2013). Simulation-assisted burn disaster planning. *Burns, 39*(6), 1122–1130.

Patterson, E. S., & Stephens, R. J. (2012). A cognitive systems engineering perspective on shared cognition: Coping with complexity. In E. Salas, S. Fiore, & M. Letsky (Eds.), *Theories of team cognition: Cross disciplinary perspectives* (pp. 173–207). New York, NY: Taylor and Francis Group, LLC.

Pfaff, M. S. (2008). *Effects of mood and stress on group communication and performance in a simulated task environment* (Unpublished doctoral dissertation). University Park, PA: The Pennsylvania State University.

Rafferty, L. A., Stanton, N. A., & Walker, G. H. (2010). The famous five factors in teamwork: a case study of fratricide. *Ergonomics, 53*(10), 1187–204.

Randall, K. R., Resick, C. J., & DeChurch, L. A. (2011). Building team adaptive capacity: The roles of sensegiving and team composition. *The Journal of Applied Psychology, 96*(3), 525–40.

Raser, J. C. (1969). *Simulations and society: An exploration of scientific gaming.* Boston, MA: Allyn & Bacon.

Rentsch, J. R., & Klimoski, R. J. (2001). Why do "great minds" think alike?: Antecedents of team member schema agreement. *Journal of Organizational Behavior, 22*(2), 107–120.

Resick, C. J., Dickson, M. W., Mitchelson, J. K., Allison, L. K., & Clark, M. A. (2010). Team composition, cognition, and effectiveness: Examining mental model similarity and accuracy. *Group Dynamics: Theory, Research, and Practice, 14*(2), 174–191.

Resick, C. J., Murase, T., Randall, K. R., & DeChurch, L. A. (2014). Information elaboration and team performance: Examining the psychological origins and environmental contingencies. *Organizational Behavior and Human Decision Processes, 124*(2), 165–176.

Reuveni, Y., & Vashdi, D. R. (2015). Innovation in multidisciplinary teams: The moderating role of transformational leadership in the relationship between professional heterogeneity and shared mental models. *European Journal of Work and Organizational Psychology, 24*(5), 678–692.

Rousseau, R., Tremblay, S., Banbury, S., Breton, R., & Guitouni, A. (2010). The role of metacognition in the relationship between objective and subjective measures of situation awareness. *Theoretical Issues in Ergonomics Science, 11*, 119–130.

Salmon, P., Stanton, N., Walker, G., & Green, D. (2006). Situation awareness measurement: A review of applicability for C4i environments. *Applied Ergonomics, 37*(2), 225–238.

Salmon, P. M., Stanton, N. A., Walker, G. H., Jenkins, D., Ladva, D., Rafferty, L., & Young, M. (2009). Measuring situation awareness in complex systems: Comparison of measures study. *International Journal of Industrial Ergonomics, 39*(3), 490–500.

Santos, E., Rosen, J., Kim, K. J., Yu, F., Li, D., Guo, Y., ... Katona, L. B. (2012). Reasoning about intentions in complex organizational behaviors: Intentions in surgical handoffs. In E. Salas, S. Fiore, & M. Letsky (Eds.), *Theories of team cognition: Cross disciplinary perspectives* (pp. 51–85). New York, NY: Taylor and Francis Group, LLC.

Schreiber, M., & Engelmann, T. (2010). Knowledge and information awareness for initiating transactive memory system processes of computer-supported ad hoc groups. *Computers in Human Behavior, 26*, 1701–1709.

Smart, P. R., Sycara, K., & Tang, Y. (2014). Using cognitive architectures to study issues in team cognition in a complex task environment. In *SPIE defense, security, and sensing: next generation analyst II* (p. 14). Bellingham, WA: Society of Photo-Optical Instrumentation Engineers (SPIE).

Smith-Jentsch, K. A., Campbell, G. E., Milanovich, D. M., & Reynolds, A. M. (2001). Measuring teamwork mental models to support training needs assessment, development, and evaluation: Two empirical studies. *Journal of Organizational Behavior, 22*, 179–194.

Sohrab, S. G., Waller, M. J., & Kaplan, S. (2015). Exploring the hidden-profile paradigm: A literature review and analysis. *Small Group Research, 46*(5), 489–535.

Stanton, N. A., Salmon, P. M., Walker, G. H., & Jenkins, D. P. (2010). Is situation awareness all in the mind? *Theoretical Issues in Ergonomics Science, 11*(1–2), 29–40.

Stasser, G. (1992). Information salience and the discovery of hidden profiles by decision-making groups: A thought experiment. *Organizational Behavior and Human Decision Processes, 52*, 156–181.

Stasser, G., & Stewart, D. (1992). Discovery of hidden profiles by decision-making groups: Solving a problem versus making a judgment. *Journal of Personality and Social Psychology, 63*, 426–434.

Stout, R. J., Cannon-Bowers, J. A., Salas, E., & Milanovich, D. M. (1999). Planning, shared mental models, and coordinated performance: An empirical link is established. *Human Factors, 41*, 61–71.

Terrell, I. S., McNeese, M. D., & Jefferson, T. (2004). Exploring cognitive work within a 911 dispatch center: Using complementary knowledge elicitation techniques. In *Proceedings of the human factors and ergonomics society 48th annual meeting* (Vol. 48, No. 3, pp. 605–609). Thousand Oaks, CA: SAGE Publications.

Tyworth, M., Giacobe, N. A., Mancuso, V. F., McNeese, M. D., & Hall, D. L. (2013). A human-in-the-loop approach to understanding situation awareness in cyber defense analysis. *ICST Transactions on Security Safety*, 2, e6.

Uitdewilligen, S., Waller, M. J., & Zijlstra, F. R. H. (2010). Team cognition and adaptability in dynamic settings: A review of pertinent work. In G. P. Hodgkinson & J. K. Ford (Eds.), *International review of industrial and organizational psychology* (pp. 293–353). Chichester, UK: Wiley.

Waller, M. J., Gupta, N., & Giambatista, R. C. (2004). Effects of adaptive behaviors and shared mental models on control crew performance. *Management Science, 50*(11), 1534–1544.

Wellens, A. R. (1993). Group situation awareness and distributed decision making: From military to civilian applications. In N. J. Castellan Jr. (Ed.), *Individual and group decision making: Current issues* (pp. 267–291). Hillsdale, NJ: Lawrence Erlbaum Associates, Inc.

Wildman, J., Salas, E., & Scott, C. P. R. (2013). Measuring cognition in teams: A cross-domain review. *Human Factors, 54*(1), 84–111.

Wittenbaum, G. M., Hollingshead, A. B., & Botero, I. C. (2004). From cooperative to motivated information sharing in groups: Moving beyond the hidden profile paradigm. *Communication Monographs, 71*, 286–310.

13 Scaled-World Simulations in Attention Allocation Research
Methodological Issues and Directions for Future Work

Dev Minotra, Murat Dikmen, Anson Ho,
Michael D. McNeese, and Catherine Burns

CONTENTS

Introduction 262
The Dual-Task Attention Research Testbed 263
    Incorporation of Dual-Task Attention Research Testbed in Cybersecurity 266
    The Dual-Task Attention Research Testbed Paradigm in the
    Living Laboratory Framework 267
The AIDL Simulator for Financial Event Monitoring 269
    Task Load 272
    Predictability 272
    Interruption 272
Discussion 273
Review Questions 274
Acknowledgments 275
References 275

ADVANCED ORGANIZER

This chapter focuses on scaled-world simulations for studying cognitive processes in complex event monitoring tasks in work domains such as cybersecurity and financial event monitoring. It presents a task paradigm named the dual-task attention research testbed (DART) and explains its applicability to study performance during nonroutine critical events (events that operators are not experienced with). This chapter then presents a simulator of a financial event monitoring task (i.e., advanced interface design lab (AIDL) simulator) that was applied to better

(Continued)

ADVANCED ORGANIZER (Continued)

understand the *disruptiveness* of interruptions. The disruptiveness of interruptions can be dependent on task load and the predictability of events monitored, and this chapter explains how these were manipulated using AIDL simulator. Advanced knowledge about what makes interruptions more disruptive can be useful in designing cognitive aids that screen for or defer interruptions for operators engaged in time-critical tasks.

In many types of work environments, operators are required to attend to nonroutine critical events that are often unanticipated by operators. Such events if left undetected can lead to accidents or costly system failures. In human factors and safety science, a lot of research has been done to explain factors underlying performance during unexpected critical events and to develop methods to design interfaces to support operator performance during such events. However, a little research has been done to examine performance during critical events in complex task environments associated with high levels of uncertainty. In this chapter, we describe our efforts along scaled-world simulations supporting experimental studies on the effects of workload and information predictability on performance in event monitoring tasks. The DART is being presented as a task paradigm applicable in many event monitoring tasks. Research inspired by the living laboratory framework may incorporate DART for experiments on event monitoring tasks that involve unexpected critical events. We also present the AIDL simulator for financial event monitoring that is used to examine the *disruptiveness* of interruptions under different levels of predictability. While explaining the benefits of using these task paradigms, we spell out a number of caveats associated with their usage. Directions for future experimental research are discussed.

INTRODUCTION

Scaled-world simulations are tools developed for human factors research and performance testing, as they afford examining cognitive processes and the effectiveness of displays and cognitive aids in the context of task environments inspired by their real-world counterparts. They capture real-world constraints in order to produce generalizable effects associated with display design and other factors. In this chapter, we focus on two areas that require more attention in human factors: responses to critical events and cognitive processing of predictive information.

Cognitive work analysis and ecological interface design are the approaches that support operator performance and situation awareness (Endsley, 1995) during unanticipated critical events (Burns & Hajdukiewicz, 2013; Vicente, 1999). Previous research has shown that these approaches can be applied to support operator performance in work domains with higher levels of uncertainty in comparison to well-structured work domains such as aviation and industrial process control. Recent human factors research in domains associated with uncertainty such as cybersecurity (Minotra, 2012) and financial event monitoring (Minotra, Dikmen, Ho, & Burns, submitted) has incorporated scaled-world simulations to answer research questions pertaining to the effects of task load, information predictability, and the incorporation of cognitive aids. Human factors research with work domains that are complex and uncertain in nature requires several phases and iterations of work involving data collection, theory building, experimentation, and theory validation. The living laboratory framework (McNeese, 1996) prescribes the organization of this process and explains interrelationships between these phases. Unlike task paradigms in traditional psychology aimed at isolating cognitive processes in task environments that may not afford generalizability, the DART paradigm and scaled-world simulations discussed in this chapter are aimed at identifying effects generalizable to real-world tasks. The development of scaled-world simulations is an integral part of the living laboratory framework. Experimental work with such simulations is aimed at informing the design of prototypes that may be incorporated in real-world environments (McNeese, 1996). In the following section, we present DART designed to address the issue of dynamic prioritization in complex and uncertain work domains. In the next section, we present the Advanced Interface Design Lab (AIDL) simulator for financial event monitoring that has been used to study the disruptiveness of interruptions (Minotra et al., under review). Then, we provide guidelines for experimental research with scaled-world simulations and provide directions for future research that stem from current work with the aforementioned scaled-world simulations.

THE DUAL-TASK ATTENTION RESEARCH TESTBED

The DART paradigm is aimed at conducting experimental research on the effects of displays (particularly predictive aids) on responses to critical events that inherently involve dynamic prioritization on the fly. There is more work that needs to be done in human factors to explain cognitive processes underlying dynamic prioritization in time-critical environments (Cummings & Mitchell, 2005; Parasuraman & Rovira, 2005). There is also more work required to understand the effect of predictive aids including workload previews (Minotra, 2012). Both issues are pertinent to complex and dynamic tasks in cybersecurity.

In the DART paradigm, the operator's task involves monitoring a complex system (e.g., large organizational computer network) for events and responding to events based on rule knowledge. The system consists of subsystems including a primary subsystem and a secondary subsystem. The operator receives most events pertaining to the status of the system in the primary channel, which pertains to the primary subsystem. Events in the secondary channel are given secondary priority, and they pertain to the secondary subsystem. Rarely the operator is presented with an unexpected critical event, which is presented through the secondary channel, and its priority level is higher than every concurrent event in the primary channel. This is intended to capture the essence of uncertain environments. Consider a scenario where a cyberanalyst may identify an anomalous pattern of attacks directed to a computer in the primary subsystem that is insusceptible to such attacks. A few hours later, if the same pattern of attacks is directed to a computer in the secondary subsystem that is susceptible to such attacks, it would be inferred that the secondary computer should be given primary importance because it is being used as a *back door* to attack primary subsystems. In such scenarios, operators are required to reprioritize subtasks dynamically, and there is no automated alerting system or alarm available to detect and alert operators to do the same.

Why to use the DART task paradigm for researching critical events? Subtask reprioritization or dynamic prioritization in such scenarios is a challenge, and there is a need to examine cognitive processes supporting reprioritization in such scenarios. Unexpected critical events in complex systems have similar constraints present in DART—unexpected critical events are rare, and displays of data pertaining to unexpected critical events are often not presented to operators proximally with displays of data pertaining to routine events (Wickens, 2000). Moreover, existing paradigms in experimental psychology and human factors do not capture the constraints present in the DART task paradigm. A description of DART is provided in Figure 13.1—it shows the progression of the overall threat level or priority of each subsystem over time The rules of engagement for this task can be described as follows: If priority levels for events in the primary channel are greater than or equal to that of the secondary channel, events in the primary channel should be given priority by the operator in all such situations. After the operator completes responding to events in the primary channel, events in the secondary channel can be attended. However, rarely would a *critical event* be presented to the operator— in this condition, event priorities in the secondary channel exceed that of events in the primary channel. Such critical events are supposed to be presented rarely to operators in the DART paradigm.

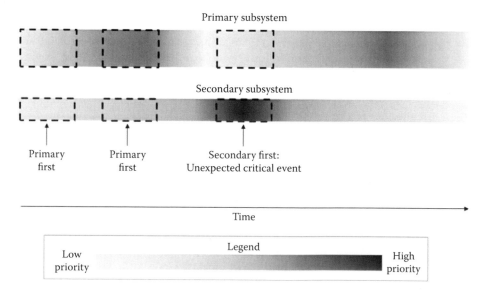

FIGURE 13.1   Description of a typical DART scenario in which threat levels or priorities change with time.

Although the DART paradigm was inspired by the cybersecurity domain, it can be applied to other event monitoring tasks. Responses to unexpected critical events involve three phases that include detection, diagnosis, and compensation. The DART paradigm was aimed at capturing constraints in cognitive processes involved in the detection of critical events in event monitoring tasks; however, the paradigm can be extended to more complex task paradigms that may capture constraints in the diagnoses and compensation of critical events as well. It should be noted that the detection of critical events in DART-based tasks should involve some top-down attentional processing, and that detection should not merely involve complying with some external auditory alert or other form of bottom-up cueing. The DART task paradigm would be useful for studying the cognitive and perceptual processes associated with change blindness, inattentional blindness, cognitive tunneling, dynamic reprioritization, and the cognitive processing of predictive information. A number of measures can be incorporated into the DART task paradigm. Response times in detection, diagnosis, and compensation of critical events can be measured. In addition, response times on other events should be collected and compared with response times on critical events. However, critical events in DART can go undetected for many participants. Thus aggregate data on response times should be accompanied with data on the percentage of timely detection of critical events. In addition, it may also be useful to consider aggregate data on the accuracies in diagnosis and compensation of critical events. Measures pertaining to diagnosis and compensation would especially be useful, if interfaces are designed to support such capabilities in tasks following the DART structure. We anticipate that other researchers may be interested in incorporating the DART paradigm in their experimental studies. In order to better facilitate such studies, we would like to list a few caveats in the use of this paradigm:

- Keeping the number of unexpected critical events low would be in the benefit of the experimenter. Truly surprising critical events should not exceed one critical event. Low-frequency critical events that are not as surprising should also be very small in number in order to avoid learning effects.
- Experimental simulations should be long enough in duration if the effects of sustained attention in the secondary channel need to be taken into consideration. Depending on how participants are trained, they would be able to expect and detect *unexpected critical events* in single-shot studies that do not run for more than 5 minutes.
- We recommend using a few different levels of task load.
- We recommend using an eye tracker and looking for drifts in visual scanning patterns over time. Do these drifts differ based on conditions of task load or the display being used?
- We recommend using two or more levels of training time if required, especially by varying the training time dedicated to unexpected critical events.

For more information on running experiments with critical events, please refer to Minotra, Dikmen, Burns, and McNeese (2015). There are a number of other task paradigms reported in the literature in psychology, human computer interaction, and human factors. The task-switching paradigm (Monsell, 2003), interruptions paradigm

(Hodgetts & Jones, 2006), and the prospective memory paradigm (Dismukes, 2008) have some similarities to the DART task paradigm; however, they do not offer the same structure inherent in the DART framework for examining performance during nonroutine critical events.

INCORPORATION OF DUAL-TASK ATTENTION RESEARCH TESTBED IN CYBERSECURITY

In order to conduct research on situation awareness issues in cybersecurity, the multidisciplinary initiatives in naturalistic decision systems (MINDS) group at Penn State developed a simulation framework referred to as idsNETS informed by previous literature and qualitative research on cybersecurity; the MINDS group used the NeoCITIES simulation framework as a basis for developing idsNETs (Mancuso, Minotra, Giacobe, McNeese, & Tyworth, 2012). NeoCITIES is a simulation environment developed by the MINDS group at Penn State to study team decision making in emergency response (Hamilton et al., 2010). Minotra (2012) extended the idsNETS framework to incorporate the DART task paradigm and named the simulation NETS-DART. NETS-DART was developed to examine the effectiveness of a predictive aid or workload-preview on the ability to detect and respond to critical events in cyber security. Figure 13.2 provides a description of the NETS-DART interface. Panel A is used for selecting a location within a given subsystem. Once a location is selected, events on that location are made visible. Locations that are highlighted in dark gray contain event(s). Panel B provides descriptions of events in the selected location. Upon reading an event description, the

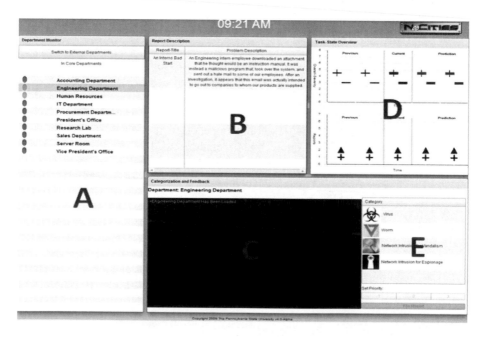

FIGURE 13.2 The interface for NETS-DART used in Minotra (2012). (From N.A. Giacobe et al., Capturing human cognition in cyber-security simulations with NETS, *2013 IEEE International Conference on Intelligence and Security Informatics*, June 4–7, 2013.)

participant would respond to the event on Panel E to indicate appropriate categorizations and level of urgency associated with the event. Panel C provides feedback pertaining to responses made. Panel D is used to compare average threat levels of the subsystems—it provides historical information, current information, and future information.

The incorporation of DART in the aforementioned cyber security simulation can be better described in Figures 13.3 and 13.4—these provide descriptions of scenarios used in the experimental study reported in Minotra (2012). Figure 13.3 represents a high-task load scenario without critical events, whereas Figure 13.4 represents a high-task load scenario with critical events. As it can be seen in Figure 13.4, three critical events were presented in the secondary channel (i.e., 1C, 3C, and 6C) and their priority levels begin to increase over time after they are dispatched. Two more scenarios of moderate task load (not represented here) were also part of the experiment reported in Minotra (2012).

THE DUAL-TASK ATTENTION RESEARCH TESTBED PARADIGM IN THE LIVING LABORATORY FRAMEWORK

NETS-DART is a product of research conducted within the living laboratory framework. This scaled-world simulation inspired by the cyber security domain was designed to test the effectiveness of cognitive aids and task load in the prioritization and detection of unexpected critical events in event monitoring. An experimental study was conducted by Minotra (2012) to test the effectiveness of a predictive aid in this task paradigm and the results of this study are informative to the design and testing of predictive aids. More generally, effects observed in lab experiments incorporating DART-based simulations would inform design and testing processes of cognitive aids aimed at improving anticipation for unexpected critical events and

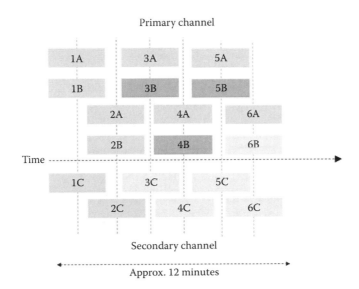

FIGURE 13.3 Description of high-task load scenario without critical events, in Minotra (2012). Each block is an event. Events with darker color represent higher priority events.

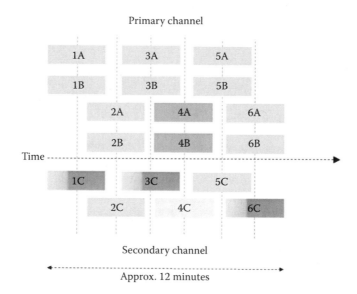

FIGURE 13.4 Description of high-task load scenario with critical events, in Minotra (2012). Each block is an event. Events with darker color represent higher priority events. The critical events are labeled as 1C, 3C, and 6C—these transition in priority after they are presented to the participant.

reprioritization of subtasks. Experimental results may inform guidelines and principles for designing and testing cognitive aids and interfaces. These guidelines and principles would be applicable in the next phase of the living laboratory framework, which involves the design of prototypes. Future research with the DART paradigm may involve field studies with cyber analysts where an emphasis will be placed in gathering data on anticipatory behavior, dynamic prioritization, cognitive tunneling, and the need for cognitive aids to support performance within these cognitive processes and challenges. This would result in the development of more advanced scaled-world simulations incorporating the aforementioned task characteristics extracted from field studies. More specific constraints may be added to the DART paradigm possibly leading to more advanced versions of the task paradigm.

MIDWAY BREATHER

We described the DART paradigm in the context of a cybersecurity monitoring task. This presents a simple task structure to examine performance during nonroutine critical events. Many traditional psychology tasks applied in studies of task switching, interruptions, and prospective memory are similar to DART but are limited in generalizability to real-world tasks. The caveats and recommendations provided can be helpful in designing experiments with DART-based tasks.

THE AIDL SIMULATOR FOR FINANCIAL EVENT MONITORING

We developed the AIDL simulator for financial event monitoring to study the effects of displays, task load, and other factors on performance in a financial event monitoring task that would involve processing differing levels of predictable information. A lot of work on projection (level 3 SA) and predictive displays has been done in the well-structured task environments that often involve visual-spatial information processing (e.g., air traffic control). However, a little work has been done in human factors to examine predictability of information in quantitative data, textual data, and graphical data. The financial event monitoring task addresses this gap as it also associated with uncertain environments.

Predictability is the extent to which an event in the future can be predicted accurately. Within a financial market, there are three main factors through which we can manipulate the change in predictability of an event: volatility, trend, and look-ahead time (LAT). Volatility, also known as variance, is the extent to which a financial product's price fluctuates over time. A larger variance or volatility represents a market value that fluctuates dramatically. Trends are the general momentum in which the market is heading. LAT is the amount of time needed for a user to determine if an event will occur (Magill, 1997; Thomas et al., 2003). Volatility, trend, and LAT can be manipulated to affect the predictability of an event as perceived by a participant.

The AIDL simulator was built to support financial event monitoring tasks, which involve receiving orders to manage a portfolio and taking advantage of windows of opportunity in a simulated financial market. The interface design of the AIDL simulation, the task, and display formats was informed by informal interactions with domain experts in financial trading software and a survey of financial trading platforms. The simulator display consists of four main panels commonly seen in financial trading platforms: portfolio summary, market data, transaction panel, and order request.

The AIDL simulator was first used to study interruptions in a task consisting of windows of opportunity or events that vary in predictability. Interruptions in time-critical and dynamic environments such as financial trading, health care, command and control, and nuclear power plants can provide time-sensitive information. However, failure to recover from an interruption may have serious consequences. Interruptions have a negative effect on individual and team performance (Cellier & Eyrolle, 1992; Kirmeyer, 1988; Van Bergen, 1968). For example, interruptions in health care have shown to negatively affect patient safety (Westbrook et al., 2010; Westbrook et al., 2010). In these environments of high levels of collaboration and multitasking (Cooke, Gorman, Pedersen, & Bell, 2007), workers are particularly prone to interruptions (Jett & George, 2003). However currently, there are few studies observing the role of interruptions in a financial trading environment. Minotra et al. (2016) explore the use of this simulator in a study to investigate the relationship between interruption *disruptiveness* and the predictability of financial events.

In using information from the four panels, users have the ability to buy and sell shares of several financial products manually in addition to capabilities in algorithmic trading of shares. The financial event monitoring task as described in Minotra et al. (2016) can be divided into three basic information processing segments pertaining to this task. Figure 13.5 highlights the interface components associated with

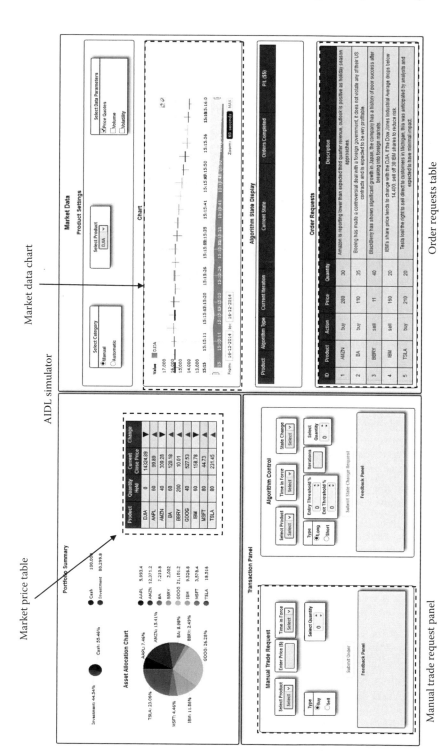

FIGURE 13.5 Layout of the interface of the AIDL simulator for financial event monitoring.

these three information processing segments. The subtasks associated these interface components can be described as follows:

1. *Review order requests*: At the beginning of any scenario, multiple orders will appear in the order request table. The participant should scan this interface component and read through all order request descriptions. An example of an order request is *Buy 40 shares of apple inc. (AAPL) under $40.00 and sell them for over $40.8.*
2. *Monitor for windows of opportunity*: After completing the above subtask, the participant should monitor share prices in the financial market to identify events or windows of opportunity that match the requirements in the order requests. A window of opportunity refers to a time window in which the market price of a particular product matches the prices in the order request. Participants are instructed to concurrently monitor for several potential windows of opportunity that can appear in any sequence. In other words, the sequence of these windows of opportunity need not match the sequence of the order requests presented in the order request panel. Windows of opportunity should be identified with information presented over time in the market price table. All windows of opportunity can be detected in the market price; however, some windows of opportunity can be better detected through graphical cues on the market data chart.
3. *Execute trade request*: Upon detecting a window of opportunity, the participant should complete the order, which will involve completing several form elements in the manual trade request panel.

The three subtasks described earlier may ideally follow the sequence, "1, 2, 3"; however, participants may occasionally revisit the order request panel for strengthening of memory elements associated with task goals. A complete scenario in this task will involve multiple order requests and would thereby involve several cycles of traversals through the three subtasks.

A typical experimental setting involves participants completing several scenarios. The scenarios consist of monitoring the market prices and completing the subtasks described. Order requests may or may not be associated with a window of opportunity, which can be fulfilled by participants. Order requests that have an associated window of opportunity disappear after a fixed amount of time, regardless of whether a participant took advantage of it or not. Order requests without a window of opportunity stay until the end of the scenario. Participants can place manual trade requests at any time in the scenario; however, the correct action is to place such manual trade requests when there is a window of opportunity for a particular product. The simulator allows experimenters to control three important characteristics of the financial trading task, namely task load, expectancies (predictability), and the timing and nature of the interruption. These characteristics are explained in the following sections.

Task Load

Task load is manipulated by the number of order requests presented in any given scenario. As the number of order requests increases, the mental demand of the task increases. More order requests require more visual information processing, more reliance on memory, and greater division of attention. Since an order request disappears after a window of opportunity has passed, the number of order requests at any given time is not a fixed number. The task load can be kept stable, if needed, by presenting a new order request after an existing one disappears. Another way to manipulate the task load is to present order requests of different nature. Apart from regular orders, which require the participant to buy or sell shares of one product if the corresponding price of that product reaches a target, the participants can also receive order requests involving two or more products or indices that require the participant to monitor two or more price or index movements with the help of the market price table and/or the market data chart.

Predictability

In the simulated task environment, market prices constantly change, and there is an uncertainty of whether a critical event will happen (i.e., market price match the price in the order request), or when it will happen. In real-world environments, some of these events can be predicted beforehand, whereas others can be surprising. This aspect is represented in the simulator by implementing gradual or sudden changes in market prices. Specifically, gradual price changes implied a trend toward a target window of opportunity, and sudden changes are represented in abrupt increase or decrease of prices.

Aforementioned, there are three variables that may affect predictability of a window of opportunity: volatility, trend, and LAT. Predictability of a particular event is further modified by manipulating different aspects of the trend such as the duration of the trend and the extent to which the prices vary during the trend (noise). If a product's price shows signs of a linear trend with low variation, it is assumed to reach the target eventually, thus have high predictability. On the other hand, if the product's price has no observable trend, yet it reaches the target abruptly, it has low predictability.

The implementation of predictability adds an interesting layer of cognitive demand. On one hand, detecting a predictable trend and being able to project that an event will occur in the near future allow participants to prioritize the tasks and allocate attention to the necessary elements on the display. On the other hand, detecting a trend itself becomes a cognitive task. It involves sampling information from the environment, and making a decision, at some point, that a task-related event might occur in a particular product.

Interruption

The simulator affords the ability to place interruptions in a scenario. The interruption task in a recent study involves reading comprehension, involving finance-related news articles (Minotra et al., 2016). The interruptions can be presented at any time during a scenario. Some of the interruptions precede windows of opportunity, whereas others do not. The scenario in the simulator does not have to be paused during the interruption. Thereby, the simulator affords the ability to study situation awareness recovery (Gartenberg, Breslow, McCurry, & Trafton, 2013), wherein events of interest can be placed during the interruption.

Overall, the AIDL simulator addresses several important cognitive processes and provides a unique opportunity to study human behavior in complex financial trading tasks. By incorporating predictability, in addition to task load and interruptions, the simulator captures the richness of these environments. A missing piece in this picture is algorithmic trading, which has now become prevalent in financial markets. As with any automation, new problems would emerge in terms of misuse and problems with trust calibration. Although the simulator has the capability to conduct algorithmic trading, we have not yet conducted experiments that use this feature.

DISCUSSION

The authors of this chapter have many years of experience running experiments with simulations. With the experienced gained, we have identified several guidelines to better develop scaled-world simulations and design scenarios and experiments. Guidelines to design experiments consisting of critical events are addressed in Minotra et al. (2015).

The scaled-world simulations reported in this chapter are intended to deeply examine cognitive processes. The identification of generalizable effects would be informative in the design of cognitive aids and displays. Although traditional psychology tasks for examining microcognitive processes can provide insights into task performance, such tasks do not represent real-world environmental constraints. More specifically, the consideration of effects associated with task load, prioritization, sustained attention, divided attention, cognitive tunneling, learning, and information integration requires the use of complex scaled-world simulations. Moreover, processes like task prioritization cannot be fully studied with microcognitive tasks (i.e., traditional tasks in psychology experiments such as tower of Hanoi) as it is difficult to elicit responses influenced by beliefs associated with urgencies in simulated tasks. This difficulty may also be experienced by researchers that incorporate high fidelity simulations in their studies.

Our experience with scaled-world simulations suggests that response times are not sufficient to understand macrocognitive processes. Data from measures associated with situation awareness, communication, task-switching frequency, recovery of situation awareness after interruptions, verbal protocols, eye tracking, and facial expressions can be highly useful. It may also be useful to run experimental studies involving microcognitive tasks in addition to those with scaled-world simulations, if feasible. This would provide comparison across experiments that differ in scale but would involve cognitive processes similar in nature. Experimental data from the microcognitive task would help better interpret results from the experiment involving the scaled-world simulation. Future work with scaled-world simulations would involve more fieldwork in order to capture constraints present in real-world tasks. Qualitative methods of data collection (such as verbal think-alouds or interviews) should go hand in hand with quantitative experimental studies as factors that contribute to accidents and human error are not necessarily attributed to a single operator or team. It is important to understand the entire socio-technical system to uncover factors underlying system failures and accidents (Rasmussen, 1997; Vicente & Christoffersen, 2006). However, the methods used may be limited by the scope and objectives of the projects.

This chapter reported the DART paradigm, which offers a simple task structure that can be incorporated in scaled-world simulations involving nonroutine critical events. It can represent certain types of task environments consisting of uncertainty where the detection of unexpected critical events requires more than just perceptual processing of external cues but also requires some top-down attentional processing that would inherently involve visual scanning, information integration, and remembering the importance of scanning for important events in a secondary information channel. Based on our experience with running experiments, unexpected critical events can be difficult to detect by participants in a DART-based task. Although this may be associated with cognitive tunneling within the primary information channel, a number of other factors may underlie the difficulty that participants face in being able to detect it. In our recent experiment with the AIDL simulator for financial event monitoring (Minotra et al., under review), participants were required to use trend information from a secondary channel to identify a unique and predictable window of opportunity. Such a window of opportunity was presented occasionally to participants and a large number of participants failed to detect such windows of opportunity. Moreover, unpredictable and abrupt windows of opportunity presented in the primary information channel were detected more often than these windows of opportunity in the secondary information channel associated with larger trend durations preceding them (thereby potentially being more predictable as standalone events). In our experiment with NETS-DART (Minotra, 2012), participants presented with a predictive aid demonstrated a decrement in sustained attention over time in the detection of unexpected critical events in a high-workload scenario. The difficulty in being able to detect unexpected critical events in DART encourages the use of this paradigm in experimental studies. The unexpectedness of critical events may also be associated with the accuracy and depth of mental models in operators. Future research with the DART paradigm should take such individual differences into consideration as it would generate more reliable results in testing ecological displays and other displays intended to support performance during critical events.

REVIEW QUESTIONS

1. Presenting nonroutine critical events in a laboratory task can be challenging because participants can expect such nonroutine critical events even though they are designed to be unexpected. This can confound experimental data and sometimes result in ceiling effects in performance. Name the three experimental design guidelines provided in this chapter that could mitigate this risk.

2. Interruptions can be programmed in scenarios used with the AIDL simulator. During an interruption, scenarios can be active; in other words, the scenario is not paused, and events can transpire during the interruption. Why was this capability built into the AIDL simulator?
3. Response time data can be limited in the information they provide researchers to test a given hypothesis. What are the other types of data that can complement response time data?

ACKNOWLEDGMENTS

The part of this research pertaining to cybersecurity was supported by the U.S. Army Research Office (ARO), MURI Grant *Computer Aided Human Centric Cyber Situation Awareness* W911-NF-09-1-0525. The part of this research pertaining to financial trading was supported by an natural sciences and engineering research council of Canada (NSERC) Collaborative Research and Development Grant (CRDPJ 445965-12), Quantica Trading Inc., and Ontario Centres of Excellence.

REFERENCES

Burns, C. M., & Hajdukiewicz, J. (2004). *Ecological interface design.* Boca Raton, FL: CRC Press.

Cellier J. M., & Eyrolle, H. (1992). Interference between switched tasks. *Ergonomics, 35*(1), 25–36.

Cooke, N. J., Gorman, J., Pedersen, H., & Bell, B. (2007). Distributed mission environments: Effects of geographic dispersion on team cognition and performance. In S. Fiore & E. Salas (Eds.), *Toward a science of distributed learning* (pp. 147–167). Washington, DC: American Psychological Association.

Cummings, M. L., & Mitchell, P. J. (2005). *Managing multiple UAVs through a timeline display.* Proceedings of the American Institute of Aeronautics and Astronautics Infotech@ Aerospace. Arlington, VA.

Dismukes, R. K. (2008). Prospective memory in aviation and everyday settings In M. Kliegel, M. A. McDaniel, G. O. Einstein (Eds.), Prospective Memory: Cognitive, Neuroscience, Developmental, and Applied Perspectives, Lawrence Erlbaum Associates, New York, pp. 411–431.

Endsley, M. R. (1995). Toward a theory of situation awareness in dynamic systems. *Human Factors: The Journal of the Human Factors and Ergonomics Society, 37*(1), 32–64.

Gartenberg, D., Breslow, L., McCurry, J. M., & Trafton, J. G. (2013). Situation awareness recovery. *Human Factors: The Journal of the Human Factors and Ergonomics Society, 56,* 710–721.

Giacobe, N. A., Mcneese, M. D., Mancuso, V., & Minotra, D. (2013, June 4–7). *Capturing human cognition in cyber-security simulations with NETS.* 2013 IEEE International Conference on Intelligence and Security Informatics. Seattle, WA.

Hamilton, K., Mancuso, V., Minotra, D., Hoult, R., Mohammed, S., Parr, A., … McNeese, M. (2010). Using the NeoCITIES 3.1 simulation to study and measure team cognition. In *Proceedings of the Human Factors and Ergonomics Society Annual Meeting, 54*(4), 433–437.

Hodgetts, H. M., & Jones, D. M. (2006). Interruption of the tower of London task: support for a goal-activation approach. *Journal of Experimental Psychology: General, 135*(1), 103.

Jett, Q. R., & George, J. M., (2003). Work interrupted: A closer look at the role of interruptions in organizational life. *Academy of Management Review, 28*(3), 494–507.

Kirmeyer, S. L. (1988). Coping with competing demands: Interruptions and the type A pattern, *Journal of Applied Psychology*, *73*(4), 621–629.

Magill, S. A. N. (1997). Trajectory predictability and frequency of conflict-avoiding action. In CEAS 10th International Aerospace Conference, Amsterdam, the Netherlands.

Mancuso, V. F., Minotra, D., Giacobe, N., McNeese, M., & Tyworth, M. (2012). idsNETS: An experimental platform to study situation awareness for intrusion detection analysts. In *IEEE international multi-disciplinary conference on cognitive methods in situation awareness and decision support* (CogSIMA), New Orleans, LA, (pp. 73–79).

McNeese, M. D. (1996). Collaborative systems research: Establishing ecological approaches through the living laboratory. In *Proceedings of the human factors and ergonomics society annual meeting*. Los Angeles, CA, (Vol. 40, No. 15, pp. 767–771).

Minotra, D. (2012). *The effect of a workload-preview on task-prioritization and task-performance* (Doctoral dissertation). University Park, PA: The Pennsylvania State University.

Minotra, D., & Burns, C. M. (2015). Finding common ground situation awareness and cognitive work analysis. *Journal of Cognitive Engineering and Decision Making*, *9*(1), 87–89.

Minotra, D., Dikmen, M., Burns, C. M., & McNeese, M. D. (2015). Guidelines and caveats for manipulating expectancies in experiments involving human participants. In *Proceedings of the human factors and ergonomics society annual meeting* (Vol. 59, No. 1, pp. 1778–1782).

Minotra, D., Dikmen, M., Ho, A., & Burns, C. M. (under review). Examining the relationship between predictability and disruptiveness of interruptions: Implications for interruption management systems. *International Journal of Human-Computer Studies*.

Monsell, S. (2003). Task switching. *Trends in Cognitive Sciences*, *7*(3), 134–140.

Parasuraman, R., & Rovira, E. (2005). *Workload modeling and workload management: Recent theoretical developments* (No. ARL-CR-0562). Fairfax, VA: Department of Psychology, George Mason University.

Rasmussen, J. (1997). Risk management in a dynamic society: A modelling problem. *Safety science*, *27*(2), 183–213.

Thomas, L. C., Wickens, C. D., & Rantanen, E. M. (2003). Imperfect automation in aviation traffic alerts: A review of conflict detection algorithms and their implications for human factors research. Proceedings of the Human Factors and Ergonomics Society Annual Meeting, *47*, 344–348.

Van Bergen, A. (1968). *Task interruption*. Amsterdam: North-Holland Publishing Company.

Vicente, K. J. (1999). *Cognitive work analysis: Toward safe, productive, and healthy computer-based work*. Lawrence Erlbaum Associates. Mahwah, New Jersey.

Vicente, K. J., & Christoffersen, K. (2006). The Walkerton E. coli outbreak: A test of Rasmussen's framework for risk management in a dynamic society. *Theoretical Issues in Ergonomics Science*, *7*(2), 93–112.

Wickens, C. D. (2000). The trade-off of design for routine and unexpected performance: Implications of situation awareness. In M. Endsley & D. Garland (Eds.), Situation Awareness Analysis and Measurement, Lawrence Erlbaum Associates, Mahwah, NJ, pp. 211–225.

Westbrook, J. I., Coiera, E., Dunsmuir, W. T., Brown, B. M., Kelk, N., Paoloni, R., & Tran, C. (2010). The impact of interruptions on clinical task completion. *Quality and Safety in Health Care*, *19*(4), 284–289.

Westbrook, J. I., Woods, A., Rob, M. I., Dunsmuir, W. T., & Day, R. O. (2010). Association of interruptions with an increased risk and severity of medication administration errors. *Archives of Internal Medicine*, *170*(8), 683–690.

14 How Multiplayer Video Games Can Help Prepare Individuals for Some of the World's Most Stressful Jobs

Barton K. Pursel and Chris Stubbs

CONTENTS

Introduction 278
Obstacles to Team Performance in Crisis Environments 280
    Prior to the Raid 281
    During the Raid 282
    After the Raid 284
Synergies between WoW Raids and NeoCITIES 286
Group Support Systems 287
Raiding Example: C'Thun 290
Similarities and Differences between Raid and Crisis Management
Environments 291
Applicability in the Twenty-First Century Workplace 292
Conclusion 294
Review Questions 294
References 294

ADVANCED ORGANIZER

This chapter takes the living laboratory framework, and uses it to draw similarities between the skills developed while playing the video game *World of Warcraft* (*WoW*), and similar skills that are needed to work effectively in large, crisis management scenarios. Although the reasons people engage in these two disparate activities are radically different, as well as the implications of

(Continued)

ADVANCED ORGANIZER (Continued)

success or failure in each environment, the skills required to be successful in a collaborative, team-based environment in both these contexts are eerily similar. After reading this chapter, you will be able to

- Identify and understand specific game mechanics included in the game WoW.
- Identify the commons skills required for success in both a WoW raid environment and a complex crisis management environment.
- Describe how playing WoW, particularly while participating in raids, can help individuals to build specific skills required to succeed in complex, team-based environments.
- Describe the relationship between cognitive engineering and team cognition with WoW.
- Identify the specific skills that WoW players learn by playing the game, and how these skills are applicable in the workplace.

INTRODUCTION

As video games continue to grow in terms of popularity and reach, they present an interesting experimental laboratory, allowing researchers to observe and explore players in closed, formal systems. Of specific interest for social science research are the massively multiplayer online games (MMOGs). Ultima Online, released by Origin Entertainment in 1997, arguably represents the first large-scale, successful MMOG, and ushered in a genre that continues to flourish. The biggest success in the genre is undoubtedly WoW released in 2004, and at its peak registering over 12 million subscribers worldwide. With so many players, and having been in the market for over a decade, we will use WoW as the focus of this chapter, exploring ways in which players are engaged in activities that build skills that go well beyond the game. Although WoW represents a specific example, it is important to note that MMOGs share a variety of common design conventions, such as social and collaborative elements, making our analysis of Warcraft broadly generalizable to other games in the genre.

This chapter will focus specifically on how the living laboratory framework (McNeese & Pfaff, 2012) can help us explore phenomenon taking place within WoW. But before taking on this task, it is worth examining some of the other research related to MMOGs. For our purposes, we will focus on research that explores MMOGs as a learning environment. Broad research in MMOGs often aims to identify various learning affordances in these spaces, exploring how specific gameplay mechanics might impact learning (Voulgari, Komis, & Sampson, 2014). Several researchers used MMOGs to explore leadership, examining how leadership skills from MMOGs might translate to virtual teams (Mysirlaki & Paraskeva, 2012) and how leadership skills from MMOGs might translate to different types of organizations (Mendoza, 2014).

Many MMOGs, including WoW, are built around collaboration and cooperation for those players engaging in the most challenging aspects of the game, which represents an opportunity to use these environments to study collaboration. Childress and Braswell (2006) explored cooperative learning in MMOGs and outlined ways in which cooperative learning that took place within MMOGs can be leveraged in other education and training opportunities. Similar research was conducted around collaboration in MMOGs, and how we can extrapolate game elements that foster collaboration into work environments (Camilleri, Busuttil, & Montebello, 2011). MMOGs often support the creation of in-game groups, such as guilds or clans, so some studies examine the formal game systems that support these groups, and how they can be leveraged for community building efforts outside of games (Papargyris & Poulymenakou, 2005).

One research team explored MMOGs and metacognition, examining what meta-cognitive strategies gamers use within games, and how those strategies can also be successfully applied outside of games (Kim, Park, & Baek, 2009). Examining WoW from this perspective, we can consider the game as a cognitive system, one that can be leveraged not only for its design intent (entertainment), but also as a learning laboratory of sorts, where players can gain skills that apply to both work and life in general.

The remainder of this chapter will examine the behaviors and skills developed in WoW, will discuss how those behaviors and skills develop, and will discuss how similar they are to skills necessary to successfully operate in a crisis management environment. But before moving on further, a few key terms are necessary to help readers unfamiliar with WoW (or other MMOGs) understand various aspects of the game (Table 14.1).

TABLE 14.1
Specific Terms That Describe Various Aspects of WoW

Term	Description
Character	The avatar that a player controls. In most MMOGs, players also have the ability to choose a class for their character.
Class	A broad, self-selected characterization that players choose for themselves. One's class has a direct impact on their character's abilities in the game and thus their role within group contexts. Roles typically fall into healer (keep others alive), tank (absorb a lot of damage), and DPS (damage per second, doing a lot of damage to enemies). Specializations further customize a class.
Dragon kill points	A measurement system, allocating points to players who participate in raids. Points are awarded for attending raids, and for defeating challenging monsters or enemies. Points are then used by players to purchase the best gear (weapons, armor) in the game.

(Continued)

TABLE 14.1 (*Continued*)
Specific Terms That Describe Various Aspects of WoW

Term	Description
Dungeon or instance	A segment of the game that groups of players enter together and are subsequently inhabited only by that group (as opposed to the rest of the MMOG world, which is shared among all players). The colloquialism *instance* refers to the fact that each group, though participating in their own unique space, is engaging in a copy, or instance, of the same content. For example if the Dungeon is *The Haunted Mansion*, instances would exist for each group of players entering the Haunted Mansion. Dungeons are a synchronous experience, forcing all participating members of the group to be online and working together at the same time.
Guild	A collection of players who come together to form a longstanding group. The group is recognized as a guild in the game. Being part of a guild often helps players coordinate and participate in large, in-game encounters (such as raids).
Raid	The most difficult content in the game, requiring anywhere from 10 to 40 players to complete. Raids are considered a type of dungeon or instance.
Raid leader	The player(s) doing the majority of organizing and leading other players on a raid.
Specialization (or spec)	The specific talents (or abilities) that each player chooses for their character class are collectively referred to as their *spec*.

OBSTACLES TO TEAM PERFORMANCE IN CRISIS ENVIRONMENTS

A number of barriers exist that teams need to overcome in order to be successful, especially in areas such as crisis management and response. McNeese and Pfaff (2012) identify several of these challenges, such as time sensitivity, knowing that without immediate action people can be injured or even killed. Information overload can also be a problem, as crisis responders often need to synthesize diverse information quickly, and either act on salient information or articulate that information to another individual who can act on it. With too much information, or without a good method to visualize or filter the information, individuals in crisis management can be overwhelmed quickly. Sometimes members of crisis response teams may not completely understand the role that each team plays in the overall crisis response ecosystem, leading to duplication of efforts, lack of action, or inefficiencies that can cost valuable time, resources, or lives. Crisis management personnel also work in an environment that is energized by tension, emotion, conflict, and stress. Team members may come from a wide variety of backgrounds, with potentially significant variance in age, gender, cultures, and geographic locations.

Interestingly, we find that these same challenges are also faced by individuals in an environment that, on the surface, could not be further from an emergency response situation: raids in MMOGs. During raids in WoW, groups of geographically distributed players must coordinate and choreograph their efforts precisely, weaving together a variety of unique skillsets in order to accomplish a single, mutually shared goal (typically, a raid encounter involves vanquishing a particularly powerful opponent).

For those not familiar with MMOG raiding, the comparison to emergency response situations might seem trivial. Online games are, after all, designed to be a form of entertainment. But upon deeper inspection, we find that raids require a degree of organization, coordination, and leadership that is, in a broad sense, applicable in a variety of nonentertainment contexts. Although we will discuss a specific raid example later in the chapter, we will begin by articulating, at a more general level, the efforts required to support, execute, and sustain raiding practices in MMOGs. As you will see, a great deal of preparation goes into planning a raid, executing the encounter, and conducting the specific activities that take place after a raid is completed.

PRIOR TO THE RAID

Long before a raid takes place, a group of people must be organized to do the raiding. For a number of reasons that we will discuss, it is difficult, if not impossible to organize regular raids by pulling together groups of random players within the game. As such, semipermanent teams (known in MMOGs as guilds) form within the game, sometimes composed of smaller groups of players who know each other in real life. Guilds typically represent groups of players who have mutually shared interests, shared gameplay availability, or similar levels of dedication to group activities. To insure an appropriate balance of players with the right skillsets, often times a process, which mirrors recruitment in many real life organizations, is put in place to identify and invite new members into the guild. High performing guilds may have players request to join them based on their reputation for excellence, guild culture, or common raiding hours, for example. For less established guilds, leadership may have to engage in active recruitment efforts to find players with the right skillsets, availability, interest level, and *fit*. For any guild, vetting potential applicants becomes a key element of sustaining long-term success. In some cases, players may be interviewed by members of the guild, forced to audition, or invited to join on a probationary period, where their performance will be evaluated before determining long-term admittance.

Although some of the players in a guild may have personal relationships outside of the game, often guildmates do not know each other prior to joining a guild, nor will they ever meet each other face to face. Guilds are rarely homogeneous in composition, often including members who span a variety of ages, ethnicities, socioeconomic classes, geographic locations, and cultures. Although the game provides a common purpose to bring players together, these differences can sometimes lead to relationship challenges, which must be overcome.

After a guild of sufficient size and ability has formed, executing a raid must be planned. Engaging in a raid involves a significant coordination of effort among what amounts to volunteers, to insure that sufficient numbers of people are available at a given time and that they possess the necessary skills and abilities required to succeed. This can involve the appointment of subteam leaders who are responsible for communicating with, training, and building cohesion among players of a common class. A raid leader must also be appointed. This individual is responsible for disseminating strategy, communicating with the team, coordinating and instructing players during the raid, distributing rewards during successful encounters, and motivating players during failures. Success in each raid encounter involves the development and articulation of an overall strategy based around the design of the encounter and the unique composition of the raid party. This variability means that even replaying the exact same encounter from week to week may require a drastically different approach. Failure to communicate this to participants may lead to a lack of preparedness, and thus failure, frustration, and tension.

For all but the most elite raiding guilds, personnel management can be a complex and time consuming issue. To succeed in large raids (25 people or more), guilds will often require multiple individuals of the same class in order to be successful. But success also involves ranking the *best* players within each class and assigning them to specific roles. This hierarchy, which must be transparent to the group, requires careful handling to avoid alienating players. If not handled carefully, this *ranking* of players can also reduce the likelihood that top performers will share valuable information with up and comers. MMOGs are played by individuals who have their own aspirations, but success at the highest level requires a successful team.

DURING THE RAID

After months, sometimes even years of preparation, a new guild may be ready to raid. For the first guilds to engage in a specific raid, the process is a step into the unknown, with no roadmap for success, and only the skills and experiences of its players to draw on. The first guilds that experience success in specific raids often share an analysis of each encounter through a variety of websites, assisting other guilds that plan on engaging in the same raid encounter in the future. Once a raid begins, nearly every participant will be expected to execute a variety of tasks in perfect harmony with their teammates. The failure of a single individual to respond correctly to the raid environment, sometimes with only seconds of notice, can result in the failure of the entire group. Though this lacks the life or death stakes of an emergency response environment, the failure of a single individual could mean hours, days, weeks, or months of progress lost, and the virtual death of many avatars in the game world. Because of this, raids demand an incredible degree of interdependence among guildmates, which can turn into tension and stress if things fail to go as planned.

Choreographing your efforts in a game may sound simple, but players in raid environments are inundated with an enormous amount of information in a variety of formats. Visuals on screen, which represent *live action*, can display hundreds of unique pieces of data at any given moment. Data are then accompanied by one, sometimes two or more simultaneous voice chats between players, audio feedback from the game and text-based communication channels. The sheer volume of data being presented, coupled with the need for near immediate analysis and response demands that players must learn (and often be coached) on how best to process so much at once. The development and use of sophisticated user interface (UI) modifications, which allow users to customize their display preferences, can be an essential element to success (Figure 14.1).

In addition to processing huge amounts of data in a timely manner, players in raid environments almost always deal with the unexpected. Game designers intentionally include varying degrees of randomness in every encounter, which force even the most well-prepared team and raid leaders to adapt on the fly (sometimes even changing roles in the middle of an encounter). Successful raiding often demands that every player fills a specific role, but when things fail to go as planned, flexibility, versatility, and constant communication by both leadership and each member of the team, can mean the difference between success and failure.

The designers behind these raid encounters, maybe unknowingly, have embedded hallmarks of cognitive systems engineering in these raid encounters. When viewing a raid encounter as a system, they are designed for flexibility and adaptability (see introduction chapter, McNeese & Forster, where flexibility and adaptability are discussed in relation to integrated living laboratory concepts). When the plan to defeat

FIGURE 14.1 A World of Warcraft raid.

a raid encounter starts to unravel, creativity and ingenuity on the part of the players can lead to new, improvised strategies, sometimes strategies the designers never considered, that can still lead to success. The raid is flexible and adaptable, and accounts for player ingenuity.

Success in raid environments almost never comes during your first attempt. Even the best raiding guilds will endure moments of failure on a given night. During these moments, raid leaders must be motivators, disciplinarians, and coaches, while simultaneously correcting mistakes in productive ways, keeping everyone sharp and keeping the mood positive. Succinct, timely communication that takes into consideration the personalities of everyone involved is crucial. But, especially in times of difficulty, the raid leader must serve as more than a coordinator and strategist; they must be the leader of the team. Failure can often lead to second-guessing of one's efforts. When there are too many cooks in the kitchen, maintaining cohesion is almost impossible. Raids, indeed entire guilds, have fallen apart because of poor leadership during these moments. A raid leader is always prepared and confident in the strategy they have implemented and have the confidence of the rest of the guild, even during challenging times. But they must also have humility to learn from other members of the group, when and where it is appropriate.

AFTER THE RAID

Once a raid is over, whether it was a success or a failure, there are lessons to be learned from the experience. The swaths of data that were used for in-game decision-making, in high performing raid guilds, are aggregated, analyzed, and synthesized into meaningful reflections, which must then be communicated back to the guild. This can be a challenging exercise for a guild leadership team. As in any organization, success in an individual encounter might be motivational, but long-term success demands constant reflection and improvement; a topic which can be dismissed as extraneous in the wake of victory. In failure, players, who are often frustrated, rarely want to discuss their faults. Raid improvement, like performance management in the business world, must be handled carefully to maximize the quality of the group without alienating individuals. Some third party websites, such as World of Logs (http://www.worldoflogs.com/) can be linked to a guild's performance, and track real time data related to each player's performance (Figure 14.2).

If a raid (or any of the individual encounters which comprise a raid) was successful, a small number of random, highly coveted rewards will also be granted to the team. These rewards, typically in the form of new equipment for characters, are not only valued for the status they carry with them, but often because they bring bonuses which improve an individual's in-game performance. This introduces a unique challenge for guild leadership. The most efficient route to maximizing team performance would be to distribute this *loot* to the individuals who perform at the highest level and attend events with the greatest frequency. But, large raids are composed of more than just top performers, and self-interest dictates that those players, who are essential to success,

FIGURE 14.2   A log from a World of Warcraft raid.

are not likely to participate if they will be passed over for rewards. Because there are almost always fewer rewards than participants, a strategy for equitable distribution, which benefits the guild but also keeps all players motivated, happy, and feeling that they are treated fairly, is essential. Moreover, this system must be upheld consistently and articulated clearly to all members of the guild. *Dragon kill points* (DKP) is a commonly used system for reward distribution in MMOGs that will be explained later in the chapter. Fewer aspects of raid management create more problems than reward management (Figure 14.3).

Though it is fraught with peril, the distribution of rewards in success is always a preferable problem to the prospect of days, weeks, or even months of repeated poor performance; an experience which may not be unusual for a raiding guild. Even the most cohesive bonds among teams can fray after consistent failure; a problem that typically leads to attrition, particularly among top performers whose services would be well received by other guilds. Managing frustration, stress, and tension, while finding ways to keep individuals motivated in the face of failure, is a constant challenge for any leader.

FIGURE 14.3   Distributing loot.

SYNERGIES BETWEEN WoW RAIDS AND NeoCITIES

While collaborating with coauthors and editors of this book, parallels were identified between the design of WoW raids, and the NeoCITIES scaled-world simulation (McNeese et al., 2005), specifically while examining the use of NeoCITIES as a simulation environment for command and control centers (Hellar & McNeese, 2010). During a NeoCITIES simulation, participants take on a unique role in an emergency response scenario, and have access to a limited pool of resources. Although NeoCITIES directly addresses emergency response contexts and requires certain skills to mitigate emergencies, such as problem identification, event prioritization, and resource allocation, raids in WoW demand the same skills in order for players to succeed. The NeoCITIES simulation also has a reward structure in place in the form of points that illustrate how well someone performed during an instance of the simulation. These points are a form of feedback for participants that not only provide feedback to individuals, but also can prompt team reflection and discussion about what went right or wrong in a given simulation. In WoW, the reward system of loot tends to also act as a feedback mechanism, in that players who know how to master a character class, and the mechanics of an encounter, are rewarded with loot that carries with it visual indicators to one's avatar. This often leads to others in the raiding community seeking these individuals out, to discuss and reflect on successful strategies for specific encounters.

A takeaway based on these synergies between the two environments is scale; NeoCITIES is a small, scaled-world simulation environment, whereas WoW is a multiplayer game boasting millions of users. Blizzard announced in 2014 that 100,000,000 unique players have spent time playing WoW (Entertainment, 2014). Although not all these players experienced raids, it is safe to assume that at least a quarter to a third of these players did raid. This means that a huge number of people have had experiences similar to those described earlier. The challenge is how to prompt these WoW players to reflect on their experiences, drawing out parallels to the workforce, and helping players reapply these skills at work and in other aspects of their lives.

GROUP SUPPORT SYSTEMS

So what are all the differing technologies that allow for raiding guilds to be successful in these complex, challenging raid environments? Similar to technologies that support distributed cognition in crisis response, Blizzard (the company that created WoW), as well as the players themselves, created a wide variety of systems and practices to help build distributed cognition for players who participate in raids. The first set of support systems are those created by Blizzard, and found within the game software. Players can form guilds within the game, allowing players to formally align with other players under a single guild name. This chapter focuses on raiding guilds, where players are often motivated to be the first guild to successfully defeat a specific raid encounter. For players who may not be involved in a guild, Blizzard later implemented a raid finder tool that can help players find others across the game world that want to raid a specific dungeon. Finally, the game also supports voice chat, allowing players from anywhere in the world to communicate via voice with one another. A small speaker icon appears above a character's name when a player is speaking through the in-game voice software. The most challenging raid encounters are nearly impossible without voice communication, due to all the decisions and coordination necessary at a moment's notice.

In addition to in-game systems that support group cognition, many external websites assist raiders in achieving in-game success. One such website is generally called a *DKP* website. These websites are often used to track individual guild member participation and contributions to raids. For example, when someone participates in a night of raiding, he might be awarded five DKP. As members of a guild accumulate DKP, they can spend it on weapons and armor that are obtained through defeating raid bosses. In essence, DKP is a form of currency, created by players to help quantify participation and contribution that can then be spent to improve a character's gear. For a guild with a long history of raiding, a DKP website becomes an artifact of the guild, illustrating how quickly they progress through raid encounters, the players who are the biggest contributors, and how a guild allocates the best gear in the game to its players.

A very unique aspect of WoW is that anyone who plays the game has the ability to modify, or mod, the user interface of the game through a scripting language called lua. Some of these mods act as a bridge, allowing external data to be pulled into the game, and used for specific purposes within the WoW user interface framework. For example, one mod allowed for players to see what weapons and armor could be found in different dungeons, to help players make decisions on what they wanted to do during any given play session in the game (Figures 14.4 and 14.5).

FIGURE 14.4   User-modified World of Warcraft UI.

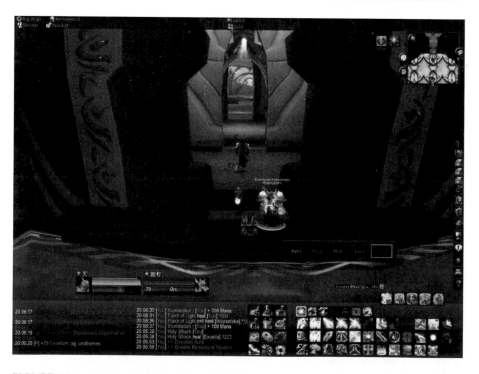

FIGURE 14.5   Original World of Warcraft UI.

A more common example of this phenomenon is smartphones. Consider the flexibility afforded to you by your smartphone. Although every phone ships with a common interface, individual users have the ability to tailor that interface to suit their specific needs. Downloading different apps, moving apps around, or adjusting notification settings are very simple examples of *modding* your phone's UI to provide you with the information you want, when and how you want it.

The modding community provides a great deal of value to Blizzard in two ways. One, the most popular mods created by players often end up being adopted or re-created by Blizzard and become part of the core user interface functionality of WoW, available to all players. By having a UI framework that is flexible and allows for players to create mods, Blizzard created an ecosystem where the players do most of the heavy lifting, in terms of both creating and adopting different mods. Blizzard can then pick the best mods to integrate fully into the game. This eliminates some of the guesswork involved in research and development at Blizzard, in terms of coming up with mods the company *thinks* are necessary in the game; instead, the players take care of this aspect of mod development. The second area of value for Blizzard is that the mod community provides a talent pool of sorts, for future Blizzard employees. As some players become competent at both understanding player needs and coding specific mods to fulfill those needs, these players make ideal candidates for potential employment at Blizzard.

When examining some of these phenomena of the WoW modding community through a cognitive engineering lens, we can consider the aspect of end users taking over various aspects of interface design and development akin to a user- or social-based approach to cognitive engineering. Also, the way in which the WoW interface evolves and adapts, through both the community and Blizzard employees, is representative of cognitive engineering (see Chapter 1).

Turning back to crisis management scenarios, what would these tools and systems look like if the creators of the software had flexible APIs, that were easy to adopt with a scripting language and allowed for the users of these software systems to customize how they received, worked with, and communicated different types of critical information? Obviously challenges exist with an approach like this, as the software that is designed to help manage crisis scenarios needs to account for a wide variety of variables when compared to game software, and often contains very sensitive information. But there is still value in examining the WoW ecosystem, in this case the way in which players can use lua to mod the interface, and try to understand what a similar modding ecosystem can enable in crisis management software. What Blizzard and other MMOG game developers understand is that there is no single user interface standard for how users can best interpret and process large volumes of data. By allowing individuals to customize their displays to suit their individual needs and play styles, game developers allow users to focus not just on consuming data, but on arranging, visualizing, and using that data to maximize performance.

This adaptability is an area of exploration in human–computer interaction (HCI), as it is important for an interface to be flexible as to meet end user needs, as well as flexible in the situations and types of data an interface can accommodate. This problem is compounded in emergency response situations, where oftentimes multiple different technology systems exist, and must somehow communicate with one another (Flentge, Weber, Behring, & Ziegert, 2008).

RAIDING EXAMPLE: C'THUN

A specific example of a raid in WoW from years ago helps exemplify some of the skills players are learning while playing. In January 2006, Blizzard released a raid called the temple of Ahn'Qiraj, with a final boss encounter that was the most difficult in the game at the time. In a team of 40, players entered the temple and faced numerous encounters before coming to the end of the dungeon to face C'Thun. It took the best players of WoW 113 days from the time the temple opened, until the first guild defeated C'Thun.

To provide more contexts, three temporal aspects of this raid were in place at the time. First, every raid was on a seven-day lockout period. What this means is that the raid would *save* a guild's progress for seven days, then reset (which happened every Tuesday morning). What this means is that if a guild defeated the first three encounters in the temple of Ahn'Qiraj in one week of raiding, on Tuesday the instance would reset, and those three encounters would need to be defeated again. From a design standpoint, this method is in place so that guilds can accrue better and better gear each week, so when they finally get to C'Thun, they are best equipped to successfully defeat him. The second temporal element to this raid is the number of nights a guild would enter the temple, in an effort to make a progress. Some guilds raided every night per week, whereas others may raid only a single night. At the time of this raid's release, most top-tier guilds were raiding every night. The third temporal aspect deals with the length of a raid, which also fluctuates between guilds. To make progress in the temple of Ahn'Qiraj, a raiding guild would likely spend at least an hour a night in the temple. Again, top-tier guilds at the time sometimes reported spending upwards of 5–8 hours of raiding in a single day. This reinforces the point that this raid, specifically the final encounter in C'Thun, was brutally difficult, and required an immense amount of dedication among raiding guilds, and also coordination, planning, and execution from those players who eventually defeated C'Thun. See Mohammed et al. (this volume, Chapter 9) for more information on temporal coordination.

As Steinkuehler and Duncan (2008) point out, a great deal of scientific thinking happens on a variety of public and private WoW forums, where players debate, often supported with data, how to optimize the performance of a specific class. Similarly, players display a great deal of scientific inquiry when facing a new challenge like C'Thun. C'Thun was a fight broken down into two complex phases, each requiring a precise amount of coordination, timing, and communication that set the stage for an intense 20-minute raid encounter. The first phase deals with destroying C'Thun's eye, which sporadically shoots a green beam at a member of the raid. One of the challenges of this phase is the green beam can *chain* from player to player, increasing in the amount of damage a player takes with each chain. Without precise coordination and positioning of raid members, one bad green beam means the death of a dozen or more players at once. As many guilds were unable to get past this first phase, a member of the community created a user interface modification to show distance from other players, specifically alerting a player that she is too close to another player, which causes the green beam to chain, or bounce, to others. By using this mod, players were better able to coordinate and monitor their distance between one another, thus eliminating the *chain* effect of the green beam, and keeping players alive to continue the battle. The second challenging aspect of this phase dealt with various smaller enemies appearing at random time intervals. These smaller enemies required different classes of the raid to perform different actions, all of which involved motion (making it incredibly hard to keep the required distance apart from other members of the raid). If each member of the raid did not perform his or her responsibilities in a timely manner, the group was quickly overrun by enemies and defeated. To this end, another mod was developed, that alerted players, through countdown bars, when specific enemies were likely to appear. This gave each class an early warning, so they could prepare to move about quickly and perform their specific roles to successfully complete the phase.

Phase one ended when C'Thun's eye is defeated, and then the rest of his body emerges. During this phase, the many small enemies continue to appear at specific intervals, whereas C'Thun's body is invulnerable. C'Thun randomly ingests players during this phase, placing up to three players in his stomach. Within the stomach, players need to defeat two enemies in order to make C'Thun vulnerable by the rest of the raid. Certain classes, particularly those that can do a lot of damage per second (DPS), are ideal for being ingested, as they can quickly defeat the enemies in the stomach. Again, as in phase one, players knowing their class, and specifically what to do in each phase, are critical to success.

Although these are the high-level strategies used to defeat C'Thun, each class also had unique strategies that needed to be executed for the high-level strategies to succeed. Players tried many differing strategies throughout the 113 days it took the playerbase to first defeat C'Thun. This represents a great deal of hypothesis generation about how C'Thun's mechanics operate, how to best deal with those mechanics, and constant testing and iteration. Players needed to have access to specific UI mods in order to coordinate things like distance and timing, as well as have a critical understanding of their unique role in the overall encounter to be successful. The approaches taken to problem solving in a context like C'Thun is very much akin to solving large, complex problems in a work environment, especially when working in teams. Brainstorming potential solutions, testing the solutions, and iterating is critical to finding the optimal path to solving a problem. The players of WoW, specifically those that excel in high-level raid environments, often have years of this type of experience that can translate nicely to certain work contexts.

SIMILARITIES AND DIFFERENCES BETWEEN RAID AND CRISIS MANAGEMENT ENVIRONMENTS

Despite their incredible complexity, raids and crisis management differ in several important areas: the most notable difference being the stakes associated with each. In crisis management, real lives often hang in the balance of the decisions an emergency response team makes. Although raid participant time is a valuable commodity that must be respected, it pales in comparison to crisis response and management situations. There is no reset button in a crisis situation, no opportunity for a do-over. The reduced stakes and opportunity for replay in raids allow individuals to hone their skills and learn from their mistakes by working with real people in simulated situations that do not put real lives on the line: a valuable learning opportunity.

One additional difference of note is in the scheduling of each raid. Crisis response, almost by definition, occurs unexpectedly. Raids, in contrast, are typically preplanned to occur at a specific time. The impact of a sudden, high-pressure situation cannot be ignored and does represent an important difference between these two types of engagements.

And yet, in considering the skill acquisition of members of a raiding guild, they align nicely with skills needed by people to be successful in a crisis response scenario. Revisiting the challenges outlined by McNeese and Pfaff (2012) in crisis response scenarios, individuals who raid are very adept at time sensitivity. They realize that reacting to fluid situations, sometimes within seconds of something unexpected happening, is critical to a guild's success in a raid. Information overload is a constant struggle for members of a raiding guild, particularly those in leadership roles. Leading 40 players in a complex, synchronous, computer-based environment requires voice communication, text communication, and monitoring an array of user interface indicators to successfully choreograph players in order to conquer an encounter. Through the strategic use of user interface modifications, players find ways to visualize important data that is critical for performing specific actions, at the right time, in raid encounters. For those who have not engaged in these raid environments, it is hard to convey the levels of tension, frustration, conflict, and stress that many members of a guild experience while raiding, sometimes all within the span a few hours. We fully understand and recognize that the tension, frustration, conflict, and stress is much more *real* in crisis response environments, though members of a raiding guild will have experiences in understanding, and managing, these types of emotions in the context of a diverse team.

The final similarity is diversity; members of a guild can come from all walks of life, including variations in age, gender, culture, ethnicity, religions, and geographic locations. These unique individuals all must band together for a common cause, and learn to work together, despite potential differences. Not only is this also the case for crisis responders, but also for any organization that leverages teams to successfully execute projects. Learning how to communicate with, motivate, coach, and critique members of such a diverse group is a key part of any guild and raid leadership team.

APPLICABILITY IN THE TWENTY-FIRST CENTURY WORKPLACE

Throughout a player's experience participating in raids, and more generally playing multiplayer computer games, a number of important skills are acquired that go beyond skills needed in crisis response contexts. Many of these skills are the same as skills required to be productive in our digital economy as an employee. By first identifying the areas that gamers are acquiring skills, organizations can then begin to leverage these skills when identifying roles and responsibilities of new employees.

Empathy refers to the understanding of other people, particularly by putting oneself in the position of someone else. Often times the participation in raids varies across age, gender, ethnicity, and geography. Because these raids are often time-intensive, the players get to know one another on a personal level outside of the game. Such things as a player having to leave his or her keyboard to deal with a sick child are common. As Lothian (2015) discovered, players often come to understand one another's contexts, which helps build empathy. Another skill acquired through raids is multitasking. To perform at a very high level, players need to be listening to the directions of the raid leader, control an avatar in the game world, and constantly monitor the different modifications of the interface, to make sure the player can execute all of her roles for the raid to be successful.

It is worth emphasizing the skills acquired by raid and guild leaders, in the context of a guild successfully progressing through raid content in WoW. The leaders of a guild often spend the most time communicating, both asynchronously and synchronously, to help the guild successfully complete a raid. Guild leaders often lead much of the planning and strategizing taking place on a guild forum, or online team space. These individuals often scour the web, looking for any and all information that will help the guild be successful. Sometimes relevant information is difficult to find. These guild leaders are honing their skills in information seeking/search, retrieval and evaluation of timely information, a skill that is very helpful when working in innovative fields that change quickly (see N. J. McNeese et al. (this volume, Chapter 6) for information on information seeking and search within cognitive systems). The ability to navigate both asynchronous and synchronous communication tools, and understand when to use each, is also a valuable skill guild leaders must understand.

Being equitable in the distribution of raid bounty is critical to a guild's long-term success. By devising and implementing these systems, guild leaders are learning about equitable treatment of others, another important skill for leaders in the workplace. Another important aspect of raid leadership is knowing how to communicate with different types of people; a single raid may have participants ranging from the 16-year old, non native English speaker that plays WoW every waking moment, to the retired professor, who shows up to a raid only sporadically. Understanding player differences and motivations helps in the way a leader approaches and frames a discussion. The same is true about how you approach an employee on the job. Leaders also must keep a positive and productive culture while on raids, making sure each guild member feels welcomed, and that the individual efforts are recognized and appreciated.

A raid leader is essentially organizing and running a 40-person team that is intensely focused on a specific task that requires collaboration. This intense concentration can sometimes last for as little as an hour or two, all the way to extended periods of time lasting up to 8 hours. We believe the skills acquired as a raid leader are critical for organizational leaders as well. The biggest challenge for these individuals is how to translate skills acquired from within WoW to the workplace.

Our final thoughts on this chapter deal with reflection, or reflective learning (Boyd & Fales, 1983), and its critical role in helping gamers take their game-based skills to the workplace (this phenomenon is broadly referred to as transferability). The ability to think back upon an experience, and gain new insights or viewpoints that can change perceptions or beliefs, is a powerful method to support learning, particularly in experiential fields (Carper, 1978). In the case of WoW, this means that players need to somehow have prompts that lead to reflecting about their raiding experiences. Perhaps this is something that seasoned employees can do, or HR professionals, when they know they have a group of employees with a shared experience of playing WoW, and want to draw out those experiences so they can be applied on the job. One of the simplest experiences all WoW players face is dealing with conflict. Nearly every raiding guild faces this when it is time to distribute rewards.

How these situations are handled speaks directly to conflict mitigation skills, though WoW players likely are not thinking consciously about this when they are in the moment. If employers had ways of prompting or facilitating reflective exercises, with WoW as the focal point, this is one method in which some of these important skills can start to translate out of the gameworld, and into the real world. For organizations to leverage the skills of gamers, particularly those with thousands of hours of time spent honing skills as described throughout this chapter, gamers need to be able to connect game-based skills to the work environment through reflection. Without a facilitated discussion or prompt, that challenges gamers to think about specific skill acquisition in games and how it can be abstracted, most gamers are unlikely to connect skills learned in the game to the work environment.

CONCLUSION

Games provide an interesting avenue for skill development that can be directly applicability to many different work environments. This chapter presented specific examples of how WoW, specifically players who engage in raids, learn valuable skills that can be applied in emergency responder scenarios. Although we spent the majority of this chapter focusing on WoW, a great deal of literature exists that illustrates a link between playing games and learning. As games continue to grow in popularity, employers should continue to think about ways in which the hundreds of hour's individuals spend playing games can be leveraged to in the workplace.

REVIEW QUESTIONS

1. What are barriers to team performance in crisis management scenarios?
2. What skills do players of WoW learn through participating in raids? How are these skills applicable to crisis response scenarios?
3. How do the players of WoW take advantage of decision support systems and user interface modifications to overcome complex and difficult raid encounters?
4. What elements of cognitive engineering and team cognition surface in players of WoW?
5. What skills do people learn while player WoW that might be directly applicable to the workplace?

REFERENCES

Boyd, E. M., & Fales, A. W. (1983). Reflective learning key to learning from experience. *Journal of Humanistic Psychology, 23*(2), 99–117.
Camilleri, V., Busuttil, L., & Montebello, M. (2011). Social interactive learning in multi-player games. In M. Ma, A. Oikonomou, L. Jain (Eds.), *Serious games and edutainment applications* (pp. 481–501). London: Springer.
Carper, B. A. (1978). Fundamental patterns of knowing in nursing. *Advances in Nursing Science, 1*(1), 13–24.

Childress, M. D., & Braswell, R. (2006). Using massively multiplayer online role-playing games for online learning. *Distance Education, 27*(2), 187–196.

Flentge, F., Weber, S. G., Behring, A., & Ziegert, T. (2008), April, 5–10. *Designing context-aware HCI for collaborative emergency management.* International Workshop on HCI for Emergencies in Conjunction with CHI (Vol. 8). Florence, Italy.

Hellar, D. B., & McNeese, M. (2010). NeoCITIES: A simulated command and control task environment for experimental research. In *Proceedings of the human factors and ergonomics society annual meeting* (Vol. 54, pp. 1027–1031). Thousand Oaks, CA: Sage Publishing.

Kim, B., Park, H., & Baek. Y. (2009). Not just fun, but serious strategies: Using meta-cognitive strategies in game-based learning. *Computers & Education, 52*(4), 800–810.

Lothian, J. (2015). *Game features and trust belief formation: A study in MMORPG.* University Park, PA: The Pennsylvania State University.

McNeese, M. D., Bains, P., Brewer, I., Brown, C., Connors, E. S., Jefferson, T., Jones, R. E. T., & Terrell, I. (2005). The NeoCITIES simulation: Understanding the design and experimental methodology used to develop a team emergency management simulation. In *Proceedings of the human factors and ergonomics society annual meeting* (Vol. 49, pp. 591–594). Thousand Oaks, CA: Sage Publishing.

McNeese, M. D., & Pfaff, M. S. (2012). Looking at macrocognition through a multimethodological lens. *Theories of team cognition: Cross-disciplinary perspectives.* New York, NY: Routledge.

Mendoza, S. H. V. (2014). *Massively multiplayer online games as a sandbox for leadership: The relationship between in and out of game leadership behaviors.* Mailbu, CA: Pepperdine University.

Mysirlaki, S., & Paraskeva, F. (2012). Leadership in MMOGs: A field of research on virtual teams. *Electronic Journal of E Learning, 10*(2), 223–234.

Papargyris, A., & Poulymenakou, A. (2005). Learning to fly in persistent digital worlds: The case of massively multiplayer online role playing games. *ACM SIGGROUP Bulletin, 25*(1), 41–49.

Steinkuehler, C., & Duncan, S. (2008). Scientific habits of mind in virtual worlds. *Journal of Science Education and Technology, 17*(6), 530–543.

Voulgari, I., Komis, V., & Sampson, D. G. (2014). Player motivations in massively multiplayer online games. In *Advanced learning technologies (ICALT), 2014 IEEE 14th international conference on advanced learning technologies* (pp. 238–239). New York: IEEE.

World of Warcraft [Computer software]. (2004). Irving, CA: Blizzard Entertainment.

Section VI

Knowledge Capture, Design,
Integration, and Practice

15 Police Cognition and Participatory Design

Edward J. Glantz

CONTENTS

Introduction...299
Living Lab Framework 300
Police Cognition...300
Literature Review...301
Methodology 304
    Police Departments 305
        Stage I: Ethnography..306
        Stage II: Knowledge Elicitation 307
        Stage III: Scaled Worlds 308
Analysis and Results..308
Discussion 310
Review Questions...310
References...311

ADVANCED ORGANIZER

This chapter describes the use of the living lab framework to investigate cognitive work in the municipal police domain, through successive application of ethnography, cognitive task analysis, and participatory design. Police officers work in large, socially distributed problem spaces with potential high-risk outcomes, while engaging in cognitive decision making, judgment, and problem solving. Findings include development of an artifact that was implemented to enhance cognitive police effectiveness during cross-jurisdictional events such as a barricaded gunman.

INTRODUCTION

This research is intended to achieve an insight into municipal police officers' use of information systems to enhance cognitive work, by studying police officers in their work environment. Similar to emergency medical and fire responders, municipal police officers face *dynamic* work environments with large problem spaces, ill-defined problems, and high-risk outcomes. Police information systems may support sense-making and thereby improve decision making, judgment, problem solving, and perception (Glantz & McNeese, 2010).

This chapter describes research that attempts to (1) develop understanding of the police domain, including cognitive constraints; (2) analyze cognitive police tasks to reveal sources of complexity and opportunities; and (3) explore technology interventions reducing task complexity, while accommodating domain constraints. The first and second objectives reflect analytical investigations, whereas the third objective extends analysis into a design intervention. Linking analysis and design of the living lab framework enabled the researcher to first identify, and then research an exigent problem, before collaborating with officers in a participatory design setting to develop and refine a reconfigurable tool.

LIVING LAB FRAMEWORK

The living lab framework extends human factors, and in particular the cognitive systems engineering discipline. Cognitive systems have informed research and practice for many years by contributing theories, methods, and design practices (McNeese, Mancuso, McNeese, & Glantz, 2015). Human factors is the idea that technology influences humans, and that humans in turn influence performance, efficiency, and effectiveness of the technology, as well as other humans. Recently, Human factors has created multiple baseline assumptions explaining human–system interaction, including four system design paradigms (McNeese et al., 2015):

- *Technology-centered*: Technology dominant functional engineering basis
- *User-centered*: Psychology dominant task analysis, principles, and guidelines
- *Data-centered*: Experimental dominant use of data sets
- *Group-centered*: Teamwork dominant use of organizational factors to develop processes decision aids, and interfaces

Each paradigm provides an insight while also neglecting contributions from the other paradigms. The living laboratory framework, on the other hand, attempts to offer a more comprehensive alternative to human factors that addresses the design of cognitive systems. The living lab's holistic perspective provides broader supporting parameter diversity, and thus broader problem solving and design. The living laboratory framework integrates human–system perspectives to merge theory, problem, and practice, as illustrated in the following police cognition case.

POLICE COGNITION

Municipal police officers typically work independently, although collaboratively, within large, socially distributed problem spaces, occasionally involving high-risk situations. Challenges during cognitive activities in municipal policing, such as decision making, judgment, and problem solving, were investigated using the living lab framework.

Police officers need to make effective decisions, show good judgment, remember details and react quickly to evolving situations (Glantz, 2006). Police systems do not always support these cognitive activities, however. In one case, an officer collided with a vehicle, while trying to key in its license plate information to a mobile data terminal. In another case, an officer could not locate the bank with the silent alarm, in an area with many new banks located in several strip malls that lacked descriptive addresses. Another officer investigating a simple retail theft was not informed of a weapons history, and ends up wrestling for control of a gun.

In an effort to understand and improve the cognitive aspects of police work, the researcher accompanied officers described in these and other scenarios during patrols and briefings, where they could be observed and interviewed as they worked. Police work involves uncertainty, evolving scenarios, collaboration with other officers and agencies, and information derived or mediated by information systems. This combination of complex work can result in suboptimal results. Poorly designed interfaces, incomplete data access, and spotty network availability contribute to the cognitive load of officers, and limit effectiveness in the performance of duties. The absence of other tools, such as global positioning system (GPS) navigation systems, apartment floorplans, and jurisdictional location tools, also creates an unnecessary cognitive burden on officers.

Understanding policing is important to law enforcement, cognitive engineers, and researchers. Effective cognition and collaboration are important for officers to make important decisions, while confronting high-stakes situations and associated risks. Cognitive engineers are challenged with evaluating, designing, and implementing systems to support cognitive user needs, in the context of the user's work. Thus, researchers seek to better understand cognitive requirements and processes of workers in dynamic settings. Time constraints, uncertainty, changing scenarios, and risks to life and property increase the challenge to support these domain decision makers.

LITERATURE REVIEW

Methodological and theoretical foundations available to study the design of information technologies used in work include cognitive systems engineering (CSE), human–computer interaction (HCI), and computer-supported cooperative work (CSCW). CSE is a multidisciplinary field with scientific conceptual foundations from the 1980s (Hollnagel & Woods, 1983; Norman, 1986; Rasmussen, Pejtersen, & Goodstein, 1994; Woods & Roth, 1998). It has since evolved from scientific foundations to include tools and methods informing engineering design (Dowell & Long, 1998; Eggleston, 2002a). Today CSE represents both a theoretical science and practical engineering within the context of complex system design. CSE frameworks have also been developed for the analysis of work domains (McNeese, Zaff, Citera, Brown, & Whitaker, 1995; Potter, Roth, Woods, & Elm, 2000; Vicente, 1999). Each framework represents multiple techniques, methods and approaches for knowledge elicitation, knowledge capture and design support.

CSE is useful to study work environments with users in an "information-rich world with little time to make sense out of events surrounding them, make decisions, or perform timely activities" (McNeese, 2002, p. 80). CSE takes an equally user-centered and problem-based focus on system design for users in complex settings. This means the designs created are not just *usable*, but also *useful* in supporting decisions and other cognitive activities given time constraints and varying levels of information richness (Woods, 1998).

Information technology relevance, including the information that it processes, is highly dependent on a user's ability to operate and understand the technology (McNeese, 2002; Woods & Roth, 1998). CSE expands the level of analysis beyond the user, or even the computer-user dyad. The unit of analysis in CSE is a *cognitive system* (Roth, Patterson, & Mumaw, 2002; Woods & Roth, 1998). This cognitive system is a triad that consists of the user, the system, and the work context. As such, CSE is a suitable method for analysis of collaborative activities, capable of improving human–computer research that focuses on just the user (e.g., solitary cognition, problem solving, and sense-making), or the user with the user's machine (e.g., human–computer interface issues and key stroke models).

A CSE approach emphasizes environment and context, and allows for a distributed unit of analysis that measures the complex interdependencies between the user and the user's artifact, and in socially distributed work, the other users and their artifacts. This distributed unit of analysis integrates the user, technology (i.e., digital artifacts), and environment, and provides theory, methods and analysis that support identification of system design principles, issues, and tradeoffs. CSE studies can answer questions such as, "Who needs to communicate with whom, when and how?" (e.g., work patterns analysis), "What information needs to be shared?" (e.g., artifact features), and "Who has what information and when?" (e.g., system design).

The intent of CSE is to design work methods, including information technologies, that are user-centric and where the artifact is part of, and contributes to, improving cognitive work (Woods, 1998). To achieve productive work, and reduce fewer errors, it is important that designs carefully consider work in its various forms, as well as subsequent relationships stemming from new design interventions.

In order to analyze work, and complex system designs supporting work, it is useful to distinguish between intrinsic work constraints, and current work practices. Intrinsic work constraints are *independent* of any artifacts currently in use. Intrinsic constraints "delimit the actions that are required to get the job done, not the actions that are required to get the job done with a particular device" (Vicente, 1999, p. 96). Intrinsic constraints originate in the environment, the organization, with the workers, and with the system itself to impact each other and shape the work context. Examples of environmental constraints include mountain location in an air traffic control and laws of physics for a power plant, whereas organizational constraints include training and technology adoption practices. Worker constraints include cognitive and ecological elements such as goals and context, whereas system constraints include issues of usability and usefulness, among others.

Current work practices depend on current artifacts, represented by actions workers use to complete tasks with available tools. Functional actions occur at the intersection between intrinsic work constraints and current work practices, and represent effectively supported work. Workarounds and unexplored possibilities represent ineffectively supported current work, but also signal opportunities. Design shortcomings are highlighted in workarounds, or *overhead* actions that workers perform to compensate for poor technology design; unexplored possibilities represent another inefficiency stemming from intrinsic work constraints currently not of the system design.

The identification of functional actions, as well as workaround activities and unexplored gaps, are all important reasons to conduct field studies of work under naturalistic conditions (Hutchins, 1995; Vicente, 1999; Xiao & Milgram, 2003). New designs should not necessarily support work in its current-articulated form, but rather should attempt to overcome current system inefficiencies, while exploiting new opportunities.

Complexity in dynamic environments stems from characteristics that increase cognitive demands on workers, such as the amount of information needed and whether this information must be shared and processed by others, as well as risks to the worker and environment (Vicente, 1999). Examples of unsuccessful sociotechnical systems in dynamic environments include reactor events at Three Mile Island and Chernobyl, onboard Apollo 13 explosion, the space shuttle Challenger disaster, and grounding of the tanker Exxon Valdez (Woods, Johannesen, Cook, & Sarter, 1994).

CSE framework techniques produce design innovation by capturing domain expert knowledge. CSE combines sets of knowledge elicitation techniques to help designers understand various aspects of cognitive work and systems, such as linking work domain analysis, cognitive task analysis, and participatory design. The results contribute to suitable design interventions that enable workers to adapt to unexpected and changing job demands.

MIDWAY BREATHER

1. When we think of *work*, we often think of the tool that we visualize ourselves using, such as the personal computer in front of us. What other insights become visible by extending this view of work to include the cognitive and physical abilities of the worker using the tool, as well as the task at hand?

2. There are challenges investigating decision making among first responders. In his 1999 book *Sources of Power*, researcher Gary Klein shared flaws in his initial plan to study decision making among firefighters. This plan called for the deployment of undergraduates as stand-by observers in fire stations. The idea was that these students could quickly accompany firefighters to conduct observation of decision making. Fortunately he realized that even in large cities it would be extremely inefficient for observers to wait around for a critical incident to observe (1999, p. 8). What other methods could be used to research first responder decision making during critical incidents?

3. Researchers can be called to develop design interventions or improvements over existing systems by investigating the practices and activities of subject matter experts, such as first responders. In these cases, a sponsor usually directs the research by identifying a specific need or opportunity to investigate. As an alternative, do you believe there could be supplemental value from periodic undirected research? That is to say, can previously unknown ideas and concepts arise from general observations and data review?

METHODOLOGY

Using the living lab framework, the researcher began by conducting an ethnography with domain experts, such as police officers and dispatchers, followed by knowledge elicitation using cognitive task analysis techniques, and concluded with participatory design to develop and refine a suitable tool artifact.

Municipal policing offers several human-system opportunities for human factors investigation such as geospatial mapping, devices permitting officers to swipe card data from drivers' licenses, and body-worn camera devices. In the first stage of the research, ethnographic methods, including observations and interviews, permitted the researcher to develop an understanding of the interplay between cognition, work, and technology within the police domain, while simultaneously identifying those activities that constrain the cognitive abilities of officers. During the ethnography, the technologies most critical during exigent situations, such as officers confronted by an armed gunman, became of interest for further knowledge elicitation (Glantz, 2006; Glantz & McNeese, 2010).

In the second stage of the research, knowledge was elicited from these police officers using cognitive task analysis that combined critical decision method (CDM) (Klein, Calderwood, & MacGregor, 1989) with concept mapping (Novak & Cañas, 2006). Guided by probing questions based on the recognition-primed decision (RPD) model (Klein, 1989), officers retrospectively evaluated cues and information used during armed confrontations. This naturalistic decision-making theory elicits worker judgments and decisions in high stake environments that consist of time pressure, multiple players, and ill-defined goals. The model suggests that absent time to deliberate, domain experts will use knowledge, training, and experience to quickly recognize and develop a course of action. Officers were asked to recall and graphically map a previously encountered armed confrontation, along with the actions, thoughts, and observations during the incident (Glantz, 2006; Glantz & McNeese, 2010).

Of particular interest for tool development were situations that evolved into barricaded gunman incidents. These are of lengthy duration and as such, require support from officers called into assist from other jurisdictions. Incident commanders struggle to quickly and accurately brief and position these visiting officers, as well as to inform these officers of desired information to observe and report back to the command (Glantz, 2006; Glantz & McNeese, 2010).

The third stage of the research utilized a scaled model to facilitate officer tool development. Officers played the roles of incident commanders and briefing support officers using a scaled model that provided an available perspective (e.g., Google Earth), projected onto a magnetic whiteboard, along with dry-erase markers, and a variety of magnetic shaped and colored pieces. The officers applied the tools and developed symbols to map the gunman's location, assigned officer deployments, and physical information such as road blocks, and the location of the command post. Temporal information was also included to track officer deployment, and schedule relief. The refined model was so effective that it has since been implemented in a mobile command post available to the research jurisdictions (Glantz, 2006; Glantz & McNeese, 2010).

This research expanded the advanced knowledge and design acquisition methodology (AKADAM) framework (McNeese et al., 1995; Zaff, McNeese, &

Snyder, 1993) to include participatory design. Refer to Chapter 1, in this volume, for more information on AKADAM and its formative relationship to the living lab framework. The ethnography stage included observation of officers and dispatchers while at work to create an analysis of the current cognitive work domain by creating a Vicente-style abstraction hierarchy diagram. This was followed by a task analysis stage where officers created concept maps to organize and represent knowledge of armed confrontation and barricaded gunman scenarios. These maps permitted officers to graphically tell their stories firsthand by linking concepts, enclosed in circles or boxes, with linking words or phrases.

The AKADAM order of methods, similar to work-centered design (Eggleston, 2002b), is important: ethnography, knowledge elicitation, and then design. Ethnography helps identify physical and environmental constraints in the work domain prior to addressing cognitive constraints during knowledge elicitation. Ethnography permits the researcher to become both familiar with the work domain and workers, and provides the foundation for subsequent knowledge elicitation, and participatory design stages. As previously mentioned, the analytical information from the first two research stages, ethnography and knowledge elicitation, informs and guides participatory design in the third research stage. Participatory design is actively developed in a living lab scaled world that models or simulates the real world, with components in the scaled world having meaning in the real world. For example, dry-erase magnets were used to mark officer deployment locations.

POLICE DEPARTMENTS

This research sampled officers from three neighboring municipal police departments. Each department is funded and managed by different municipalities. These departments cooperate in a county-based law enforcement consortium where they work closely together, are dispatched by a common 911 center, share radio frequencies and computing technology, and are *cross sworn* and thus able to support each other when needed. Municipal Police Department (MPD)-A was the largest department in this study with 63 officers, serving the borough as well as two adjoining townships. As such, most of the ethnography was conducted with MPD-A, the most active of these three departments, based on investigated incidents. Working three full shifts every day of the week, MPD-A provides an opportunity for observing a wide variety of police work during varying shifts, workloads and activities in an area that includes retail shopping, bars, restaurants, residential and university student housing (Glantz, 2006; Glantz & McNeese, 2010).

MPD-B and MPD-C have 18 and 15 sworn officers respectively, and serve neighboring townships that include residential and student housing, rural farmland and developed commercial sectors. The three police departments combine to cover approximately 130 square miles, and a total population of 81,183.

Department size relates to the number of residents, type and complexity of work, as well as workload (as reflected in crime statistics). The majority of the approximately 1,200 police departments in Pennsylvania have fewer than thirty officers, including MPD-B (18 officers) and MPD-C (15 officers) (Glantz, 2006; Glantz & McNeese, 2010).

The police track criminal investigations based on seriousness of offense:

- *Part I*: Serious offenses including murder, rape, aggravated assault and arson
- *Part II*: Lesser offenses including assaults, forgery, drug abuse, drunkenness, and disorderly conduct.

These three municipal police departments are close to state averages for occurrence of the more serious Part I crimes per officer. The Part II crimes per officer rate are higher than the state average, due to an active student population. This is especially true in the central and southern borough area of MPD-A adjacent to the university (Glantz, 2006; Glantz & McNeese, 2010).

Stage I: Ethnography

The first analytical research stage used ethnographic methods such as observation to identify cognitive domain constraints of officers in naturalistic settings. The second analytical stage used cognitive task analysis to elicit knowledge from police experts to reveal cognitive task complexity during performance of critical incidents.

In the first stage, ethnographic techniques of domain training and observation provided insight into how officers perform their cognitive activities. This includes the ecological and cognitive constraints impacting cognitive police work. Domain training included six weeks of Pennsylvania Act 235 training and certification that included classroom and range instruction, as well as psychological and physical assessments. Observation included 46 total hours of immersive fieldwork such as dispatch center observation. Thirty-five of the hours were dedicated to riding with officers from the three jurisdictions during different shifts, attending briefings, and conducting interviews. Of primary interest was observation and analysis of current police information system use.

For approximately seven months, the researcher conducted police ride-along ranging from two to five hours each, with eleven police officers, during different shifts. Since incidents frequently begin with a citizen's 911 call to dispatchers, and could require more than one shift to complete, ride-along were often conducted in overlapping research *segments*. Segments could begin or end at the dispatch center, and continue the ride-along over two shifts, including the in-between briefing shifts.

The goal during this stage was mostly to observe officers and dispatchers while they worked. Only a few unstructured questions were used to gain insight into why or how something was being done. For example, dispatchers would radio and *push* incident information, such as street address, to the officer's vehicle information system. One officer explained that he had learned from experience to reboot the car's information system when crossing a certain geographical boundary, to keep the system connected to dispatch. Also, dispatchers and police officers seemed to know each other mostly through the shared radio contact. Dispatchers creatively used a mapping system to locate a possible accident location, and rarely learned of the outcomes of incidents they processed. Official police communications were limited to routine briefings, or over the radio. Unofficial information exchanges required some effort, relying upon cell phones, or simply pulling cars adjacent to another for a brief discussion.

Outcomes from the first stage included work domain analysis of intrinsic police work constraints and current work practices, with an emphasis on currently unexplored possibilities and workaround activities. This included early-stage *design seeds*, or an attempt to identify a leverage point, or generalized area of cognitive police work not currently supported by technology (Patterson, Woods, Tinapple, & Roth, 2001). Design leverage points in the ethnography stage focused design seed identification during the knowledge elicitation stage. Design seeds represent specific cognitive tasks not already effectively supported by current artifacts. A design seed represents a preliminary notion of a tool, and arises from identification of a leverage point. Design seeds can then be used to form the basis for prototype development.

Stage II: Knowledge Elicitation

The second analytical stage elicited knowledge in task analysis by using a version of the CDM (Klein et al., 1989). CDM sessions were conducted between August and November 2005, using individual interview and concept mapping techniques. These sessions included 10 officers from MPD-B and MPD-C. The CDM cognitive task analysis elicited cues, decision requirements, and decision points during the retelling of a critical incident case, often involving an armed gunman. Probes were used to reveal decision requirements for each decision point by these police officers solving real world problems.

The primary difference between the interview questions during ride-alongs and those developed for elicitation during cognitive task analysis was focus level, as well as specificity. Whereas the ride-alongs were primarily intended to generally observe the work domain environment, the cognitive task analysis was primarily to investigate specific cognitive domain activities by having the respondent describe activity, actions, thought processes, and decision-making approaches. The first four CDM questions used asked the respondent describe a critical incident, its timeline, the decision-making heuristics, and provide a post hoc analysis of decision-making effectiveness. A fifth question was added to have the officer use concept mapping to envision technology enhancements to improve decision-making effectiveness, seeding ideas for scaled-model participatory design in the third stage.

Ten officers were selected for cognitive task analysis and knowledge elicitation based on work experience and specialized training. Training included 12 or more years of experience, whereas experience included multiple positions, or highly skilled backgrounds, such as an *observation-sniper*. The elicitation and follow-up sessions totaled 59 field research hours, and generated 26 CDM procedures, or situation assessment records (SAR). These included 61 decision points during a variety of critical events.

The CDM was adapted to elicit cognitive task knowledge, as well as benefit from graphically retelling the story using concept mapping. Officers would graphically describe what they saw, knew, felt, and believed, as the critical incident evolved. For example, one officer found himself in a narrow hallway facing an armed gunman, with relatives of the gunman unaware of the emerging danger in the room immediately behind him. The officer knew that he had to continue facing the gunman, while using the constrained space of the hallway to radio for backup, and request the relatives exit the building. Without looking, the officer knew what to expect as the department's response formed outside. Years of experience and training permitted this officer to use situational awareness to calmly deescalate a situation that otherwise could have ended violently.

Having the respondent concept map directly permitted the researcher to probe, and be more actively engaged. The time to transcribe interview notes was not eliminated, however, but only shifted to transcribing the concept maps as needed. The researcher and respondent simultaneously shared in the representation of knowledge, with immediate benefits for interpretation and discussion. Using this technique, the police expert represents his or her perspective directly, and with minimal interpretive bias by the researcher.

Stage III: Scaled Worlds

The third and final stage used insight from the previous stages to create a *scaled model* prototype capable of enhancing cognitive police work. The scaled model simulated a mobile police command center that are used as on-scene headquarters providing command, control, and communications during a critical event, such as a barricaded gunman. This stage combined previously obtained domain and task knowledge in a participatory design setting with police officers. In participatory design, the researcher and domain experts work together to jointly innovate concepts and ideas (Greenbaum & Kyng, 1991; Kyng, 1994).

Mobile police command centers are quickly deployed in response to critical incidents. A barricaded gunman incident, for example, may not resolve quickly, and may require mobile capabilities to integrate and orient officers from other jurisdictions.

In this case, the researcher provided tools to the officers who would then decide what to utilize, and in what manner. Tools included a 4' × 5' magnetic whiteboard and easel, a laptop and portable projector, aerial photo jpg files, and dry-erase markers and magnets of different colors and widths. A flip chart and digital and video cameras were used to document the session. The officers pretended to use the scene to orient and deploy officers, track scheduled redeployments, locate barricades, and of course, locate the barricaded gunman. By combining the computer with the whiteboard, the benefits of both media (i.e., map and whiteboard) could be exploited.

As in other participatory design, codesign, or Scandinavian design settings, the *users* of the system actively played a role in the configuration and use of the technology.

ANALYSIS AND RESULTS

The living lab framework provided a useful design method sequencing work domain analysis, cognitive task analysis, and participatory design. The design output is a technical artifact capable of supporting cognitive police work during cross-jurisdictional incident command events, such as a barricaded gunman. Figure 15.1 reveals the deepening understanding, as issues and opportunities in cognitive police work emerged through successive living lab methodologies, beginning with ethnography.

Ethnography techniques afforded the researcher an important personal understanding of the cognitive police domain. Clearly, there are benefits for any researcher

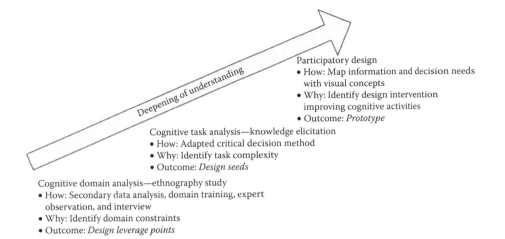

FIGURE 15.1 Living lab framework provided progressive design refinement, linking cognitive analysis through to cognitive design.

to begin directed or undirected research with a general ethnography. If nothing else, this serves to strengthen the researcher's understanding of the cognitive domain under investigation. Understanding can be gained by simply observing workers as they perform tasks in their environment with their tools. In this research, ethnography revealed a design leverage point around the need for officers to patrol individually, yet quickly coalesce and work as a team in response to critical situations. These critical situations guided the design seed development during the subsequent knowledge elicitation stage.

Knowledge elicitation with domain experts focused on knowledge and behaviors during critical incidents. One-on-one sessions with experienced police officers creatively combined the CDM with concept mapping. Officers acted as scribes as they drew their critical incident experience in the form of a concept map on a sketch pad, guided by questions by the researcher. Sketches revealed the chronology of the events, physical surroundings, and thinking behind actions. For example, one officer was face-to-face with a distraught man holding a gun outside the man's trailer-home. The officer described not being able to risk turning from the armed man to *see* what his support officers were doing. Instead, he cognitively *knew* what they were doing. This knowledge permitted the officer to continue undistracted his efforts to deescalate the situation, without removing his gaze from the armed man.

During these critical incident interviews, a design seed emerged to cognitively support incident command centers during extended events. Incident command events of any duration require officer support from other jurisdictions. These officers arrive, unfortunately, with little or no familiarity with the geography, topology, or infrastructure where they are being deployed. Designing an artifact to alleviate this lack of situational awareness guided the participatory design stage of the research.

The approach to participatory design in this research was practical, as opposed to theoretical or conceptual. Satellite imagery of the police jurisdiction readily existed on the web, similar to Google maps. The idea was to quickly simulate a proof of concept by combining projection of these maps on a magnetic whiteboard. Officers simulated an actual incident command situation along with dry-erase markers, and various shaped and colored magnets. From this exercise, it was clear that even in this simple form, there was immediate benefit to including these artifacts in the mobile vehicle used by law enforcement as an incident command.

As a result, the artifact was actually deployed in its proof of concept form. Officers arriving to the incident command could become situationally aware, and be clearly instructed on where to deploy, and what area to observe. Relieved officers could more clearly report where they were, and what they observed. Tables on the whiteboard could record and track individual officer deployment and relief schedules as well. Future research could further develop and test the design, perhaps even to develop a mobile application for situations with quickly responding officers prior to incident command deployment.

There should be no surprise that even though undirected, the living lab framework was able to develop and deploy a needed design artifact. It is particularly interesting that this design was relatively simple, and yet previously unidentified. More interesting, perhaps, is that this was not the only design opportunity to emerge. Others such as data incompatibility between police vehicle systems and dispatch, labor intensive applications for police, and ergonomic problems for system use in police vehicles also appeared.

DISCUSSION

It is important that researchers and designers become immersed in the work domain within they wish to work, and the ethnography stage of the living lab framework provides this opportunity. During ethnography, the researcher became familar with the duties of a municipal police officer. During this stage, the researcher discovered several design leverage points, while becoming aware that the most valued mobile police technology was probably the radio, and the least the computer. From the literature, we learned that community policing pushes police work out of the office and into the field and that now, routine office police work is done from the rather cramped space of the front car seat. This is not so onerous during the day, but at night the dome light used to transcribe notes also reduces situational awareness. Since the officer's car serves as office, time in the station with other officers is reduced to briefings at shift beginning. As a result, police officers experience reduced opportunities to informally share best practices through *water cooler* knowledge exchanges. As such, the living lab was perhaps the most appropriate framework to investigate this distributed work domain.

REVIEW QUESTIONS

1. The research described in this paper was undirected, but still identified and developed an intervention to quickly orient police during barricaded gunman situations. Is this the same methodology beginning with ethnography appropriate for sponsor-directed research of a specific opportunity, such as deploying video cameras in police cars?

2. What are the benefits of using qualitative research, such as ethnographies, before attempting quantitative research, such as survey questions?

3. How does the living lab framework improve upon development of design interventions that do not include a qualitative investigation, or generation of scaled-world representation of the problem space?

REFERENCES

Dowell, J., & Long, J. (1998). Conception of the cognitive engineering design problem. *Ergonomics, 41*(2), 4.

Eggleston, R. G. (2002a). Cognitive systems engineering at 20-something: Where do we stand? In M. D. McNeese & M. A. Vidulich (Eds.), *Cognitive systems engineering in military aviation environments: Avoiding cogminutia fragmentosa!* (pp. 15–78). Wright-Patterson Air Force Base, OH: Human Systems Information Analysis Center Press.

Eggleston, R. G. (2002b). *Work-centered design: A cognitive engineering approach to system design*. Paper presented at the Proceedings of the Human Factors and Ergonomics Society 47th Annual Meeting, Denver, CO. Retrieved from http://www.ise.ncsu.edu/nsf_itr/794B/papers/Eggleston_2003_HFES.pdf

Glantz, E. J. (2006). *Challenges supporting cognitive activities in dynamic work environments: Application to policing* (Ph.D. Doctoral). University Park, PA: The Pennsylvania State University. Retrieved from http://ezaccess.libraries.psu.edu/login?url=http://search.proquest.com.ezaccess.libraries.psu.edu/docview/59989813?accountid=13158 (01)

Glantz, E. J., & McNeese, M. D. (2010). A cognitive systems engineering approach to support municipal police cognition. In D. B. Kaber & G. Boy (Eds.), *Advances in cognitive ergonomics* (pp. 76–86). Boca Raton, FL: CRC Press.

Greenbaum, J. M., & Kyng, M. (1991). *Design at work: Cooperative design of computer systems*. Mahwah, NJ: Lawrence Erlbaum Associates.

Hollnagel, E., & Woods, D. D. (1983). Cognitive systems engineering: New wine in new bottles. *International Journal of Man-Machine Studies, 18*(6), 583–600.

Hutchins, E. (1995). *Cognition in the wild*. Cambridge, MA: MIT Press.

Klein, G. A. (1989). Recognition-primed decisions. In W. B. Rouse (Ed.), *Advances in man-machine systems research* (pp. 47–92). Greenwich, CT: JAI Press.

Klein, G. A. (1999). *Sources of power*. Cambridge, MA: MIT Press.

Klein, G. A., Calderwood, R., & MacGregor, D. (1989). Critical decision method for eliciting knowledge. *IEEE Transactions on Systems, Man, and Cybernetics, 19*(3), 462–472.

Kyng, M. (1994). *Scandinavian design: Users in product development*. Paper presented at the Proceedings of the SIGCHI Conference on Human Factors in Computing Systems: Celebrating Interdependence, Boston, MA.

McNeese, M. D. (2002). Discovering how cognitive systems should be engineered for aviation domains: A developmental look at work, research, and practice. In M. D. McNeese & M. Vidulich (Eds.), *Cognitive systems engineering in military aviation environments: Avoiding cogminutia fragmentosa!* (pp. 77–116). Wright-Patterson Air Force Base, OH: Human Systems Information Analysis Center Press.

McNeese, M. D., Mancuso, V. F., McNeese, N. J., & Glantz, E. J. (2015). *What went wrong? What can go right? A prospectus on human factors practice*. Paper presented at the 6th International Conference on Applied Human Factors and Ergonomics (AHFE 2015), Las Vegas, NV and the Affiliated Conferences, AHFE 2015.

McNeese, M. D., Zaff, B. S., Citera, M., Brown, C. E., & Whitaker, R. (1995). AKADAM: Eliciting user knowledge to support participatory ergonomics. *The International Journal of Industrial Ergonomics, 15*(5), 345–363.

Norman, D. A. (1986). Cognitive engineering. In D. A. Norman & S. W. Draper (Eds.), *User centered system design; New perspectives on human-computer interaction* (pp. 31–61). Hillsdale, NJ: Erlbaum.

Novak, J. D., & Cañas, A. J. (2006). *The theory underlying concept maps and how to construct and use them.* Retrieved from http://cmap.ihmc.us/Publications/ResearchPapers/TheoryUnderlyingConceptMaps.pdf

Patterson, E. S., Woods, D. D., Tinapple, D., & Roth, E. M. (2001). *Using cognitive task analysis (CTA) to seed design concepts for intelligence analysts under data overload.* Paper presented at the Proceedings of the Human Factors and Ergonomics Society 45th Annual Meeting, Minneapolis, MN.

Potter, S. S., Roth, E. M., Woods, D. D., & Elm, W. C. (2000). Bootstrapping multiple converging cognitive task analysis techniques for system design. In J. M. Schraagen, S. F. Chipman, & V. L. Shalin (Eds.), *Cognitive task analysis* (pp. 317–340). Mahwah, NJ: Lawrence Erlbaum.

Rasmussen, J., Pejtersen, A. M., & Goodstein, L. P. (1994). *Cognitive systems engineering.* New York: Wiley.

Roth, E. M., Patterson, E. S., & Mumaw, R. J. (2002). Cognitive engineering: Issues in user-centered system design. In J. J. Marciniak (Ed.), *Encyclopedia of software engineering* (2nd ed., pp. 163–179). New York, NY: Wiley-Interscience, John Wiley & Sons.

Vicente, K. J. (1999). *Cognitive work analysis: Towards safe, productive & healthy computer-based work.* Mahwah, NJ: Erlbaum.

Woods, D. D. (1998). Designs are hypotheses about how artifacts shape cognition and collaboration. *Ergonomics, 41,* 168–173.

Woods, D. D., Johannesen, L. J., Cook, R. I., & Sarter, N. B. (1994). *Behind human error: Cognitive systems, computers, and hindsight.* No. CSERIAC-SOAR-94–01.

Woods, D. D., & Roth, E. M. (1998). Cognitive systems engineering. In M. Helander (Ed.), *Handbook of human-computer interaction.* Amsterdam: Elsevier.

Xiao, Y., & Milgram, P. (2003). Dare I embark on a field study? Toward an understanding of field studies. *The Qualitative Report, 8*(2), 306–313.

Zaff, B. S., McNeese, M. D., & Snyder, D. E. (1993). Capturing multiple perspectives: A user-centered approach to knowledge acquisition. *Knowledge Acquisition, 5*(1), 79–116.

16 Intermodal Interfaces

Arthur Jones

CONTENTS

An Introduction to Intermodal Interfaces 314
 Definition and Overview.. 314
 The Case against Straight Digitalization.. 315
 Relationship with Naturalistic Decision Making and Situational
 Awareness.. 317
 Application Domains 318
 Design Challenges 318
Exploration by Experiment 318
 Experiment Overview 318
 Interface Requirements and Experimental Hypotheses 321
  Primary: Quick and Accurate Collection of Real-Time Intermodal Data..........321
  Secondary: Development and Maintenance of Situational Awareness....321
 Findings... 322
  There is a Significant Effect on the Speed of Data Entry Correlated
  with the Different Interface Design.. 322
  The Validity of the Data Collected Quickly Is Maximized in Interface #2.... 322
  The Rate of Errors When Measuring Situational Awareness across
  the Three Interfaces Was Just under Statistical Significance;
  However, the Magnitude of the Errors Was Significantly Higher
  for Interface #3 324
  Throughout the Analysis, Gender Was the Most Often Significant
  Covariate; Other Covariates Can Be Correlated with Gender 324
Summary... 326
References 326

ADVANCED ORGANIZER

The term *intermodal interface* refers to a specific category of human–computer interaction (HCI) artifacts that enable users to work with data in the translation from real-time real-world activity to digital representations that can be aggregated, analyzed, or otherwise operated upon. The utility of intermodal HCI considerations is especially relevant to situations where a great deal of decision-critical information is available only in highly analog and subjective formats. Under such conditions, even high-accuracy digitization may fail to capture important nuances of information that are readily apparent to human analysts. A challenge therefore exists in developing interfaces capable of facilitating the speedy and accurate collection of data that has been initially interpreted by a human user. In this chapter, we introduce the challenges of intermodal interface design, and present an experiment which explores some of these challenges within the domain of emergency services resource assignment. The experimental design also evaluates the efficacy of designing for the dual purposes of collecting real-world data while simultaneously improving the situational awareness of the users working with the interfaces. The case is made that even rudimentary changes to interface designs can have significant impacts on their efficacies as decision-support utilities.

AN INTRODUCTION TO INTERMODAL INTERFACES

This chapter discusses intermodal interfaces, which are a class of HCI systems designed to facilitate the transformation of real-world information into computer-usable data with the intention of transforming that data in some way, so that it can be presented back to users to aid in decision making. Intermodal interfaces address a gap between technology's ability to capture data, and the need for some data to be interpreted and refined anthropologically, prior to being collected, due to subtleties of information meaning that computers cannot handle.

DEFINITION AND OVERVIEW

Note the distinction between the terms *information* and *data* in the aforementioned description. In this work, data are treated as a physical phenomenon, whether that is the movement of electrons, the rendering of magnetic fields, or the polarization of beams of light. An easy way to think about this is that technology is confined to working with data *only*. This treatment of the term *data* pushes the term *information* into the realm of cognitive systems. That is to say, information can only exist within a mind. (And we will leave it to philosophers and artificial intelligence researchers to define the term *mind*.)

With this model in place, many examples can be identified where the simple collection of data is not synonymous with the collection of information. An easy example of this is having a computer work with spoken language. Even if a system could be built, which would provide perfect transcription of speech, there would remain a question of correct interpretation. Reliable interpretation of speech is beyond the current capabilities of technology-only systems, even when reliable transcription is available. People learn from an early age to appreciate the nuances of verbal communication. Even without accompanying visual cues such as body language or spatial referencing (i.e., pointing at something), humans normally develop the ability to use tone of voice, speaking speed, pauses between words, and similar auditory artifacts to modify their interpretation of the actual spoken words. Moreover, people are able to very rapidly incorporate other auditory cues, inferring demographic information such as gender, age, ethnicity, and so on, to further contextualize our interpretations.

Intermodal interfaces form a partial bridge between information available to a human in the environment—what some refer to as the *infosphere* (Floridi 2006), and the data-handling capabilities of technology. The other major component of this bridge is one or more people tasked with the interpretive work needed to capture useful information into a digital format.

Intermodal interfaces might be cast along with all other HCI designs, but for these factors:

1. The role of intermodal interfaces is to collect information-dense data, which is difficult for technology to interpret, for purposes of decision support. This stands in contrast to the body of work dedicated to the presentation and visualization of data for decision support.
2. Consideration must be given for the anticipated decision(s) to be made in order to build-in the facilities needed to support the development of situational awareness as data is collected.

The contrasting role of intermodal interfaces remains when comparisons are made to command and control types of designs, as many of those systems are focused on collecting affective directives from a user rather than collecting situational descriptors. The extent of those directives can be defined programmatically. Although intermodal interfaces may similarly programmatically limit the scope of a user's inputs, and thus their interpretive range, the decision-support purpose is paramount.

For instance, consider information that is presented in an audio format and must be collected quickly in order to be analyzed and/or shared efficiently as part of a decision-making process. The use-case for this example, and the basis for the experiment described later, is the communications within an emergency operations center (EOC). (In the United States, EOC's are often referred to as *911-centers* in reference to the telephone number used to access them.) Many of the verbal messages coming into the EOC must be interpreted beyond the meaning of the spoken words in order to assess the situation and assign appropriate resources.

THE CASE AGAINST STRAIGHT DIGITALIZATION

Ideally, raw data are maintained with full fidelity. In the case of audio messages, this has been achieved through simple analog recording technologies, and more recently by digital means. The digital recordings of audio messages have some significant benefits over their analog predecessors: time indexing becomes trivial, archiving and duplicating recordings has become trivial, and even producing text transcripts from the original audio is becoming more accurate and efficient. However, quickly discerning the accurate meaning of the words that were spoken is still often beyond the technology. This can be due to nuanced use of vocabulary in a particular situation or other contextual factors that may not even be available to the technology. A common example of this phenomenon is when a person is being cynical. Most people are able to pick up on verbal and nonverbal cues from the speaker to inform their interpretive processing and discount the literal meaning of what is said.

In my former career as a paramedic, I learned the importance of listening to how something was said as well as what was being said. In many cases, what might be considered to be noise, was actually a critical part of the signal. One classic example presented to medical providers is a patient who is hyperventilating. A person who is hyperventilating is—literally—breathing too fast. Counterintuitively, many of these people will report that they cannot get enough air and that they are not able to breathe.[*] It occurred to me that a voice recognition system, listening to the words of a hyperventilating person, would not be able to recognize the true problem; it would have to recognize the rate of breathing as an important contextual component of the situation.[†]

There is also a concern with information retrieval speed. Although time-indexed digital recordings of audio messages may afford a user the ability to quickly jump to a specific part of an exchange, that piece of audio would still have to be heard and interpreted in real-time. Furthermore, if the message is only one in a sequence (and there might be several independent parallel sequences), the full context necessary to accurately resolve the meaning of the message might not be under consideration, leading to misinterpretations.

The importance of fast recognition of context is supported by another anecdote from my paramedic career. Emergency providers will often have their radios scanning to listen for activity by other agencies in the area. One night, my crew and I overheard a neighboring ambulance service dispatched for an automobile accident on a nearby interstate highway. We listened to the crew respond that they had arrived at the accident scene. We recognized the voice of the paramedic who was on the call. He was one of the most experienced medics in the area, and had a hand in training many of the emergency medical providers in the region. He had a reputation for being extremely calm and analytical under pressure. When he called over the radio to tell the hospital that he was on the way to them with a patient, he ended his report with the calmly spoken words "… be ready; this is really bad." Such sentiments are often felt, and occasionally voiced, but to hear this guy say

---

[*] The actual problem in this situation is an acute drop of carbon dioxide in the bloodstream; oxygen levels are usually higher than usual. The real danger is that the changes in blood chemistry due to the lack of CO_2 will cause the person to pass out and sustain an injury as they fall down.

[†] The visual cues which accompany hyperventilation would also be relevant, but would extend this example beyond the scope needed here.

something like this, no matter how calmly, told everybody listening that it was really, REALLY, bad. This led me to consider that even if speech-recognition could be extended to effectively consider attributes such as tone and speed of speech as factors in inferring the context of the words, the system could still be very inaccurate. The excited voice of a rookie saying *it is bad* is not the same as the calm voice of a veteran saying exactly the same thing.

The idea of an intermodal interface occurred to me at a time when some of my colleagues were discussing multimodal interfaces, designs which supplemented traditional mouse and keyboard inputs with voice and/or gesture recognition. In the instructions for using such interface enhancements, the need for environmental control was often discussed. For example, voice recognition was easily fouled by background noise; the sound environment had to be controlled. Similarly, hand gestures had to be contained within a fixed area monitored by the computer, with a static background. Precalibration and other methods must have been employed to allow the signal to be extracted from the noise. The focus of many of the multimodal interface projects I heard about was for directing the actions of a computer system. Voice commands and hand gestures were being used to tell a computer what to do. It seemed that these alternative input modalities would be extended beyond providing direction to include more generic data capture. From there, the idea of an influx of uncontrolled multimodal input lead into the consideration of the potential chaos coming from mass technologic misinterpretations. A stepping-stone was needed between the ambiguity of full fidelity real-world information and the organized and controlled environment of the computer's database.

RELATIONSHIP WITH NATURALISTIC DECISION MAKING AND SITUATIONAL AWARENESS

One of my main concerns with the emergency response management information systems investigated in my research was their use primarily for information capture with little, if any, real-time utilization for decision making. The realm of emergency response coordination implies an obviously unpredictable (naturalistic) environment. Any system, which behaves primarily as an information sink without providing any significant decision-making assistance, is easily seen as a burden to efficient real-time management.[*] The need for rapid intermodal data capture combined with the need for technology-supported decision making, particularly in new or atypical circumstances in emergency management is obvious. I therefore concluded that the fields of naturalistic decision making (NDM) (Lipshitz 1995, Klein 2008) and ecologic interface design (EID) (Vicente 1999) had much to offer to the considerations for any new information interface related to emergency response management. However, the underlying assumption of *good* decision-making capability is the acquisition and maintenance of accurate and timely situational awareness (Endsley 2000).

---

[*] The information captured regarding emergency incidents was most commonly used for long-term planning purposes. There was very limited (if any) utility in real-time decision making.

APPLICATION DOMAINS

Although I have concentrated on the work domains involved with emergency services communications, and operations management, the use of intermodal interfaces is appropriate in any situation wherein information is provided in a format that requires some form of anthropologic interpretation from its raw presentation.

DESIGN CHALLENGES

The anecdotes presented earlier are to demonstrate that intermodal transformations are often much more complex than a simple digitization of a single primary factor, such as the textual representation of speech. This necessary complexity is at the core of the requirement for intermodal data collection systems to involve both human and technologic components. The anthropologic parts of the system should be optimized to handle the subjective interactions of signal, noise, and context. The technologic parts can be tasked with objective data manipulation.

There are many challenges facing the designers of intermodal interfaces that extend beyond the typical ease-of-use layout considerations. The design considerations must be drawn from analysis beyond the obvious *what information is more or less likely to be encountered* to include some measures of the likely priority of the information if/when it is encountered. A simple, though dramatic, example of this consideration can be found in the activation controls for emergency ejection seats in military aircraft. One hopes to never use those controls, but when you need them, you need them immediately.

EXPLORATION BY EXPERIMENT

As a foray into intermodal interface design, an experiment was devised based on the activities that take place in a typical EOC.[*]

EXPERIMENT OVERVIEW

In this experiment, subjects were given audio messages emulating those received by a typical emergency communications center. The messages described the occurrence of various emergency events (e.g., robberies and automobile accidents)[†] within a selection of neighborhoods, and the resulting sets of activities completed by the assigned emergency services resources. The messages were scripted to report each stage of each resource (e.g., police, fire department, and emergency medical services) working through its assigned event (e.g., enroute to the incident, arriving at

---

[*] "Emergency Operations Centers" (EOC's) and "Emergency Communications Centers" (ECC's) are synonymous for the purposes of this work. Both are often referred to as "9-1-1 centers" in the United States.

[†] A set of seven different incident types were specified, each type requiring a unique combination of emergency resource types (police, fire, and medical). For example, a robbery required only a police response, whereas an automobile accident required all three services to respond.

the scene, and returning to stand by). The audio was prerendered by a text-to-speech software utility utilizing a high-resolution American English voice model. The timing between messages was controlled to provide a *slow* period of activity (messages arriving approximately every 10 seconds), which would then ramp up and sustain a period of *fast* activity (messages arriving approximately every 4 seconds). Subjects were selected to use one of three graphical user interface (GUI) variants with which to capture the activities reported to them in the audio messages. The GUI variants all had the same basic arrangement of screen elements representing neighborhoods, resources, incidents, and so on, but had some differences that were the focus of the experiment.

Interface #1 was text-only (though still utilizing GUI *point-and-click* input). Interface #2 had the same screen layout as interface #1, but had parts of the screen shaded in different colors corresponding to the different types of emergency service: police, fire, and medical.* The third interface variant was identical to interface #2, with the addition of a text-output area where transcripts of the audio messages would appear as the messages were heard. This was effectively a simulation of a high-accuracy and high-speed voice recognition system (Figures 16.1 through 16.3).

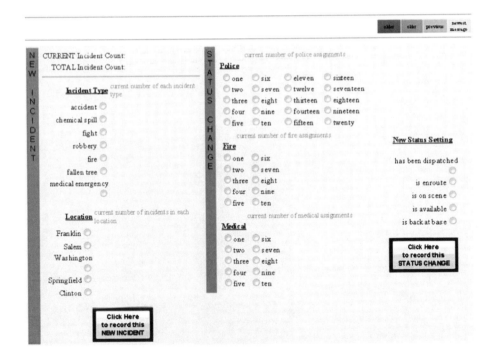

FIGURE 16.1   Screen-shot of interface #1 (no status data displayed).

---

* Any form of color blindness or other uncorrected problems with vision were disqualifications from participation, as was any problem with hearing or any experience with emergency services dispatching.

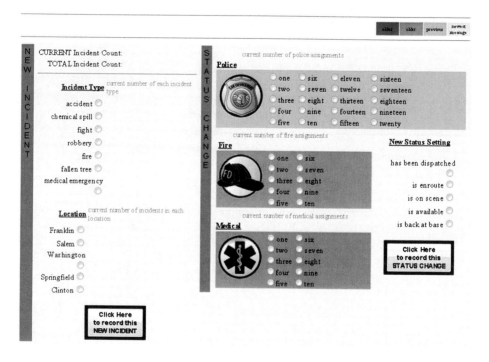

FIGURE 16.2   Screen-shot of interface #2 (no status data displayed).

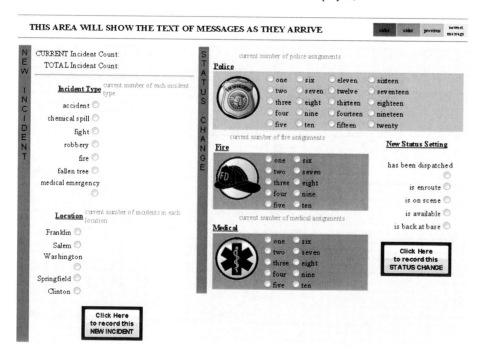

FIGURE 16.3   Screen-shot of interface #3 (no status data displayed).

INTERFACE REQUIREMENTS AND EXPERIMENTAL HYPOTHESES

Primary: Quick and Accurate Collection of Real-Time Intermodal Data

The primary purpose of the system being tested was the quick and accurate collection of real-time information arriving via vocal messages into a structured database. It was hypothesized that even the minor visual differences between the three interface variants would have an impact on the timing and accuracy of the intermodal process. The impact of the voice-to-text transcription simulation on the third interface was a point of particular interest, as it could have been used as a verification of the message that was heard, hypothesized to result in a slower response time, but better accuracy.

Secondary: Development and Maintenance of Situational Awareness

The secondary purpose of the interfaces being tested was the establishment and maintenance of situational awareness by the user. Several research programs have addressed aspects of situational awareness applied to decision making. Boyd's Observe-Orient-Decide-Act (OODA) loop framework (Boyd 1996) is one of the best-known and often cited. The stimulus-hypothesis-option-response (SHOR) model (Wohl 1981) is very similar, but there is no indication that Boyd and Wohl were aware of one another's work (Grant and Kooter 2005).

The definition of situational awareness provided by Endsley was adopted for this work. Endsley's model of situational awareness provides for a decomposition into three well-defined levels, which support a well-tested objective evaluation methodology: The situation awareness global assessment technique (SAGAT) (Endsley 1988, Endsley et al. 1998). SAGAT calls for a simulation to be paused at a random point, have the informational displays blanked, and then query the subject's recollection and understanding of the frozen situation across the three levels of situational awareness: (1) perception of elements in the current situation, (2) comprehension of the current situation, and (3) projection of the future state of the situation.

The simulation system driving the experiment was configured to deliver a partial SAGAT assessment at a random time once the fast message delivery period had been running long enough to establish a moderate level of situational complexity.[*] Only a partial SAGAT assessment was possible as the timing and sequencing of events was controlled by the simulation to be sufficiently randomized that participants could not predict even the very near-term future, which could have impacted the speed and accuracy of their data entry. Therefore, no assessment of level-3 situational awareness (projection of the future state) was made.[†]

---

[*] A quantitative proxy for the concept of "situational complexity" was established for this experiment. The contributing factors were the number of concurrent incidents within the simulation, as well as a short-term average of the message arrival rate.

[†] An additional factor that led to the exclusion of an assessment of level-3 situational awareness was the time constraints on each simulation session. Had the simulations been extended in time, allowing individual resources to work through several cycles of activity, a subject might legitimately be able to make a determination of how much longer an incident was likely to last or perhaps which would be the next resource to clear their assignment and become available for another.

Within the three interface designs being tested in this experiment, indicators were placed to show the number of incidents per neighborhood, number of committed resources, number of current incident types, and so on. These indicators would be updated by the simulation, and so always contain accurate information, even if it had not been properly recorded by the subject. On the first interface type, which was the nongraphical variant, the situational information would be displayed as simple numbers. On the other two interface variants, the information would be displayed as numbers supplemented by small bar graphs. The experimental hypothesis was that the bar graphs would support better situational awareness, but that the text transcriptions of the audio messages, which were visible only on, interface #3 might counteract this effect.

FINDINGS

Undergraduate students were the primary source of subjects, although about 20% of the pool were over the age of 24. Seventy-three percent of the subjects were male. Eighty-three percent were in science/technology/engineering/math (STEM) majors. In all, 112 subjects' simulations were used for analysis, generating 6664 data entry events for timing and accuracy analysis, and 1662 situational awareness assessment data points.

Each of the 112 subjects received 72 audio messages, which would optimally result in 8064 data entry events. Of the 6664 entries (82.6% of the optimal number) that were recorded, only 5339 (80.1%) could be matched to an audio message delivered within the prior 30 seconds, and so could be used for timing and accuracy assessment.

There is a Significant Effect on the Speed of Data Entry Correlated with the Different Interface Design

A series of analyses of variance (ANOVA's) comparing the timings across the three interface variants at the slow message arrival rate showed that the time needed to enter data was significantly reduced with the interface supplements of color-coded backgrounds and icons, as well as the text transcriptions. ($p = 0.0494$) However, the significant difference in timing was not evident when the messages arrival rate was increased. ($p = 0.6076$, Figure 16.4).

The Validity of the Data Collected Quickly Is Maximized in Interface #2

Accuracy of the data entries were measured in two distinct ways: entries that could be matched to audio messages delivered in 30 seconds prior to the entry event and errant entry events that could not be matched. A gross analysis comparing the accurate and errant reporting rates across the three interface variants was run, finding a slightly improved accuracy rate for the interface, which included text transcriptions. All other comparisons were flat, with accuracy rates in the mid-50%'s and errant reports in the upper 20%'s. The remaining percentages for each interface represent audio messages, which were not reported at all (Figures 16.5 and 16.6).[*]

---

[*] The slightly improved accuracy of interface #3 was not statistically significant in this gross (all data per interface, without timings broken out) analysis. Other tests of significance were not assessed on these values in favor of the more detailed analysis to follow, with the message timings being included.

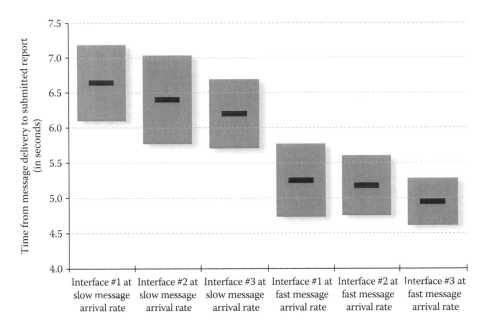

FIGURE 16.4 Session aggregated response timing (in seconds) across experiment conditions; mean values are surrounded by 95% confidence intervals.

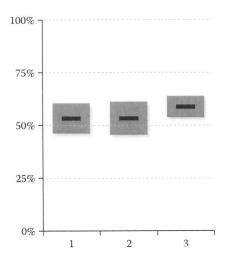

FIGURE 16.5 Accurate reporting rates (based on announcements) by interface.

The percentage of *active errors* was used as the primary metric of accuracy. Active errors were defined to be the data entries, which could not be matched with an audio message that had been delivered in 30 seconds prior to the entry event. This metric excluded errors of omission, might have been explained simply by the pace of the arriving messages.

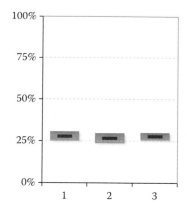

FIGURE 16.6   Errant reporting rates (based on announcements) by interface.

The percentage of invalid reports during the *slow* period was minimized (though not significantly so) with the third interface variant, whereas in the *fast* period, errors were minimized (this time with statistical significance: $p = 0.0049$) using the second interface variant. The difference between the second and third interface designs is the addition of a text transcription of the audio messages. A possible interpretation is that during the slow period, participants were able to verify the messages using both the audio and the text, whereas at the fast rate of message arrival they struggled to continue to use both, a strategy which backfired and caused more errors (Figure 16.7).[*]

The Rate of Errors When Measuring Situational Awareness across the Three Interfaces Was Just under Statistical Significance; However, the Magnitude of the Errors Was Significantly Higher for Interface #3

The SAGAT measures across the three interface variants did not show a significant difference in the number of accurate SA-Level-1 reports ($p = 0.0580$). However, the magnitude of errors, when they were made (roughly 25% of the time) was highly significant ($p < 0.0001$). Interface #3, with the transcriptions of the audio messages showed a near doubling of the average difference between the reported values and the actual values (Figure 16.8).

Throughout the Analysis, Gender Was the Most Often Significant Covariate; Other Covariates Can Be Correlated with Gender

Throughout the analysis of the collected data, the most common significant covariate was gender. Other covariates that appeared are all commonly correlated with

---

[*] Covariate analysis showed gender to be a highly significant factor ($p = 0.0002$). Subsequent analysis showed that both genders were more accurate with the addition of graphical cues to the interface, but the improvement was much more dramatic for males. However, the addition of the text transcriptions in interface #3 negated the improvement for women, nearly doubling the error rate.

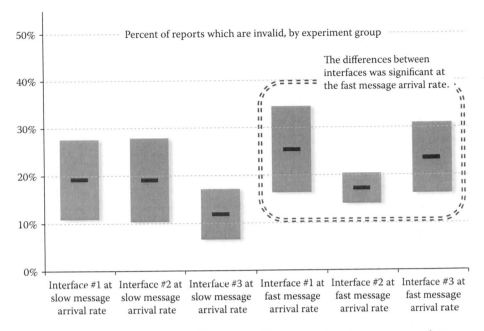

FIGURE 16.7 Percent of reports which are invalid, by experiment group; mean values are surrounded by 95% confidence intervals; the *fast period* measures are significantly different.

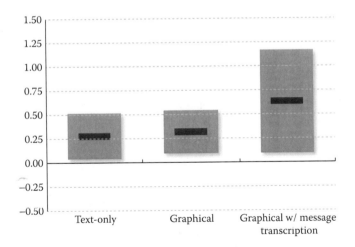

FIGURE 16.8 Mean values surrounded by 95% confidence intervals for level-1 situation awareness error magnitudes (Reported values minus actual values) across interface variants.

gender: time spent playing video games, cognitive style (a measure of brain-side dominance), and so on. However, the size and composition of this study's sample group (112 valid participants, 27% female, and 77% STEM majors) call any would-be conclusions regarding gender into question.

SUMMARY

Intermodal interfaces are part of the stepping-stone process that allows real-world meaningful information to be collected, transmitted, stored, analyzed, or otherwise used to make actionable decisions. In work, domains where the timeliness of such activities is significant, such as in emergency response command and control, the speed and accuracy of this data collection has been shown to improve experimentally by interface design variations that may seem inconsequential at first glance. Furthermore, careful selection of interface components can be used to facilitate better situation awareness, which logically thereby should lead to better decision making. The development of intermodal interfaces should be grounded in work domain analysis with consideration of the informational demands of NDM. These requirements bring intermodal interfaces squarely within the realm of cognitive systems.

REFERENCES

Boyd, J. R. (1996). *The essence of winning and losing.* (unpublished lecture notes).

Endsley, M. R. (1988). *Situation awareness global assessment technique (SAGAT).* Aerospace and Electronics Conference 1988, NAECON 1988, Proceedings of the IEEE 1988 National. Dayton Convention Center, Dayton, OH.

Endsley, M. R. (2000). Theoretical underpinnings of situation awareness: A critical review. In M. R. Endsley & G. D. J. Mahwah (Eds.), *Situation awareness analysis and measurement.* Mahwah, NJ: Lawrence Erlbaum Associates.

Endsley, M. R., Selcon, S. J., Hardiman, T. D., & Croft, D. G. (1998). A comparative analysis of sagat and sart for evaluations of situation awareness. *Proceedings of the Human Factors and Ergonomics Society Annual Meeting, 42*(1), 82–86.

Floridi, L. (2006). A look into the future impact of ICT on our lives. *The Information Society, 23*(1), 59–64.

Grant, T., & Kooter, B. (2005). *Comparing OODA & other models as operational view C2 architecture.* Proceedings of the 10th International Command and Control Research Technology Symposium. Washington, DC.

Klein, G. (2008). Naturalistic decision making. *Human Factors: The Journal of the Human Factors and Ergonomics Society, 50*(3), 456–460.

Lipshitz, R. (1995). Converging themes in the study of decision making in realistic settings. In G. Klein, J. Orasanu, R. Calderwood, & C. E. Zsambok (Eds.), *Decision making in action: Models and methods* (pp. 103–137). Norwood, NJ: Ablex Publishing.

Vicente, K. J. (1999). *Ecological interface design: Supporting operator adaptation, continuous learning, & distributed, collaborative work.* Conference on Human Centered Processes–HCP'1999, Brest, France.

Wohl, J. G. (1981). Force management decision requirements for air force tactical command and control. *IEEE Transactions on Systems, Man, and Cybernetics, 11*(9), 618–639.

17 Bridging Theory and Practice through the Living Laboratory

The Intersection of User-Centered Design and Agile Software Development

D. Benjamin Hellar

CONTENTS

Introduction 328
Living Laboratory and User Experience 328
    Identifying the Problem 329
    From Knowledge Elicitation to Sense-Making 330
    Mitigating Risk through Personas 331
    Building the Product Vision 332
    Mock-Ups and Prototypes 333
Living Laboratory and Agile Development 333
    Sprint Planning 336
    Development 336
    Sprint Review 337
    Evaluating User Feedback 338
Conclusion 338
Review Questions 339
References 339

ADVANCED ORGANIZERS

This chapter describes the challenges and benefits of applying the living laboratory framework to the software development process common in government work. The profession of user experience (UX) is discussed as it relates to the discipline of cognitive systems engineering. This chapter continues with a deeper look into the tools and techniques of the UX professional and how they relate to the living laboratory framework. Next, the chapter presents the agile methodology for software development and specifically details the steps of the agile scrum framework. Finally, it concludes with a short discussion on the benefits of an integration of the living laboratory with agile software methods.

INTRODUCTION

In the context of government work, one of the current buzzwords in the industry is *Big Data*. *Big data* is a general term used to describe the set of problems that arise when humans try to perform sense-making and decision making on vast quantities of data (Van Dijck, 2014). In addition to the basic throughput and performance problems of system engineering, what makes *Big Data* such a ripe problem for human factors is determining the critical insight that is necessary for decision makers to achieve situational awareness in their domain.

In the domain of intelligence community, the problems of *Big Data* are exacerbated by two factors. First, the collection of intelligence is constantly growing—data are collected continuously from a variety of disparate sources and methods. Second the requirements for solutions are tightly coupled to their missions, which are, by their nature, highly specific with custom parameters (Hall & McMullen, 2004). Together these factors mean that building a solution for *Big Data* problems, requires an approach that is both agile to the changing technologies and methods of data collection, and tailored to the specific needs of the problem domain.

This chapter explains how the living laboratory framework can be used as a cognitive systems engineering approach to solving *Big Data* problems and how it works in tandem with agile software development methodologies. This chapter will show how the living laboratory framework enables cognitive system engineers to define, bound and decompose problems in order to develop innovative solutions. These solutions aim to provide the critical insight that enables decision makers to view the most salient aspects of a situation. Mastery of this domain results in the ability to provide a timely, actionable picture to users helping them to do their job faster, and to the best of their ability. In the domain of the intelligence community, this can mean the difference between life and death.

LIVING LABORATORY AND USER EXPERIENCE

At the heart of the living laboratory framework is the idea of problem-centered learning. Cognitive systems engineers use the living lab framework to study problems that emerge from practice and represent issues that regular workers face in their workflow (McNeese, 1996). The living laboratory framework provides cognitive systems

TABLE 17.1

A Brief Description of the Living Laboratory Framework

Method	Description
Ethnographic study	Observation of the problem domain, analysis of the user's workflow and environment
Knowledge elicitation	Interviews with subject matter experts to further understand the problem domain
Scaled worlds	The initial concept of a workflow that simplifies the task environment for empirical validation
Configurable prototypes	The instantiation of the workflow concept deployed into the original problem domain

engineers the methods for understanding the problem space. The methods of the Living Laboratory Framework are briefly described in Table 17.1, with more detailed descriptions found in Chapter 1.

In modern software design, cognitive systems engineering and the use of the living laboratory framework is performed by the UX professional. In its relation to cognitive systems engineering, UX is a subset of the applied skills with focus on usability engineering and human-centered design. A UX engineer functions as a member of a product development team, whose role is to consider the holistic UX of a product, from its inception as a concept workflow, to its final graphical design and usability (Garrett, 2010).

The UX engineer does not work alone, but works in tandem with other members of a product development team including but not limited to software developers and testers, communications and marketing personnel, and various managerial stakeholders. The UX professional is responsible for designing the product interface and working with the engineering team to implement the specifications. In this way, a UX professional is often referred to a *team of one*, serving multiple roles, and acting as a liaison from the customer and stakeholders to the software development team (Buley, 2013).

Although the scope of work is narrower compared to the larger domain of cognitive systems engineering, the application of the living laboratory framework is still relevant to the process of the UX professional. Modern software development requires a collaborative effort between stakeholders, users, UX professionals, and software developers that is aptly described through application of the living laboratory framework. Table 17.2 summarizes the areas of the living laboratory framework as seen through the lens of a UX engineer working in a modern software development environment (Table 17.2).

The remainder of this chapter will explore each of the main areas of the living laboratory framework as it pertains to the applied work of UX professionals in a software development team.

IDENTIFYING THE PROBLEM

In the living lab framework, discovery and investigation of the problem space occurs within ethnographic study. Cognitive systems engineers are meant to carefully observe and study the ways in which people work in their environment

TABLE 17.2

The Living Laboratory Framework Applied to UX Workflow

Living Laboratory Method	UX Workflow Equivalent	Description
Ethnographic study	Persona development	Developing profiles of users to better understand their individual use cases
Knowledge elicitation	Product vision	Vetting the needs of the stakeholders and defining the ideal end state
Scaled worlds	Mock-ups	The initial concept of the workflow and the *dreamvision* design of the product
Configurable prototypes	Agile software development	Agile software implementations based on the mock-ups

(Woods & Hollnagel, 2006). Analysis of this research identifies key problems that can be further refined through knowledge elicitation with subject matter experts.

In government work, especially those associated with the military, the hierarchical nature of the organization dictates that problems are identified at the top and pushed downward (Ender, 2006). Although grassroots discovery and skunkworks efforts for reporting problems do exist, in the end, the acknowledgment of the problem occurs separate from the actual user environment. In software development, products begin development once the business contracts are signed, regardless of whether the problems identified are well defined or not.

This presents a dual-faced challenge to the UX professional on the software development team: Not only must they design a solution that addresses the problem as contracted, but they must also reinvestigate the problem space to verify the nature and validity of the original problem. This challenge also illustrates the key difference between the theoretical study of the living laboratory and its application in the government work—for researchers, the study occurs *before* the potential solutions have been identified. For UX professional, the earliest opportunity to study the problem space occurs *during* the development of a software-based solution.

Given additional access restrictions to users based on the sensitivity of their work, many UX professionals in the government space may only see user feedback *after* the product is deployed, or even worse, some will never receive feedback at all (Buie & Murray, 2012). Ideally once user feedback is available for a developed software solution, the cycle of the living laboratory can be restarted providing further development and understanding of the problem space. To compensate for these barriers, UX professionals rely heavily on the knowledge elicitation performed during stakeholder analysis to achieve understanding of the problem space.

FROM KNOWLEDGE ELICITATION TO SENSE-MAKING

In the living laboratory framework, knowledge elicitation follows ethnographic study as a means for the cognitive systems engineers to better understand the problem space. In this phase, users and subject matter experts are consulted to provide context to the researchers collected knowledge. Techniques such as structured interviews

(Hove & Anda, 2005), concept mapping (Davies, 2011), and scenario development (Obendorf & Finck, 2008) are commonly used to define the issues and use cases that are relevant to the user workflows. This stage enables the cognitive systems engineer to understand the context of user actions in the environment, which will then enable them to hypothesize new methods for optimizing and improving user workflow.

From the perspective of the UX professional, the research performed in this stage is focused on understanding the requirements that have been elicited by project stakeholders. They often start a project on the back foot without the benefit of the ethnographic study to inform their understanding of the context from which these requirements have been defined. To make matters more complicated, given the weft and weave of government hierarchy, the requirements for these projects are constantly evolving, as the needs of the stakeholders change (Vartiainen & Hyrkkänen, 2010). The operating environment of the end users is likewise often dynamic and unpredictable.

In government work, access to end users is limited and often outright prohibited due to security, privacy, or other reasons that negate direct physical or virtual access. Therefore, the elements that inform the sense-making of the UX professional range from the stakeholder that initiated the project, to individual past experiences, to subject matter experts. Each of these elements is fraught with their own inherent biases and issues that need to be identified in the UX process: Mainly, without proper access to users all developed workflows and resultant interfaces risk not being relevant to end users.

As a result, the sense-making timeline that occurs during proper cognitive systems engineering, the time lapse between ethnographic study and knowledge elicitation is dramatically condensed for the UX professional. This problem is exacerbated in business and government work that places an increased focus on cutting costs with shorter term deliverables (Ghani, 2016). To compensate, UX professionals have developed their own strategies and techniques for maximizing the value of human factors analysis, while minimizing the time needed to design effective workflow solutions: developing personas and defining the product's DreamVision.

MITIGATING RISK THROUGH PERSONAS

Cognitive systems engineers develop a theory of practice as a result of the ethnographic study and the knowledge elicitation stages of the living laboratory framework (McNeese, Perusich, & Rentsch, 2000). This theory of practice is a hypothesis of how to improve or evaluate user workflow grounded by the knowledge gathered in the previous stages. This theory then serves as guidance in the further construction and modeling of the subsequent scaled-worlds simulation.

Similarly, UX professionals build theories and models called *personas* that represent who the users are and how they benefit from the various workflows. A persona is a profile representation of a user that provides insights into the user's behavior, motivation, attitudes, challenges, and goals. Personas are portrayed as a *slice of life* for an individual user, but they are designed to represent an aggregate of your intended user population (Ward, 2010). This allows for the UX professionals to empathize with the user to better understand the *why* behind their actions, choices, and decisions. A typical project will have multiple personas that represent the different demographics of the user population (LeRouge, Ma, Sneha, & Tolle, 2013). This is a hermeneutical process that requires iteration and adaptation. As the UX professionals understanding of the user community evolves, new personas are created and old personas are updated or discarded as needed.

This model of the user community, the collection of personas, is also referred to as a persona matrix. Personas matrices enable UX professionals the ability to compare personas against the list of stakeholder's product requirements, features, and user stories. In practice, persona matrices are excellent tools that can be used to evaluate and synchronize the priorities of the stakeholders with the UX professionals and the rest of the software development team (Gualtieri, 2009). Based on the feedback from a project stakeholder, the UX professional can pivot their design to focus on the primary user and use cases of the system, while keeping in mind the secondary users and use cases that may become important later on. This important foundation enables the UX professional to have grounding needed to build further theories of user workflow and produce a coherent product vision.

Overall the persona development process provides the same primary insight that is gained by the cognitive systems engineer in their sense-making process—context. Understanding the context of a user's decision-making process allows for the UX professional to understand the user workflow and ultimately design better methods to improve the user's workflow.

BUILDING THE PRODUCT VISION

In addition to understand the needs of the users through personas, UX professionals must also understand and vet the needs of the stakeholders through defining the product vision. The product vision is a focus on the end state—the goals and concepts that the stakeholders are trying to reach that best match the needs of the users (Pichler, 2009). This vision is not bound by specific technology solutions or current task constraints as it represents the ideal *dream* state of the user workflow. In this author's practice of the UX profession, I refer to this colloquially as the DreamVision.

The DreamVision often takes the form of high-level storyboards, marketing slick sheets, or concept diagrams that aim to create a shared understanding of the solution space among the primary stakeholders (Sutcliffe, 2016). These documents also serve the stakeholders to get buy-in from their managers or other dependent organizations. Socializing the DreamVision is also important to other actors of the software development team, such as testers, tech-writers, system administrators; as it deemphasizes the focus on specific technologies and helps the team better understand how the product will provide a meaningful interaction to end users (Moffett, 2014).

In the age of modern software development, where there is an emphasis on short-term deliverables, the DreamVision provides the UX professional a method to consider the larger implications of their design and not be stuck in a frame by frame mindset. Similar to how the persona matrix helps UX professionals understand the impact of features on the user base, the DreamVision helps the UX professional understand the impact of a feature on the overall design of the user workflow. A focus on the DreamVision contextualizes new features that are developed, minimizing the needs to redesign the product every time a new feature is conceptualized, requested, or demanded.

MOCK-UPS AND PROTOTYPES

In the living laboratory framework, a scaled-world's simulation represents the development of the theory of practice. These scaled-world simulations become synthetic task environments that allow cognitive systems engineers to empirically test their theories within an actual context of use (McNeese et al., 2005).

For UX professionals, mock-ups and storyboards are designed to model individual tasks and workflows that can be used to gauge validity early in the development process. The mock-ups represent the UX professional's theory of how a user's tasks should be as understood through personas and defined in the DreamVision.

For the UX professional, mock-ups can be developed in either low fidelity or high fidelity, depending on scope of the problem and the precision of the original concept (Peres et al., 2014). A low-fidelity mock-up is a rough sketch of a feature that focuses on workflow and information architecture and allows the UX professionals the ability to explore multiple options quickly and easily with little risk for wasted time. High-fidelity mock-ups, in contrast, are precise images that present a clear picture of the intended task structure and allow the UX professional to focus on the usability and accessibility of the product.

One of the benefits of creating mock-ups is the ability to gauge the validity of the suggested tasks and workflow prior to software implementation. A common technique used by UX professionals in this process is known as *paper prototyping*. In paper prototyping, the mock-ups are printed on paper, and users (or stakeholders) are asked to *click* through the interface as if they were using an actual input device (Hartson & Pyla, 2012).

Paper prototyping provides insights into the expectations of users, and identify early usability problems and fundamental issues in an information architecture (Hartson & Pyla, 2012). Techniques like paper prototyping also save time and money for the development team, as not all mock-ups make it to the software development phase, where once implemented, it can be costly to reverse course.

MIDWAY BREATHER: STOP AND REFLECT

- A UX professional relies on an interdisciplinary skill set comprised of knowledge of human factors, graphics design, front-end software development, psychology, and qualitative research methods.
- UX professionals and cognitive systems engineers both share a goal of developing human-centered solutions. Through this lens, the tools and techniques of the UX professional can be seen as analogues to the phases of the living laboratory framework.

LIVING LABORATORY AND AGILE DEVELOPMENT

In the living laboratory framework, the configurable prototype phase contains the development of the artifact. However as a research methodology, the living laboratory framework does not provide details or guidelines into how development should be done, other than that the resultant output should be flexible and reconfigurable to the demands of the users and researchers. In this way, development can be considered a *black box* that models of task environments are passed into and outputted are useable and accessible prototypes that meets the needs of users.

In practice, engineering a system that is flexible and responsive requires a methodology that is equally flexible to the demands of its users and its stakeholders. Traditional waterfall development methodologies have been criticized for being too slow as the time from formal requirements gathering to implementation can take weeks or months (Abrahamsson, Salo, Ronkainen, & Warsta, 2002). This speed of development then causes further downstream issues with user feedback, where the software remains static until a new build is completed.

In comparison, *Agile* software development is a relatively new set of methodologies that emphasize rapid iterative development with collaborative, cross-functional, and self-organizing teams. Agile development contains a common set of practices for delivering solutions that evolve over time. The following list is a summary of these principles from the agile Manifesto.org website, the original creators of the agile methodology:

The 12 principles of agile development are as follows (Beck et al., 2001):

1. Our highest priority is to satisfy the customer through early and continuous delivery of valuable software.
2. Welcome changing requirements, even late in development. Agile processes harness change for the customer's competitive advantage.
3. Deliver working software frequently, from a couple of weeks to a couple of months, with a preference to the shorter timescale.
4. Business people and developers must work together daily throughout the project.
5. Build projects around motivated individuals. Give them the environment and support they need, and trust them to get the job done.
6. The most efficient and effective method of conveying information to and within a development team is face-to-face conversation.
7. Working software is the primary measure of progress.
8. Agile processes promote sustainable development. The sponsors, developers, and users should be able to maintain a constant pace indefinitely.
9. Continuous attention to technical excellence and good design enhances agility.
10. Simplicity—the art of maximizing the amount of work not done—is essential.
11. The best architectures, requirements, and designs emerge from self-organizing teams.
12. At regular intervals, the team reflects on how to become more effective, then tunes and adjusts its behavior accordingly.

Agile software development methodologies are an appropriate fit for engineering reconfigurable prototypes within the living laboratory framework. Agile development is a proponent of integrated collaborative team, wherein a UX professionals work side-by-side with developer teammates to craft designs that are technologically feasible and meet customer goals (Jurca, Hellmann, & Maurer, 2014). With agile development, feature development cycles are short, allowing teams to remain flexible and work on features that are important as problems arise and needs change (Chin, 2004). Agile also emphasizes communication, and provides a stable framework in which to engage the team on what can and should be accomplished.

From the perspective of a system's engineer, the living laboratory framework fuels the requirement process necessary for development! In agile methodologies, the living laboratory represents the investigation and understanding of the problem space. This is where UX engineer follows the aforementioned described steps to understand the user's context and create a vision of the product and mock-ups of the intended workflow. Software developers take these mock-ups and scale them into actual products through the agile methods of sprint planning—development—and sprint review.

Figure 17.1 illustrates this intersection between the living laboratory framework and agile software development methodologies.

The living laboratory framework, together with agile methodologies provides a user-centered approach to software development. Seen a whole, the living laboratory provides the necessary understanding of the problem space, with methods to examine the user community and construct a vision of the product. Agile methodologies take the resultant workflow schematics and provide the flexibility and rigor in the solution space to ensure deployed software is received into the hands of the user community. In turn, these solutions can be studied and evaluated to further fuel the living laboratory framework.

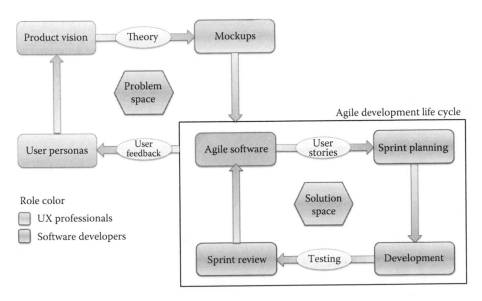

FIGURE 17.1   Where agile development intersects the living laboratory framework.

The remainder of this chapter describes the agile development life cycle in depth following the terminology consistent with the *scrum* methodology of agile development. Scrum is a specific agile methodology that describes each iteration of software development as a *sprint* with individual features described as *user stories* (Schwaber, 2004). Each sprint contains a planning phase where user stories are reviewed and prioritized, a development phase where the software developers create a robust and accessible system, and a review phase where the state of the project is evaluated against the vision of the product.

SPRINT PLANNING

Each sprint begins with a planning stage, where stakeholders consider the impact that individual features (user stories) will have on the user base, while keeping in mind their relative development costs (in terms of manpower and time). The goal of this phase is to reach a common understanding between the project stakeholders, UX professionals, and software developers on what is the minimum viable product (MVP) for this sprint.

The MVP is a user-centered approach to considering which of the proposed features will have the most impact on the user community (Moogk, 2012). To ensure everyone's expectations remain properly managed, the stakeholders define the criteria for when the MVP should be considered complete. These criteria can range from vague and informal, *ready when the feature works* to specific and qualitative *ready when page load performance is less than 3 seconds.*

For software developers, these success criteria inform the type, and amount of testing that will be performed during the development phase (Myers et al., 2011). Testing procedures can vary from feature to feature based on the precision and time-to-deployment demanded by the stakeholders. For UX professionals, these success criteria inform the intended scope of the user workflow. This allows the UX professional time to adjust their mock-ups to better fit the constraints of the current deployed solution, while keeping in mind the larger product vision.

This constant balancing act, between what is best for the users right now, and what is needed in the long term is the ultimate challenge in planning for agile software methods. Software developers need to be able to select the appropriate technology to fit the current problem, while engineering the architecture to scale to system demands in the future. Similarly, the UX professional wrestles with how to optimize the usability for the MVP, while considering the larger DreamVision. When balanced poorly, systems can be short-sighted, and can require significant reengineering.

DEVELOPMENT

In an agile methodology, software development is a collaborative effort that requires coordination between project stakeholders, UX professionals, and software developers. To facilitate better communication, the software developers are collocated and embedded with other operational support members including the UX professionals, technical writers, and system administrators (Jurca et al., 2014). Being in close, daily communication reduces or eliminates the delays often associated with changing requirements. For this reason, one of the tenants of agile software development is the concept of the *daily standup.*

The *daily standup* is a regular meeting that occurs between the members of the team who are engaged with completing tasks for this project (Cohn, 2009). The purpose of this meeting is to synchronize tasking across members of the team, and report any issues or delays that occur for any internal or external reason. The daily stand-up and other informal communication pathways provide a proactive means for resolving issues as they occur, as opposed to *after* they happen. For the UX professional, these meetings provide a method to receive early feedback on interface mockups, eliminating misunderstandings, and improving the design interactions.

During the development process, software developers are able to validate that work is complete by leveraging test-driven development (Beck, 2003). The success criteria defined in sprint planning are converted into unit, integration, functional, and performance tests. These validation tests become key parts of the solution to ensure that stakeholder's needs are met for security purposes, data retrieval, performance, scaling, and monitoring.

Features are pushed to a live production environment, where users are able to access the system as soon as they are complete. During planning and development, communication between the stakeholders and the development team determines when features can be deployed. This ensures that users get their functionality in a timely manner and never have to wait for critical fixes.

Depending on the needs of the stakeholder, a development sprint is complete when either the MVP has been pushed to production or a set amount of time has passed. Some MVPs, depending on their complexity, can take several weeks to complete; as a result checkpoints are needed along the way to verify progress is on target and other are priorities are being met. For small features, that are deployed as they are completed, the development team continues development by pulling feature requests from a backlog of prioritized user stories. Regardless of the size of the features, it is important for the team to periodically review their progress towards the vision of the product and review the efficacies of their processes.

SPRINT REVIEW

Depending on the agile methodologies employed, this review can occur in one of two different meetings—the sprint review and the sprint retrospective. A sprint review is held with the stakeholders of the project, including managers, subject matter experts, and any available power users of the system. The purpose of the sprint review is to demonstrate progress towards the MVP, and receive feedback that could adjust the course of future development (Cohn, 2009).

In contrast, a sprint retrospective is held privately with members of the integrated development team, including the software developers, UX professionals, technical writers, testers, and system administrators. In a sprint retrospective, each member of the team asks themselves the questions of "what went well, what did not? And what can we do to improve?" (Cohn, 2009).

For the UX professionals, these questions facilitate the introspection to evaluate the current user interface in respect to the overall DreamVision of the product.

Do the current mock-ups address the needs and wants of the user community? Can the mock-ups be adjusted to better match the realities of development?

For software developers, the sprint retrospective allows them to evaluate and manage technical risk—sticking with current technologies might seem like the prudent choice, but newer technologies might be available that gives the team more advantage in the long term. The development team needs to research and test the boundaries of new technologies, and gain insights on where the appropriate use is—and where it is not.

Finally, the purpose of the sprint review and sprint retrospective is for team needs to hold each other accountable for the success of the program (Cohn, 2009). UX professionals and software developers need to function as a single cohesive team to succeed. There is no room for cowboy developers or dictatorial UX designers. Adjustments to team attitude, and composition should be made, when team members are found under performing or not cooperating with the rest of the team.

EVALUATING USER FEEDBACK

Once the product is in the hands of the end user, methods for evaluating user feedback can be found in both the living laboratory framework and in agile software development methodologies. First, from the development perspective, once a system is deployed to an operational environment, the performance is continually monitored by the combined development team. If performance in a live environment does not match the expectations set initially by the stakeholders in sprint planning, then development priorities will shift until the demands are met.

System developers and administrators also employ quantitative feedback mechanisms such as in-app metrics and data analytics to help validate the relevance of introduced features. The collected analytics and metrics are valuable to the entire team as a form of limited market research. This research paints a picture of the critical path that users and system agents follow in accessing your system. As a whole, the analytics are able to tell you how your system is accessed, for how long, and by whom (Jansen, 2009). The only question these analytics are not able to answer is *why*?

For that answer, the living laboratory perspective provides insight into user behavior. Once users begin to access the system and provide feedback, the UX professional is able to apply a variety of qualitative research techniques. As outlined in the ethnographic study and knowledge elicitation portion of the living lab, UX professionals can evaluate the qualitative feedback through techniques such as observations, structured interviews, surveys, and focus groups.

CONCLUSION

The living laboratory framework, combined with agile methodologies, provides a user-centered approach to software development. The living laboratory framework excels at providing an in-depth understanding of the problem space. Although the individual notes are different for the UX professional compared to the cognitive systems engineer, the overall rhythm and beat is the same—study the user—understand the user—hypothesize on ways to improve the work of the user.

When it comes to the mission critical needs of government work, rapid development and iteration is the key to solving big problems. Agile software development methodologies provide the answers to the black box question of the living laboratory framework—tools and techniques to ensure that the problem space of the user community is addressed in a timely and efficient manner. In turn, the technical requirements introduced through agile development impact the vision of the product as not all ideas are feasible. Being able to adjust the ideal vision of a product, while considering the realities that occur in software development, are what separate those who build products from whose who build *solutions*!

REVIEW QUESTIONS

1. Describe how the profession of UX relates to the discipline of cognitive systems engineering?
2. Describe the challenges and limitations that UX professionals face when working in the government space?
3. How does persona development help the UX professional mitigate risk?
4. Describe the purpose of defining the product vision aka DreamVision?
5. Define the iterative steps of the agile scrum methodology?
6. Where does the living laboratory intersect with agile software development? How does the living laboratory help feed the agile development process?

REFERENCES

Abrahamsson, P., Salo, O., Ronkainen, J., & Warsta, J. (2002). Agile software development methods: Review and analysis. VTT Publications 478. pg 107.

Beck, K. (2003). *Test-driven development: By example.* Boston, MA: Addison-Wesley Professional.

Beck, K., Beedle, M., Van Bennekum, A., Cockburn, A., Cunningham, W., Fowler, M., … Kern, J. (2001). Manifesto for agile software development. http://agilemanifesto.org/

Buie, E., & Murray, D. (2012). *Usability in government systems: User experience design for citizens and public servants.* Burlington, MA: Elsevier.

Buley, L. (2013). *The user experience team of one.* New York, NY: Rosenfield Media.

Chin, G. (2004). *Agile project management: How to succeed in the face of changing project requirements.* New York, NY: AMACOM Division of American Managementt Association.

Cohn, M. (2009). *Succeeding with agile: Software development using scrum.* Upper Saddle River, NJ: Addison-Wesley Professional.

Davies, M. (2011). Concept mapping, mind mapping and argument mapping: What are the differences and do they matter? *Higher Education, 62(3)*, 279–301.

Ender, T. R. (2006). *A top-down, hierarchical, system-of-systems approach to the design of an air defense weapon* (PhD Thesis). Atlanta, GA: Georgia Institute of Technology.

Garrett, J. J. (2010). *Elements of user experience: User-centered design for the web and beyond.* Boston, MA: Pearson Education.

Ghani, I. (2016). *Emerging innovations in agile software development.* Boston, MA: IGI Global.

Gualtieri, M., Manning, H., Gilpin, M., Rymer, J. R., D'Silva, D., & Wallis, Y. (2009). *Best Practices in User Experience (UX) Design.* Cambridge: Forrester Research, Inc.

Hall, D. L., & McMullen, S. A. H. (2004). *Mathematical techniques in multisensor data fusion* (2nd ed.). Norwood, MA: Artech House.

Hartson, R., & Pyla, P. S. (2012). *The UX book: Process and guidelines for ensuring a quality user experience.* Boston, MA: Elsevier.

Hove, S. E., & Anda, B. (2005). Experiences from conducting semi-structured interviews in empirical software engineering research. *Proceedings of 11th IEEE International software metrics symposium (METRICS'05)* (10 pp.). Washington, DC: IEEE Computer Society.

Jansen, B. J. (2009). Understanding user-web interactions via web analytics. *Synthesis Lectures on Information Concepts, Retrieval, and Services, 1*(1), 1–102.

Jurca, G., Hellmann, T. D., & Maurer, F. (2014). Integrating agile and user-centered design: A systematic mapping and review of evaluation and validation studies of agile-UX. In *Agile conference (AGILE), 2014* (pp. 24–32). Washington, DC: IEEE Computer Society.

LeRouge, C., Ma, J., Sneha, S., & Tolle, K. (2013). User profiles and personas in the design and development of consumer health technologies. *International Journal of Medical Informatics, 82*(11), e251–e268.

McNeese, M. D. (1996). An ecological perspective applied to multi-operator systems. In *Human Factors in Organizational Design and Management*, V. O. Brown Jr. & H. W. Hendrick (Eds.) (pp. 365–370). Amsterdam, the Netherlands: Elsevier Science B.V.

McNeese, M. D., Bains, P., Brewer, I., Brown, C., Connors, E. S., Jefferson Jr, T., ... & Terrell Jr, I. (2005, September). The NeoCITIES simulation: Understanding the design and experimental methodology used to develop a team emergency management simulation. In *Proceedings of the Human Factors and Ergonomics Society Annual Meeting* (Vol. 49, No. 3, pp. 591–594). Los Angeles, CA: SAGE Publications.

McNeese, M. D., Perusich, K., & Rentsch, J. R. (2000, July). Advancing socio-technical systems design via the living laboratory. In *Proceedings of the Human Factors and Ergonomics Society Annual Meeting* (Vol. 44, No. 12, pp. 2–610). Los Angeles, CA: SAGE Publications.

Moffett, J. (2014). *Bridging UX and web development: Better results through team integration.* Waltham, MA: Morgan Kaufmann.

Moogk, D. R. (2012). Minimum viable product and the importance of experimentation in technology startups. *Technology Innovation Management Review, 2*(3), 23.

Myers, G. J., Sandler, C., & Badgett, T. (2011). *The art of software testing.* Hoboken, NJ: John Wiley & Sons.

Obendorf, H., & Finck, M. (2008, April). Scenario-based usability engineering techniques in agile development processes. *CHI'08 Extended abstracts on human factors in computing systems* (pp. 2159–2166). New York, NY: ACM.

Peres, A. L., Da Silva, T. S., Silva, F. S., Soares, F. F., De Carvalho, C. R. M., & Meira, S. R. D. L. (2014, July). AGILEUX Model: Towards a reference model on integrating UX in developing software using agile methodologies. In *Agile conference (AGILE), 2014* (pp. 61–63). Boston, MA: IEEE.

Pichler, R. (2009, January 9). *The Product Vision.* Retrieved from the Scrum Alliance: https://www.scrumalliance.org/community/articles/2009/january/the-product-vision

Schwaber, K. (2004). *Agile project management with scrum.* Redmond, WA: Microsoft press.

Sutcliffe, A. (2016). Designing for user experience and engagement. In *Why engagement matters,* H. O'Brien & P. Cairns (Eds.) (pp. 105–126). Boston, MA: Springer International Publishing.

Van Dijck, J. (2014). Datafication, dataism and dataveillance: Big data between scientific paradigm and ideology. *Surveillance & Society, 12*(2), 197.

Vartiainen, M., & Hyrkkänen, U. (2010). Changing requirements and mental workload factors in mobile multi-locational work. *New Technology, Work and Employment, 25*(2), 117–135.

Ward, J. L. (2010). Persona development and use, or, how to make imaginary people work for you.

Woods, D. D., & Hollnagel, E. (2006). *Joint cognitive systems: Patterns in cognitive systems engineering.* Boca Raton, FL: CRC Press.

18 Incorporating Human Systems Engineering in Advanced Military Technology Development

Patrick L. Craven, Patrice D. Tremoulet, and Susan Harkness Regli

CONTENTS

Background 342
Introduction 342
Human Systems Engineering for Technology Development 344
Embedding Human Systems within Systems Engineering 346
    Origin of IDEAS 346
    What Is IDEAS? 347
        Needs Analysis 348
        Requirements Generation 350
        Design and Engineering 350
        Interface Review 351
        Implementation 353
        Evaluation 353
Real-World Application of IDEAS 354
    A Challenging Experience 354
    Successful Application 354
    Using IDEAS with Software Development Methods 355
Conclusion 356
Review Questions 357
References 357

ADVANCED ORGANIZERS

This chapter reviews the work performed by a human systems engineering group that endeavored to incorporate user-centered design best practices into military advanced technology research and development. It begins with an overview of human systems engineering, followed by a discussion of its role in technology research and development (R&D). It then introduces a model for including human systems engineering in new technology development that has much in common with living lab framework, and shares two real-world examples of applying the model with a multidisciplinary systems development team.

BACKGROUND

This chapter relates how and why the interaction design and engineering for advanced systems (IDEAS) methodology was created, and shares lessons learned through applying this methodology in a technology R&D environment. In addition, IDEAS will be compared to the living lab framework in an effort to identify commonalities and best practices.

The authors worked in an R&D lab whose primary customer was the Department of Defense—most commonly the Defense Advanced Research Projects Agency (DARPA) and the military service labs (Office of Naval Research, Air Force Research Labs, and Army Research Labs). Working with DARPA requires creativity because its primary purpose is to prevent strategic surprise through advanced research. DARPA was created during the Cold War in response to Sputnik; the United States did not want to be caught off-guard again (Jacobsen, 2015). Current projects include both explorations of futuristic technologies and development of near-term solutions to challenges encountered during the Iraq War. Some of the work that DARPA funds is directed toward creating concept demonstrations that illustrate the potential for modern commercial products such as smart phones by demonstrating technological feasibility, while spawning novel human–technology interaction paradigms.

Human systems engineering is an important part of many of defense projects. In fact, the military has recently renewed its interest in topics such as human–systems integration and human–autonomy interaction (Dahm, 2010). Despite explicit calls for improved human-centered technology, there are still many cases where these experts are brought in at the tail end of projects, to try to fix designs that were developed without a solid understanding of the users and their work context.

INTRODUCTION

Human factors (or ergonomics) has been defined as

> The scientific discipline concerned with the understanding of interactions among humans and other elements of a system, and the profession that applies theory, principles, data and methods to design in order to optimize human well-being and overall system performance. (International Ergonomics Association)

This field is generally considered to have originated during World War II (Wickens & Hollands, 2000), though it has its roots in the industrial revolution in the early 1900s (Boff, 2006). Prior to this time, humans were trained to fit the machine instead of designing machines to fit humans (Gilbreth & Gilbreth, 1917; Gilbreth, 1914; Taylor, 1911). The early pioneers who looked at human factors in equipment design included researchers under the leadership of Sir. Frederick Bartlett at the Applied Psychology Unit of Cambridge University (Bartlett, 1943; Craik, 1940). With the start of World War II, members of this group switched from studying pilot selection and training to the development of flight instrument design. Many advances in human factors were due to military necessity, such as the widespread adoption of airplanes in combat, which created a need for methods to rapidly select and train qualified pilots (Meister, 1999). The greatest impetus for a change in design philosophy came from the vast number of men and women needed to fight the war, which is estimated at 16.1 million American and 1.9 billion worldwide (World War II Foundation). This large number of warfighters made it impractical to select individuals for specific jobs, and instead there was a shift toward designing for people's capabilities. At the same time, technological advances first outpaced the ability of people to adapt and compensate for poor designs, as evidenced by highly trained pilots experiencing crashes due to problems with control configurations (Fitts & Jones, 1947a) and instrument displays (Fitts & Jones, 1947b).

Military R&D played a pivotal role in technology development, which also fueled commercial development in personal computing. DARPA's investment in network technologies enabled the creation of the Internet (Perry, Blumenthal, & Hinden, 1988), as well as advances in graphics, artificial intelligence, timesharing, and massively parallel processing (Norberg, 1996). This infusion of research paved the way for advancements in commercial capabilities, which were included in new cutting edge defense technologies. The result was widespread use of commercial off-the-shelf (COTS) computers and technology across the military (DoD ESI/Navy Enterprise Licensing Agreements Team, 2014). Computing technology became a critical component of the military's forces, and systems were developed with increasingly complex technological capabilities. However, complexity can lead to confusion, which can have catastrophic results, as evidenced by the U.S. Navy's accidental shooting of Iran Air Flight 665 (New York, 1988) and a similarly tragic outcome for Air France Flight 447 (Echo, 2012).

The definition of human factors emphasizes the importance of studying not just the human but also the *interaction* among humans and systems (Dul et al., 2012; Russ et al., 2013). In the early 1980s with the advent of more widespread computational devices and more users interacting with them regularly, the field of *cognitive systems engineering* (CSE) was pioneered by researchers such as Norman (1981), Hollnagel and Woods (1983). This new field focused on the design and engineering of a *cognitive system*, as opposed to *systems engineering* with a cognitive tilt (E Hollnagel, 2016). In developing or engineering new technologies it is important to note that CSE accepts that the whole may be different than the sum of its parts (E Hollnagel, 2016). This field more strongly focuses on the cognitive rather than physical or mechanical aspects of human factors. It has been defined as the "study of cognitive work and the application of this knowledge to the design and development of technology" (Endsley, Hoffman, Kaber, & Roth, 2007).

Although the authors' work predominately focused on cognitive function, it also included the broader effects of physiological and affective influences. It is for that reason that the work described in this chapter can be most accurately described as *human systems engineering*. The goal of this engineering discipline is to provide equal consideration of the human along with the hardware and software systems as part of the overall technical and management process for systems engineering (ODASD, 2016).

HUMAN SYSTEMS ENGINEERING
FOR TECHNOLOGY DEVELOPMENT

One of the desired outcomes of human systems engineering is the prevention of errors. The consequences of usability errors can range from annoying to severe, but they are real. Three-mile Island was a catastrophe that could have been avoided through the use of a proper human systems engineering process (Meshkati, 1991).

Despite the benefits that human systems engineering can provide, it is still too often neglected. Healthcare.gov is an excellent example that affected more than 8 million users in the first four days of its ill-fated rollout. At its launch, the site attracted five times more visitors than it was designed to handle (Mullaney, 2013). The problem was exacerbated by the site design, which forced users to create an account before reviewing and comparing plans (Bryant, 2013). This laborious account creation process created a bottleneck and increased the number of simultaneous hits to the site. Having a larger number of visitors than expected can affect even the best designed websites, but the account creation process could have been easily identified with user-centered processes. As President Obama describes:

> Part of the problem with Healthcare.gov was not that we didn't have a lot of hardworking people paying attention to it, but traditionally the way you purchase IT services, software, and programs is by using the same procurement rules and specification rules that were created in the 1930s... What we know is, the best designs and best programs are iterative: You start out with, "What do you want to accomplish?" The team starts to brainstorm and think about it, and ultimately you come up with something and you test it. And that's not how we did Healthcare.gov.
>
> It's something, by the way, I should have caught, I should have anticipated: That you could not use traditional procurement mechanisms in order to build something that had never been built before and was pretty complicated. So part of what we're going to have to do is just change culture, change administrative habits, and get everybody thinking in a different way. (Obama, Barack, Pres., 2015)

Even when programs are proactive about including human systems engineering, they may fall victim to budget cuts. For example, in 2008, the federal budget experienced cuts in basic human systems engineering research in aeronautics (DeAngelis, 2008), continuing a 15-year trend toward minimizing human-based research. Although government research efforts in general have been plagued with budget cuts, anecdotal evidence suggests that human systems engineering was cut at a disproportionately higher rate than other work. The reasons for this are unclear, but decisions may be influenced by a perception that human systems engineering is a *soft* science with diminished importance compared to the *hard* science of technology creation. In addition, military and government procurement processes ensure that the end user of the technology is not the same as the individual selecting or developing it. Warfighters do not select the technology they use—rather, they are expected to learn to cope with the design of systems selected by others. In the commercial world, if a product is not usable, it is purchased less frequently by consumers, so the manufacturer has an incentive to design with the user in mind. This effect is reduced if not eliminated outright by the military procurement process.

Some have argued that in system development, there is "a long and successful record concerning the use of training to compensate for poor design" (Hancock & Hart, 2002). People may assume that systems have to be complex and thus will require highly trained operators when, in reality, time and money could be saved by investing in human systems engineering to simplify designs. This logic mirrors the pre-World War I mentality that humans are plenty in number, and that if someone is confused by a system, the person should be swapped with someone more qualified. Training soldiers is not cheap, though, and publicly available information provided by the UK's Ministry of Defense has estimated that basic training for an infantry recruit costs ~£34,000* (United Kingdom Army Secretariat, 2015). Although exact figures were not obtained for training within the United States, one can estimate that it is likely to be similar. The time spent on training should be focused on improving performance of the human system team, and not on humans trying to overcome poor design. If the design problem is fixed, it eliminates the need for training to understand a system and reallocates that time to training to perform well with a system.

Training is also unable to alleviate certain problems, such as inefficiencies. Just because someone is trained to work with an inefficient design does not suddenly make it more efficient. For example, UPS tracking numbers are 18 characters long, and due to their length, human interactions with these numbers are performed almost exclusively with bar code readers because humans have difficulty accurately transcribing sequences of this length (Johnson, 1991). In addition, user satisfaction will not improve through additional training. In fact, the opposite may be true, and users may seek to actively avoid the system altogether and seek alternative ways to meet their objectives.

There are clear benefits to ensuring that system development includes a focus on the user, including maximizing the performance achieved by the human-machine team, improving safety, and reducing inefficiencies. This understanding is gained through an iterative process (Nielsen, 1993). Both the living lab framework and IDEAS are built around the fundamental idea that good system design requires an exploration using a scientific process of theory and empirical discovery. This scientific process hinges on the systematic quantification of humans (Sauro & Lewis, 2012) as exemplified in careful measurement of behavior, cognition, and relevant environmental factors as users work with design artifacts, prototypes, and fully functional systems. Understandings of the technology, the human, and how they interact

---

* $51,411 at the time of estimation.

should evolve together. When they do not, the final product may fall short of the need it was meant to meet.

MIDWAY BREATHER

1. Think of at least two examples from your own experience that illustrate technology designs that failed to consider something about you as a user. What would you have done differently?
2. What might some of the differences be between human systems engineering in (a) the defense community, (b) consumer applications, (c) healthcare applications? What is at stake in these domains?

EMBEDDING HUMAN SYSTEMS WITHIN SYSTEMS ENGINEERING

Once the decision to include human systems engineering is made, one must choose a method to incorporate it. The living lab framework (McNeese, Zaff, Citera, Brown, & Whitaker, 1995), which has provided a common theme for the chapters in this book, is an excellent iterative approach to system design. However, in a collaborative environment where there are individuals with many different backgrounds, it can be challenging for some team members to understand the value of the living lab framework since it was (intentionally) designed from a human systems perspective. In particular, it does not clearly identify points of intersection between more general engineering processes and human factors specific ones. The IDEAS process (Regli & Tremoulet, 2007) creates a shared engineering representation that attends to these intersections.

ORIGIN OF IDEAS

The idea of integrating user-centered design activities into product development is an industry's best practice that has been adopted in a variety of different domains. IDEAS is a refinement of this practice, which specifies how user-centered design activities can be seamlessly integrated into applied technology R&D efforts. IDEAS was originally developed during a proposal effort when an interdisciplinary group with expertise in human systems engineering recognized the value of articulating a process for ensuring that user needs, capacities, and limitations would be taken into account when conducting research aimed at developing new technologies for defense applications. Part of the motivation was to develop a proprietary process that could be used in future proposal efforts but which would be rooted in proven user-centered design best practices. However, it also served as a helpful tool for educating peers inside the company. Because customers tend to value products, not process, it was advantageous to be able to point to a formal methodology that could be applied to support creation of new research products, rooted in theory and evidence. Also, as part of explaining our proprietary process, readers could be educated about why designs or visual concepts were not typically included in proposals as part of describing IDEAS, namely because the needs analysis had not been done nor had the requirements been defined yet.

The creation of IDEAS was led by two individuals who came to human systems engineering from different directions. The first individual had a solid grounding in theory and HCI research methods. She brought extensive skills designing and conducting user research, including experiment design for human in the loop evaluations. The second individual had a history of bringing human factors engineers and documentation and training experts together to found a user advocacy group at a major telecommunications company. She was also an expert in needs analysis, writing requirements and prototyping, as well as educating engineers and management about how user-centered design activities fit into systems engineering of a production development process. The two worked together to create similar materials showing how IDEAS could be applied on behalf of technology R&D efforts. Ironically, a couple of years after IDEAS was initially laid out, the company's Engineering and Technology office funded the group to create materials explaining how IDEAS could be integrated into a Model Based Systems Engineering approach to product development.

As IDEAS was applied, it became possible to contrast programs which included user-centered design activities from the outset to those where *build user interface* was left until a second or third phase. The latter course of action led to less useful, less usable final products and/or huge development rework costs to implement what users needed. It was emphasized that applying IDEAS involves facilitation, a translator between engineers/scientists and domain subject matter experts (SMEs). It is not good enough to simply extract user needs and write requirements, it is also important to expose SMEs to what is technologically possible and bring any ideas that this sparks back to project leadership and customers to determine if project goals may need to be updated.

Inside the company and with some external customers, the biggest hurdle was educating (some of) them on the value and necessity of human systems engineering activities when the final product of a research effort is a prototype and not a commercial product. Several customers understood the need for IDEAS or something like it; but one sat through a 20-minute presentation nodding and smiling and said "this is great, I'm so glad the company has a group working on this, but MY program is developing technology that is SO novel and SO exciting that I do not have to worry about the UI—the users will want my technology no matter what the GUI looks like so it can be built at the end of the project." Unfortunately, he was not a rare exception, as many as 10%–20% of program managers for advanced technology research projects held the same view. On the other hand, for customers who already understood the need for human systems engineering, it was quite easy to get them to buy-in to investing in IDEAS activities.

What Is IDEAS?

Although the *I* in IDEAS stands for *Interaction*, it could just as easily have stood for *Iterative* as the cyclical nature of the process is critical. User-centered methods are designed around a discovery process, and these processes require an opportunity for

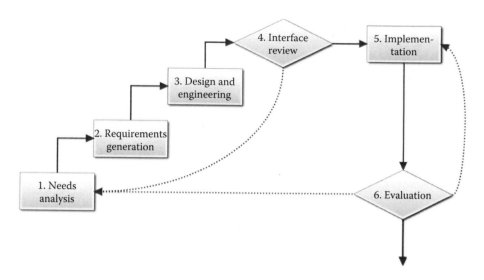

FIGURE 18.1   The six steps of IDEAS.

observing the effect of changes. Because humans are complex and often unpredictable, one cannot know with certainty how they will respond to novel technological solutions. The IDEAS process starts with discovery, where domain-specific knowledge is combined with a researcher's understanding of cognition and behavior, and the cycle ends with an evaluation of a candidate solution. The full IDEAS process (Regli & Tremoulet, 2007) is comprised of six steps, which are detailed in Figure 18.1.

Needs Analysis

Needs Analysis is the first step in the IDEAS process. Its goal is to gather and generate information about the users and the contexts in which they operate. For this step, the desired outcome of is to identify users, goals, tasks, and cognitive processes.

Needs Analysis can be accomplished in a number of different ways and should be scoped for the project's size and budget. It is important to identify domain experts who can act as SMEs. Although SMEs typically have rich contextual knowledge, discussions with them are more focused if the human systems engineer (HS engineer) has developed some level domain knowledge through reading training manuals, concept of operations, or other available materials. In defense work, SMEs are often seasoned in the domain but sometimes are so senior that they do not perform the relevant tasks on a day to day basis anymore. For that reason, it is helpful to also identify representative target users who do perform those tasks day to day, but may lack the breadth of view that the SME possesses.

Using these three sources of information (existing documentation, SMEs, and target users), HS engineers apply validated techniques to develop research artifacts to aid in the next step of the process. Examples of these techniques include:

- *Ethnography:* The etymology of the word comes from Greek for ethnos (people) and *grapho* (I write), and is the systematic study of people and cultures. The technique has been popularized in anthropology, and an important aspect of ethnography is studying people within their natural environments. For an HS engineer, ethnography is an important first step in understanding the *environment* in which individuals work as well as capturing qualitative data about those individuals. It allows the HS engineer to identify important design considerations that may not otherwise have been identified. For example, the National Training Center afforded an opportunity for HS engineers to perform ethnography with Army Companies who were training in an extremely realistic setting simulating parts of the high desert of Afghanistan. The battalion was encamped in the same manner it would be overseas, and each of its companies performed patrol operations in a realistic manner. This realism helped reveal the importance of the volume of equipment that dismounted warfighters carry, the dust of the environment, the onset of fatigue and its effect on performance. All of that was critical for designing a product that would be useful for dismounted warfighters.

- *Semistructured interviews:* This technique blends the concrete questions from structured interviews with the openness of unstructured interviews. The value of being neither too structured nor too unstructured is the interviewees can all respond to a limited set of the same basic questions. The specific questions can target unknowns that the HS engineer needs to better understand. For example, one might ask "on average, how many times a day are you finding yourself waiting for this software to process a response?" Follow-up questions can be generated on the fly to gain further insight and explore the underlying issues. In addition, unbiased probe questions can be used to further expand on a structured interview response, or explore a new topic area. For example, one might ask "could you tell me about how your interactions with this software have gone?"

- *Task analysis:* A task analysis is a valuable artifact to create during the needs analysis process. It serves as a concrete representation of the HS engineer's understanding of users' goals and tasks and what actions they take to achieve them. Two approaches have been established: Hierarchical Task Analysis (HTA) (Stanton, 2006) focuses predominately on system goals, whereas (CWA) (Vicente, 1999) focuses on constraints within the system. Additionally, HTA is descriptive and focuses on how a system should operate, and CWA is formative and highlights how functions could be achieved. Finally, HTA has a more clearly-defined process. HS engineers will need to consider the particular needs of a research project to determine which technique is more appropriate. For a more thorough comparison of these two techniques, see Salmon, Jenkins, Stanton, and Walker (2010).

- *Card-sorting:* This technique was adapted from the field of psychology and is used to study how people organize concepts and knowledge (Gaffney, 2000; Spencer, 2009; Wood & Wood, 2008). It is particularly helpful for developing the information architecture of a system, which is defining the organization, structure, and labeling of content in an effective manner. For example, it may be important to understand how different system actions cluster together with a sort of family resemblance. Commonly, word processing functions like cut, copy, and paste are situated near one another because users perceive them to be similar. Microsoft® Word's menu architecture illustrates groupings of various functions in the Ribbon menu into groups of Insert, Page Layout, Review, and more.

Requirements Generation

At this stage, artifacts created during needs analysis are used to specify requirements for the system. A full description of the requirements generation process is beyond the scope of this chapter, and the generation of requirements should be the result of a collaboration between HS engineers and technical experts in the system domain. Requirements can include specifications about the function, data, environment, and usability (Foraker Labs, 2016). Broadly, requirements are grouped into *functional requirements* and *nonfunctional requirements*. Functional requirements relate to the functional aspect of software. For example, a functional requirement might be the ability to search any incoming alert messages that arrive for the user. Nonfunctional requirements address implicit aspects of software such as storage, performance, security, configuration, and so on.

Zimmerman and Grötzbach (2007) have defined a user-centered requirement framework that includes three types of requirements: (a) usability requirements, (b) workflow requirements, and (c) user interface requirements. Usability requirements are typically nonfunctional requirements that specify the criteria that the system must meet from a usability perspective. The workflow requirements are specified at a level that relates the user goals and they specify how the system should support the user to achieve those goals. Essential interaction steps, required information, and commands are specified at this level but without the level of concrete details in the user interface requirements. The user interface requirements come from a synthesis of usability and workflow requirements, and this is where the information architecture, navigation, and wireframe elements are initially defined. User interface requirements, when fully specified, begin to blend with design artifacts listed in the next section.

A defined set of requirements is more commonly associated with the waterfall and spiral software development methods. Agile methods such as scrum and extreme programming use a different process that focuses on user stories rather than rigidly defined requirements. A discussion of methods for applying a human systems engineering method within an agile software development process is described by Düchting, Zimmermann, and Nebe (2007). Briefly, the authors suggest that the brief exploration phases prior to development should be used for HS engineers to provide the development teams with a rough set of inputs to help explain the users natural work environment. If one works within an agile environment, it is important to adapt the artifacts and not insist on generating comprehensive documentation of the results, but rather emphasize lightweight artifacts and strive to quickly share essential knowledge with the team.

Design and Engineering

Design is fundamentally a creative endeavor, constrained in a beneficial way by the detailed work invested in needs analysis and requirements generation. Undoubtedly, seeds of design are sown during the earlier steps, but it is at this time that initial tangible artifacts are created. Design might start with simple sketches of concepts that started to form, or perhaps a rough wireframe of where system components might end up. The requirements help reduce the potential variability. For example, if one is designing an Android app, the built-in navigational commands and user interface guidelines serve as valuable constraints.

A designer is no longer a single specified role, but rather it has morphed into a team consisting of software engineers, HS engineers, as well as other stakeholders who share a *participatory culture* (Sanders, 2002). In this participatory design process, folks use tools such as Balsamiq,[*] proto.io,[†] or Axure[‡] to provide the ability to quickly adjust designs based on iterative testing. Design concepts can often be shared remotely through web-based platforms, and some even offer means to explore the wireframe or mock-up on the targeted technology system. One can model both the static representation as well as the interaction among system states. In general, the tangible artifacts that can be produced at this stage include both static and interactive versions of the following:

- *Wireframe:* A wireframe is a low fidelity representation of a design that serves as an initial team-wide medium for communicating about the product as well as for performing reviews of the initial design concepts. The wireframe should represent each of the pieces that will be in a fully fleshed out design, but present them in a simplified manner. For example, a logo might be needed on a homepage, but the logo has not been designed yet. The wireframe would still depict a placeholder for the logo in the specified location and of the desired size.
- *Mock-up:* A mock-up is higher fidelity than a wireframe, and serves as a representation that is closer to what a finished product may look like. The color scheme, typography, icon set, and other detailed visual representations are included at this at this stage. It is becoming increasingly popular to create mock-ups such that the effort can be leveraged for the user interface software front end. For example, the mock-up could be written as software code or created in a package that allows exporting the mock-up as usable software code.

Interface Review

Once design artifacts are created, they can be reviewed by a variety of individuals: prospective users, SMEs, and internal and external stakeholders, and an HS engineer. This step at the formative stage can provide the greatest cost-savings to an effort because design concepts have been turned into tangible products that can be reviewed. This allows the whole team to develop a shared representation of what the system will look like and how the user will interact with it, and that representation can be iteratively tested without costly implementation.

---

* https://balsamiq.com.
† https://proto.io.
‡ https://www.axure.com.

An HS engineer can perform a heuristic evaluation and examine the interface serving as a proxy for the user due to the knowledge the HS engineer obtained during the earlier needs analysis process. This perspective is coupled with human systems engineering training and experience to examine the interface and compares it to standard usability principles. The most commonly used set of principles was developed by Nielsen and Molich (1990) and subsequently modified by Nielsen (1994a). The principles are as follows:

- Visibility of system status
- Match between system and the real world
- User control and freedom
- Consistency and standards
- Error prevention
- Recognition rather than recall
- Flexibility and efficiency of use
- Aesthetic and minimalist design
- Help users recognize, diagnose, and recover from errors
- Help and documentation

The HS engineer records areas where these principles are violated and can make suggestions for how to redesign the system to address the violation. For example, terminology used in the system might not match the terminology identified during the first step of IDEAS, needs analysis. In that case, the HS engineer would identify the inconsistencies and would suggest revising the terminology. Although evaluations may be performed by a single expert, there are benefits to having multiple evaluators to examine the interface (Nielsen, 1995). Heuristic evaluators are great at identifying concerns that might cause issues for users, and when potential users are rare (e.g., an interface designed for professionals who are in short supply) or when it would not be cost effective to work with them often. In this manner, more obvious concerns may be caught by the trained eye of the evaluator.

In addition to a heuristic evaluation, formative usability testing can be employed. Because the system has not yet been implemented, it is critical to only work with individuals who recognize the limitations of the mock-ups and have appropriate expectations for the system's functionality (or lack thereof). This type of user testing benefits from having detailed explanations of what the participants can and cannot do with the design artifact. A well-fleshed out narrative can help set the stage and help approximate that kind of context in which the user would actually be working. Users can be asked to imagine parts of a scenario, and then be shown the wireframe or mock-ups and asked to share specific thoughts about it.

This type of early testing is particularly helpful for obtaining feedback on early design concepts. For example, system designers might be considering multiple ways to organize a user profile. Mock-ups of each design could be generated, and participants could be shown counterbalanced orders of the different designs and be asked to rate them using a Likert scale as well as provide open-ended feedback. HS engineers can develop a study protocol that minimizes bias in the participant's responses by using sound experimental design principles. The quantitative and qualitative data can then be analyzed to improve the design artifacts. It is important to note that in Figure 18.1, this step has a path that leads back to needs analysis. As part of the iterative process, one can use feedback from the interface review to step back and revisit Steps 1, 2, and 3 to generate improved models of the users, perhaps new system requirements, and new or revised designs that also need to be tested. This cycle can be repeated as often as needed with the understanding that changes at this stage are far more affordable than after the system is implemented.

Implementation

Once a design has gone through several iterations of interface review and it has been determined that there are diminishing returns for another cycle, the design can be implemented to create a fully functional system. In software efforts, this would be the phase where the lines of software code are written. The brunt of implementation work falls on the shoulders of software developers, but HS engineers continue to serve as consultants throughout this stage. Throughout this process, HS engineers coordinate with the software development team to ensure that the software is developed in accordance with the design.

Although the goal is to address every design issue prior to implementation, it is not uncommon for challenges to emerge during implementation that result in design changes. For example, a design could have pointed toward a particular software module needing to be used, but the software developer encounters unexpected challenges integrating it into the system. The functionality that the package provided would now need to be replicated with a different module, or the design could be altered to accommodate different functionality. The HS engineer works with the software developer and the program lead to understand the ramifications on functionality, usability, and program budget.

Evaluation

System testing can occur prior to evaluation or be included as a step in evaluation. More specifically, the overall system must be tested in an end-to-end fashion to ensure that it meets the requirements that were previously specified. If a task analysis was created during needs analysis, these tasks serve as helpful input for the creation of test plans to verify system functionality. HS engineers, who have developed a strong sense of how the intended users work, can often serve as good user proxies and can play a role in system testing. Bugs that are identified at this stage should be resolved before moving on to user testing.

There is no substitute for observing real users working with a prototype or functional system, and this is where *usability testing*, comes into play. Although usability testing can be an important tool during interface review, in summative usability testing the representative users are asked to work with a functional product while observers carefully watch and record the manner in which the users work with the system (Nielsen, 1994b). Usability testing can reveal issues that an HS engineer may have missed. An evaluator can learn a lot by carefully observing, frequently recording and analyzing the interaction, and speaking with the users afterwards.

REAL-WORLD APPLICATION OF IDEAS

A Challenging Experience

Some of the authors worked on a proposal for a program that was designed to be an enabler of net-centric warfare. The technology developed in this program would provide greatly improved connectivity and data transfer to users and systems that would not normally have reliable internet access, thus connecting commanders with enhanced information and ability to see the battlefield.

The proposal effort was very large, with a significant number of personnel working on it. In early discussions with the proposal leads, the newly minted IDEAS process was shared. There was interest in talking about the way that the method allows for rapid prototyping of design concepts. The proposal leads were enthusiastic about the in-house capabilities the human system engineers could provide, and were excited that this approach would offer a competitive advantage. As the proposal process continued, circumstances emerged that tempered the initial enthusiasm. The proposal team had more than 50 members, and the proposed design process followed a strong waterfall software development approach with very distinct phases for requirements, design, implementation, and verification. There seemed to be no obvious way to build in an iterative model that allowed for early user feedback that could serve not only to get the requirements correct, but to reduce later risk.

The proposal process itself served as a microcosm for the proposed project design process. There were frustrating months advocating for a user-centered process, but it was extraordinarily difficult to make a meaningful change in either the proposal process or the proposed waterfall design process. In retrospect, there were at least two sources of failure. First, assumptions were made about the flexibility in the design process. From a human systems perspective, the waterfall approach was difficult to work with because it did not allow for flexibility through the process which would make it difficult to use the results of the evaluation to modify the system. From the systems engineer perspective this was a tried and true method that had worked previously and that the customer was comfortable with, so there was little reason to change. It was simply perceived as either too risky or too challenging to modify the process to accommodate the flexibility that would be needed for meaningful modifications based on a user-centered approach.

One of the lessons learned from this effort was a recognition that the IDEAS steps had not been adequately layered onto the existing software design methodology. Granted, there was a fundamental mismatch of a rigid waterfall process vs. an iterative user-centered approach. Still, had the HS engineers better overlaid their process onto the other engineering processes there may have been a chance to modify the process and introduce spirals into an otherwise straight sequential process. This was a lesson that proved crucial in the next major proposal opportunity.

Successful Application

IDEAS was also applied to a proposal for creating technology that could radically shift the way in which future sailors control multiple unmanned systems launched from naval ships. For the first phase, a two evolution spiral software development model was proposed. The spiral model contains the same basic steps as a waterfall model, but includes multiple iterations. Unlike the proposal effort described earlier, the software development team was collocated and they were more attuned to user-centric approaches. In addition to colleagues who were more receptive, prior experience was used to motivate the creation of a tight integration of the human systems engineering and the software development processes.

The team worked together to super-impose the IDEAS process on the spiral software development process. The four steps in the software method are Requirements & Design, Development, Integration & Test, and Evaluation. An HS engineer conducts needs analysis, requirements generation, design & engineering, and interface review all within the requirements and design step in the spiral process. The IDEAS Implementation step overlaps with both the development and integration and test stages of the spiral process. Finally, evaluation is the final step for both the IDEAS and spiral methodology. When one process is overlaid on the other it illustrates that a significant portion of the human systems engineering steps occurs up-front prior to implementation. The primary software tasks are development and integration, and the entire team works to evaluate the systems functionality.

Overlaying the processes in this manner was remarkably helpful because it allowed the HS engineers, the software engineers, and the customer to understand how the two processes interacted and could work together. It retained the original language of each, so individuals from one camp or the other could look at the figure and still feel a sense of familiarity. This successful integration of the processes helped the team win the proposal and ultimately execute the multiyear project. Throughout execution there was a recognition of the benefits of the human systems engineering process, and this allowed the team to hone a strong collaborative effort throughout. In particular, it was noteworthy that the design process became a truly shared process where HS engineers and software engineers contributed creative input. This enriched the design artifacts and allowed the team to test multiple viable versions. Ultimately, the prototype met both the customer program manager and user expectations.

USING IDEAS WITH SOFTWARE DEVELOPMENT METHODS

As noted above, a majority of the steps in IDEAS precedes software implementation. This should feel intuitively correct to an HS engineer, who would expect to put in a lot of effort in the early stages to understand target users. IDEAS is overall an iterative process, but there are also two internal spirals embedded within the process.

The first internal spiral takes place after the interface review and reflects an iterative design process. Once an interface mock-up has been designed, it is shown to prospective users and SMEs to obtain feedback. This is sometimes also called a *formative evaluation*. The interface can take many forms: wireframe, mock-up, paper prototype, or even functional user interface software code. The defining characteristic at this stage is that the interface has been created affordably. Granted, the interface needs to be significantly fleshed out to allow the prospective users to understand the design and how it would (and would not) support their workflow. Testing multiple design concepts early allows for greater objectivity as well as minimizing the risk of a sunk-cost mentality of spending time to implement the first design. User feedback at this early stage can identify early design flaws, and because the design artifact is lightweight it can be more easily modified. Any areas of concern are identified early, and the internal spiral can start again by going back to revisit the user needs.

For example, when designing an operator control station for the second project described above, prospective users were shown interactive wireframes of a design that allowed flexible placement of interface elements, and another that preserved a rigid layout. The flexible layout worked similarly to windowed operating systems and allowed the control mechanisms to be moved based on the preference of the operators. The latter design offered consistent placement and size of control mechanisms. Due to the program's goal of blending operator roles, there were many different possible combinations of control elements. Early review of interface designs allowed for a determination of which design would ultimately be implemented (flexibility won). Although this was a difficult choice, an informed decision was made based on the user feedback and not one made purely for technological or programmatic reasons.

The second internal spiral occurs after the evaluation is conducted and the results of that evaluation require a change to the implemented system. The evaluation can often include a quality assurance usability test to determine whether the implemented system meets relevant standards. In Figure 18.1, there is an arrow pointing back to implementation, and if the usability testing revealed things requiring change then this could be immediately addressed and then retested. This type of testing is also commonly called a *summative evaluation*. This type of evaluation typically covers testing usability and functionality and can identify properly functioning software that has usability concerns as well as properly designed software that was implemented with errors or bugs. Both of these outcomes necessitate revisiting the implementation stage to correct the software.

CONCLUSION

Several themes related to the impact of human systems engineering have been discussed in this paper. First, the value of human systems engineering should not be underestimated. For the last 80+ years, the value has been demonstrated across numerous domains. The early domains usually included safety-critical areas like high performance aircraft design, weapon systems, power plants, and health care. Success in these areas has helped pave the way for the widespread use of human systems engineering in commercial and consumer applications, and the proliferation of technology systems in our everyday lives has emphasized the need for simple, easy to use, intuitive products. Successful design does not happen by accident, and has been the result of tireless effort by individuals who carefully study the interaction between humans and the systems in order to make them as efficient as possible.

A second theme is the value of including human systems engineering in advanced technology engineering efforts. When humans engineer, they create and design in an effort to solve human-relevant problems. Consequently, the human perspective plays an important role in ensuring that the engineered artifact works well for its intended audience. This holds true in advanced technology development as well, even if a technology concept is in its infancy there is considerable merit in developing an understanding of how the technology could be used. In particular, the cumulative experience of HS engineers can help avoid simple errors such as failing to account for users who are color-blind, or are left-handed, or above or below average height or weight. They can also help the team leverage or develop models for how differences in mental models or cognitive processes would lead to different interactions with the technology system.

Finally, our own experiences have shown that it is beneficial to adapt methods for the context in which they are used. This chapter focuses on our experiences performing human systems engineering within an advanced R&D environment, and other engineering environments might require a similar customization. The important point is to identify the points of intersection where a product or result from one group impacts the others. Human systems engineering is not a single process or method but rather consists of a collection of tools and techniques that can be customized for particular efforts. IDEAS was created out of a recognition of common engineering approaches in advanced technology and allowed the HS engineers in that environment to communicate those common steps with other types of engineers.

REVIEW QUESTIONS

1. Describe the key difference between human systems engineering and CSE.
2. It is claimed that training can address what deficiency? Think about your own experiences and describe at least one example where you were expected to overcome this deficiency.
3. Given that software developers shoulder most of the burden in the *implementation* stage, what role is there for a human systems engineer in this phase?

REFERENCES

Bartlett, F. (1943). Instrument controls and display. *GB MRC-RNPRC, Psychological Laboratory, Cambridge, RNP, 43*(81), 7.

Boff, K. R. (2006). Revolutions and shifting paradigms in human factors & ergonomics. *Applied Ergonomics, 37*(4), 391–399. doi:10.1016/j.apergo.2006.04.003

Bryant, N. (2013, October 21). Obama addresses healthcare website glitches. *BBC News.*

Craik, K. (1940). *The fatigue apparatus (Cambridge cockpit)(Report 119).* London: British Air Ministry, Flying Personnel Research Committee.

Dahm, W. (2010). *Technology horizons: A vision for air force science & technology during 2010–2030.* Arlington, VA: United States Air Force.

DeAngelis, T. (2008). NASA budget leaves human factors in the cold: Continued cuts prompt concerns about air safety and the loss of important basic research. *Monitor on Psychology, 39*(3), 39.

DoD ESI/Navy Enterprise Licensing Agreements Team. (2014). Best practices and lessons learned for DoD commercial off-the-shelf software licensing: Best value, best pricing, best terms and conditions for the software you need. *CHIPS.* Retrieved from http://www.doncio.navy.mil/CHIPS/ArticleDetails.aspx?ID=5009

Düchting, M., Zimmermann, D., & Nebe, K. (2007). Incorporating user centered requirement engineering into agile software development. In J. A. Jacko (Ed.), *Human-computer interaction. Interaction design and usability: 12th international conference, HCI international 2007, Beijing, China, July 22–27, 2007, Proceedings, Part I* (pp. 58–67), Berlin: Springer.

Dul, J., Bruder, R., Buckle, P., Carayon, P., Falzon, P., Marras, W. S., ... van der Doelen, B. (2012). A strategy for human factors/ergonomics: Developing the discipline and profession. *Ergonomics, 55*(4), 377–395. doi:10.1080/00140139.2012.661087

Echo, R. (2012, July 6). South Wales Echo: Air France pilots "lost control" of Flight 447. *South Wales Echo (Cardiff, Wales)*.

Endsley, M. R., Hoffman, R., Kaber, D., & Roth, E. (2007). Cognitive engineering and decision making: An overview and future course. *Journal of Cognitive Engineering and Decision Making, 1*(1), 1–21.

Fitts, P. M., & Jones, R. E. (1947a). *Analysis of factors contributing to 460 "pilot error" experiences in operating aircraft controls*. Dayton, OH: Aeromedical Laboratory, Air Material Command.

Fitts, P. M., & Jones, R. E. (1947b). *Psychological aspects of instrument display. Analysis of 270 "pilot-error" experiences in reading and interpreting aircraft instruments*. Dayton, OH: Aeromedical Laboratory, Air Material Command.

Foraker Labs. (2016). Requirements specification. Retrieved from http://www.usabilityfirst.com/about-usability/requirements-specification/

Gaffney, G. (2000). Card Sorting. Retrieved July 25, 2016, from http://infodesign.com.au/usabilityresources/cardsorting/

Gilbreth, F. B., & Gilbreth, L. M. (1917). *Applied motion study: A collection of papers on the efficient method of industrial preparedness*. New York, NY: Sturgis & Walton Company.

Gilbreth, L. M. (1914). *The psychology of management: The function of the mind in determining, teaching and installing methods of least waste*. New York, NY: Sturgis & Walton Company.

Hancock, P., & Hart, S. (2002). What can human factors/ergonomics offer? *Ergonomics in Design, 10*(1), 6–16.

Hollnagel, E. (2016). *CSE—Cognitve systems engineering: CSE: RIP*. Retrieved from http://erikhollnagel.com/ideas/cognitive-systems-engineering.html

Hollnagel, E., & Woods, D. D. (1983). Cognitive systems engineering: New wine in new bottles. *International Journal of Man-Machine Studies, 18*(6), 583–600.

International Ergonomics Association. Retrieved from http://www.iea.cc/whats/

Jacobsen, A. A. (2015). *The Pentagon's brain: An uncensored history of DARPA, America's top secret military research agency* (1st. ed.). New York, NY: Little, Brown and Company.

Johnson, G. J. (1991). A distinctiveness model of serial learning. *Psychological Review, 98*(2), 204.

McNeese, M. D., Zaff, B. S., Citera, M., Brown, C. E., & Whitaker, R. (1995). AKADAM: Eliciting user knowledge to support participatory ergonomics. *International Journal of Industrial Ergonomics, 15*(5), 345–363. doi:10.1016/0169-8141 (94)00081-D

Meister, D. (1999). *The history of human factors and ergonomics*. Mahwah, NJ: Lawrence Erlbaum Associates.

Meshkati, N. (1991). Human factors in large-scale technological systems' accidents: Three Mile Island, Bhopal, Chernobyl. *Organization & Environment, 5*(2), 133–154.

Mullaney, T. (2013, October 6). Obama adviser: Demand overwhelmed HealthCare.gov. *USA Today*.

New York Times. (1988, August 3). Report: Human error doomed Iran jet probe cites combat stress aboard *Vincennes*. *The Orlando Sentinel*, p. A1.

Nielsen, J. (1993). Iterative user-interface design. *Computer, 26*(11), 32–41.

Nielsen, J. (1994a). *Enhancing the explanatory power of usability heuristics*. Paper presented at the Proceedings of the SIGCHI Conference on Human Factors in Computing Systems, Boston, MA.

Nielsen, J. (1994b). *Usability inspection methods*. Paper presented at the Conference Companion on Human Factors in Computing Systems, Boston, MA.

Nielsen, J. (1995). How to conduct a heuristic evaluation. Retrieved from https://www.nngroup.com/articles/how-to-conduct-a-heuristic-evaluation/

Nielsen, J., & Molich, R. (1990). *Heuristic evaluation of user interfaces.* Paper presented at the Proceedings of the SIGCHI Conference on Human Factors in Computing Systems, Seattle, WA.

Norberg, A. L. (1996). Changing computing: The computing community and DARPA. *IEEE Annals of the History of Computing, 18*(2), 40–53. doi:10.1109/85.489723

Norman, D. (1981). Steps towards a cognitive engineering: System images, system friendliness, mental models. *Department of Psychology and Program in Cognitive Science (Draft),* San Diego, CA: University of California.

Obama, Barack, Pres. (2015). *President Obama: The Fast Company Interview.*

ODASD. (2016). Initiative: Human systems integration. Retrieved from http://www.acq.osd.mil/se/initiatives/init_hsi.html

Perry, D. G., Blumenthal, S. H., & Hinden, R. M. (1988). The ARPANET and the DARPA Internet. *Library Hi Tech, 6*(2), 51–62. doi:10.1108/eb047726

Regli, S. H., & Tremoulet, P. D. (2007). *The IDEAS process: Interaction design and engineering for advanced systems.* APA Division 21, Division 19, & HFES Potomac Chapter Annual Symposium on Applied Research. Fairfax, VA: George Mason University.

Russ, A. L., Fairbanks, R. J., Karsh, B.-T., Militello, L. G., Saleem, J. J., & Wears, R. L. (2013). The science of human factors: Separating fact from fiction. *BMJ Quality & Safety, 22*(10), 802–808. doi:10.1136/bmjqs-2012–001450

Salmon, P., Jenkins, D., Stanton, N., & Walker, G. (2010). Hierarchical task analysis versus cognitive work analysis: Comparison of theory, methodology and contribution to system design. *Theoretical Issues in Ergonomics Science, 11*(6), 504–531. doi:10.1080/14639220903165169

Sanders, E. B.-N. (2002). From user-centered to participatory design approaches. In J. Frascara, (Ed.), *Design and the social sciences: Making connections* (pp. 1–8). London: Taylor & Francis Group.

Sauro, J., & Lewis, J. R. (2012). *Quantifying the user experience: Practical statistics for user research.* Amsterdam: Elsevier.

Spencer, D. (2009). *Card sorting: Designing usable categories.* Brooklyn, NY: Rosenfeld Media.

Stanton, N. A. (2006). Hierarchical task analysis: Developments, applications, and extensions. *Applied Ergonomics, 37*(1), 55–79.

Taylor, F. W. (1911). *The principles of scientific management.* New York, NY: Harper & Brothers Publishers.

United Kingdom Army Secretariat. (2015). *Ref: FOI2015/00342, Army Secretariat official response to request for cost of training soldiers.* Retrieved Jun 18, 2016, from https://www.gov.uk/government/uploads/system/uploads/attachment_data/file/397258/Army_Training_Costs_Per_Recruit.pdf

Vicente, K. J. (1999). *Cognitive work analysis: Toward safe, productive, and healthy computer-based work.* Boca Raton, FL: CRC Press.

Wickens, C. D., & Hollands, J. G. (2000). *Engineering psychology and human performance* (3rd ed.). Upper Saddle River, NJ: Prentice Hall.

Wood, J. R., & Wood, L. E. (2008). Card sorting: Current practices and beyond. *Journal of Usability Studies, 4*(1), 1–6.

World War II Foundation. WWII Facts & Figures. Retrieved from http://www.wwiifoundation.org/students/wwii-facts-figures/

Zimmermann, D., & Grötzbach, L. (2007). *A requirement engineering approach to user centered design.* Paper presented at the International Conference on Human-Computer Interaction. Beijing, P.R. China.

Section VII

The Future of the
Living Laboratory

19 The Future of the Integrated Living Laboratory Framework
Innovating Systems for People's Use

Michael D. McNeese and Peter Kent Forster

CONTENTS

Introduction 363
Perspective 364
    Problem-Centered Challenges 364
    Technological Advancements 366
    Practical Considerations 367
Concluding Thoughts 368
References 369

INTRODUCTION

The chapters in the book have collectively laid out the overall history, purpose, foundation, conceptual elements, and application of the integrated living laboratory framework (LLF). Taken as a whole, these chapters represent a specific methodology for the engineering design of cognitive systems within complex real-world environments where human agents must transact, conform, and yet adapt to changes given their abilities, resources, and provisions. Through the intervention of designs that have evolved through application of multiple sources of knowledge, humans can accomplish their intention and act upon their environment for both purpose and worth. As suggested in the introductory chapter, one of the themes of our book is to *make life better.*

The connotation of a living environment is appropriate. Cognitive systems engineering must approach design not within a vacuum but within the living forces that impinge upon phenomena as humans attempt to overcome problems, issues, and constraints. Indeed, designing systems for living agent–environment transactions requires multiple levels of understanding, multiple representations and models of how *use* evolves when conditions undergo change, and having knowledge about the structure and activities as they come about within a given context. As mentioned earlier, Perkins (1986) *knowledge as design* concept is relevant to discover and integrate multiple levels of understanding and representation as activities are distributed across a context or multiple contexts. Although others have delved in cognitive systems engineering with both functional and ecologically design viewpoints (Flach & Rasmussen, 2000; Rasmussen, 1986; Sanderson, 2003; Vicente, 1999), have represented cognition in action through multiple arrays for different purposes (Woods, Johannesen, Cook, & Sarter, 1994; Zhang & Norman, 1994), designed cognitive systems with affordances distributed across internal and external states (Hutchins, 1995; Zhang & Patel, 2006), directed cognitive systems engineering toward a given field of practice (aviation systems, Flach & Rasmussen, 2000; McNeese & Vidulich, 2002; health care, Bisantz, Burns, & Fairbanks, 2014)—this particular volume of works seeks to provide a multi-methodological, integrated framework for scientists, practitioners, academics, and students.

PERSPECTIVE

Throughout the volume, the integrated LLF is viewed from different angles with specification and emphasis to a specific element of interest. This enables the presentation of chapters through a certain window of light that draws illumination to the whole. Each chapter has been designed to simultaneously draw out use within the overall framework and to enlighten the reader for a specific topical focus. The hope in writing this book has been to provide a broad end-to-end continuum in state-of-the-art applications and fields of practices that are evident around the world today. Although any volume cannot cover all areas of application, this volume does look at the LLF for some unique settings that in turn produce unique problems for which this cognitive systems engineering methodology is apropos. The following sections take a quick look at specific themes (and barriers that will exist) that we see as steps along the way of a future path for the integrated living methodology.

PROBLEM-CENTERED CHALLENGES

As elaborated throughout this book, the necessity of finding, exploring, and addressing specific problem spaces lies at the heart of learning and setups up the direction of knowledge flow toward design and then practice. As a new world emerges, new sets of problems besiege us with much calamity, stubbornness, and unpredictability. Challenges in the world today crisscross a number of dimensions and solutions to problems can impact people in different ways (some advantageous–others disadvantageous). The interconnected world demands a completely new process for business, enterprise, society, and threats to security. This complexity may result in many unintended consequences wherein just defining what a problem is and how it came about is a problem within itself. Owing to the Internet, mobile computing, cameras capturing the who and what of visible activities, ubiquity of information pervading many contexts, and being able to access so much knowledge so quickly presents unprecedented problems we have yet to fathom.

As the problem space gets more ubiquitous, complex, and presents examples of massive distributed cognition, the reliance on both theory and practice will become even more important. Strong theories will be needed to understand the landscapes of how humans solve problems and make decisions in the midst of social networks, virtual interaction, and just in time actions. The rapid access of a myriad of information sources simultaneously generating greater information seeking creates a new world with new kinds of problems to breakdown. Increasingly cognitive systems engineering must look at problems that dovetail the social with the cognitive, the perceptual with the political, and the threat with hidden knowledge. What looks to be true in defining a problem may only expose a small part of the problem space. For example, this might be explored in the ever-expanding world of cybersecurity where many operations are deceptive, hidden from the human eye, and only detectable after the fact. Likewise, designing information technology to detect terrorist activities such as the internet in recruitment strategies is not as simple as defining simple sensor fusion algorithms that identify specific words or phrases but involves deep understanding of culture, religion, and the possibility of manipulating the human psyche. As the world becomes more global yet more interconnected, the challenges in defining and exploring problems with past theoretical frameworks and models will become increasingly inadequate. Theories must keep up with advancements in the digital global economy.

Concomitantly, observing and understanding the user (via ethnography and knowledge elicitation components of the living laboratory) in praxis become more difficult when a lot of activity is embedded through the information exchanged via mobile devices (e.g., apps such as Facebook, Twitter, Pinterest, Instagram, and YouTube). Whereas in the past, ethnographic activity placed the researcher in the midst of a social context (as in collocated activity that could be seen and heard), which made it meaningful; now a lot of social activity is embedded and distributed across multiple contexts, which are often silent to the researcher (e.g., most young people communicate by text rather than the spoken word when using a phone). In our own research, we have recently explored the roles of hidden knowledge (McNeese, McNeese, Endsley, Reep, & Forster, 2016) as well as other cultural factors in joint cognitive activity. Although hidden knowledge is present when humans operate in complex activities, it may or may not be used among teams. Sharing is contingent upon the circumstances -but in contemporary society- privacy often controls access to knowledge. Therein, just observing what is happening is not enough, but a lot of hidden knowledge is within portals that are not available to researchers.

Considering the changes in culture and technology and how it impacts the instantiation of the LLF for given research projects, we are increasingly running into further barriers on the *knowledge elicitation* side of the coin. Not only is studying the context often missing the most important data (it is on the phone!) but further elucidation of mental models from individuals and teams is often coming under severe restraints either owing to privacy issues, military classification, or just people wanting to protect their own interests (legal). Recently, we concluded a major research grant in the DoD related to human-centered cybersecurity activities. For the most part, actual, real cyber analysts are off limits to academic researchers, so obtaining interviews, surveys, or concept mapping them is very difficult. Hence, one begins to wonder what one actually knows, given the constraints placed on knowledge. We understand why these limitations exist, but the old cliché about *who knows what about who knows what* is applicable. What can really be known is a deeper more esoteric effect of limited access to experts. In turn, research experience (RX) can be muddied and even shallow when it comes to user experience (UX) if only surface knowledge can be discovered and deep knowledge is invisible. There are some workarounds to limitations and constraints in this area, but again one really wonders about the veracity of knowledge elicited especially when combined with only partial views of observations. There is no substitute for experiential knowledge of true experts. Without that knowledge, the development of scaled worlds and reconfigurable prototypes will necessarily be deficient creating weak links in knowledge as design.

On the other hand, technological disruptions can offer new feasibilities/innovations for applying the LLF as we look into the future. Crowd sourcing (where the power of many can be used to acquire, enhance, and deliver socially inspired knowledge) leverages social networks to produce new forms of citizen science (Hall et al., 2010), therein developing powerful mechanisms to explore emergent problem spaces. Due to the rampant and ubiquitous use of the internet and mobile computing, there is a greater probability of video recordings of events and situations, being made available, as everyone has a cell phone and may contribute to unique forms of knowledge/perceptual acquisition and soft/hard data fusion. Although data for knowledge acquisition and observation afford on-the-fly and present new potential sources of data for scrutiny, privacy and security are still major concerns and must be considered. Notwithstanding the potential for a lot of naturalistic sensing is highly encouraging.

The next challenge is the Internet of things that epitomizes the proliferation of distributed computing. These new kinds of data supply additional layers of understanding of user(s) as they transact with information that flows across the LLF. In addition, the use of datasets in large-scale exercises provide explorations within complex situations that are often not classified. They can provide answers as to how human-environment transactions emerge under various conditions. Large-scale datasets also offer oversights about (1) the interdependencies among agents, objects, and events (2) how success or failure unfolds. Newer techniques and modeling employ machine learning or deep learning to detect important patterns in massive datasets. This can help to gain new insights from video, and other forms of sequential data that humans generate.

Technological Advancements

The previous section looks at the use of new information technologies (such as crowd sourcing, citizen science, metadata, machine learning, Internet of things, mobile-distributed computing, and big data) as the basis of acquiring new forms of knowledge. Other forms of technological sophistication and advancement are coming into their own especially as they relate to socio-cognitive support of humans. We have already seen, for example, the use of uninhabited air–land–sea vehicles (drones) support humans in simultaneously complex mission advancements. Drones and robots may employ artificial intelligence as they almost always have coupling to human agents in one fashion or another. Our early work in humane intelligence (McNeese, 1986) for the Pilot's Associate program pioneered some of the early concepts that are relevant for human–autonomous interaction and human–robotic interaction. With the prevalence of these areas coming into their own, we would forecast that this will be one of the next main application areas for the LLF.

There are a lot of unique problems to be discovered owing to the unique contexts that will utilize these innovative developments. Being able to create new simulations that have the infrastructure and information architecture to scale up interdisciplinary problems with these kinds of agent orchestrations will be a key component to extend LLF outputs. Some of our work involving teams of agents (Fan et al., 2009) that can employ cognitively inspired architectures to jointly work with humans to make joint decisions in complex domains such as command, control, communications, and intelligence are pertinent examples. Here, the LLF can produce mutual learning and prodigious outcomes in cognitive systems engineering. Augmented cognition utilizing 3D worlds for generating the combination of decision aiding, remote operations, and sensory integration should provide intriguing opportunities for new kinds of macroprototyping wherein scaled worlds and new innovations can move human–environment transactions to new levels of intelligence. Being able to inform researcher and user experience within these exotic technological worlds will be paramount for success in actual performance in practice.

PRACTICAL CONSIDERATIONS

Although people are becoming exceedingly similar in certain ways, the search for invariance across many situations, cognitive systems, and social interaction becomes daunting. Individual differences are difficult to account for in design and are becoming exceedingly stratified as diversity multiplies in so many ways. This is increasingly more evident in multinational teamwork (see Endsley, 2016) where differing cultures must put aside biases, risks, language differences, communication barriers, and prejudices in order to work together for the greater good. These challenges likely require negotiation to form joint meanings, establish new levels of situation awareness, and to develop new team mental models. In real world practice, this may be easier said than done as teamwork can easily breakdown and be subject to multiple distractions. Adaptive interfaces may also be useful kinds of reconfigurable prototypes to experiment with when it comes to cognitive systems engineering of collaborative cultural work.

As enterprise and human endeavors become inherently coupled and interdependent—within a global digital economy—distributed cognition is not just unitary but in fact entails managing several contexts simultaneously in order to produce requisite and acceptable performance. Knowledge management of these kinds of situations must address the simultaneity of interdependent information swirls where information overload is prevalent and where chaotic behavior is persistent as it unfolds in the face of uncertainty. We have referred to the dynamic back-and-forth among tasks evident in an environment as *context switching* (see Fan, Sun, Sun, McNeese, & Yen, 2006). This enables many action potentials but unfortunately can place severe limits on attention, memory, and flow of information within highly coupled and interdependent systems.

As we project trajectories into the future use of the LLF, there are a few of these kinds of areas/contexts that require the strategic engineering of cognitive systems—that may pose high risk to address but also offer tremendously high payoffs for the *information society*. As mentioned earlier in the chapter, the meshing of information security, privacy, crises, and socio-cognitive technologies will become even more prevalent in specific multilayered situations. However when coupled with terrorism and extreme events, much calamity and fear can transpire. We have seen extreme events over the last few years that represent this multilayered meshing such as the Boston marathon bombing, the San Bernardino attacks, and the Paris attacks in November 2015. The problems that emerge in these kinds of crises often are formulated through massively distributed cognition contexts that contain a team of many teams, a system of many systems, and a problem of many problems, where acute judgments have to be made with time stress or catastrophic results ensue. Preparing responses that make a difference in terms of life-saving consequences will be important to study and research. Part of the response involves development of envisioned designs of socio-cognitive technologies that can be tested within scaled-worlds representative of these dynamically multilayered situations.

These unique types of extreme events frequently have components wherein spreading activation unfolds and potentially creates new corridors of the problem space that envelop resources consisting of the use of additional information, technology, and people. Although multifaceted, large-scale emergency response exercises are valuable to get a handle on some elements of the problem space (and we have done ethnographic participant–observer studies in these kinds of events, see MacEachren, Cai, Furhman, McNeese, & Sharma, 2004), they lack the control, precision, and experimental design that is present within experiments in a scaled-world simulation). Hence, it is often difficult to layout *what* is affecting *what* for *what reason* when so many variables are unconstrained and vary with unwieldy fashion. While studying these kinds of exercises may yield insights about collaboration, reasoning and judgment, the allocation of resources, and execution within a common operational picture, this type of knowledge must be complemented by an integrated LLF to triangulate and increase reliability, validity, and resilience that comes into play while contextualizing envisioned designs within hard, wicked problems. Just generating a partial view of complex wicked problems will result in weak designs that fail when adaptation is required in uncertain, risky, and volatile environments. Hence, the simulations designed for this future trajectory must begin to look at team of teams, system of systems, and problem of problems approaches in order to get a better representation of the dynamics and nested elements of macrocontextual problem solving.

CONCLUDING THOUGHTS

As we come to the end of our book, it is our hope that the integrated LLF provides the necessary and sufficient multimethodological approach to transdisciplinary problem solving with the intent to use multisourced knowledge as the basis for consistent, reliable, and effective human-centered design to make life better. The goal of our book has been to provide specific research insights into the various elements of the LLF by stratifying the book sections to reflect the critical elements of the framework itself. Therein, each chapter represents a specific point of light of the LLF and points upward to a given element of the framework. Comprehensively then the book yields a broad bandwidth of cognitive systems engineering methods that is reflective of living systems, growth and change, that are the foundation of sustainable, resilient, ecological, and sound designs, which are necessarily human centered and problem focused.

REFERENCES

Bisantz, A. M., Burns, C. M., & Fairbanks, R. J. (2014). *Cognitive systems engineering in health care.* Boca Raton, FL: CRC Press, Taylor & Francis Group.

Endsley, T. (2016). *An examination of cultural influences on team cognition and information sharing in emergency crisis management domains: A mixed methodological approach* (Unpublished Doctoral Dissertation). University Park, PA: The Pennsylvania State University.

Fan, X., McNeese, M., Sun, B., Hanratty, T., Allender, L., & Yen, J. (2009). Human-agent collaboration for time stressed multi-context decision making. *IEEE Transactions on Systems, Man, and Cybernetics (A), 90,* 1–14.

Fan, X., Sun, B., Sun, S., McNeese, M. D., & Yen, J. (2006). RPD-enabled agents teaming with human for multi-context decision making. *Proceedings of the Fifth International Joint Conference on Autonomous Agents and Multiple Agent Systems* (AAMAS), Association for Computing Machinery (ACM), P. Stone & G. Weiss (Eds.), May 8, Hakodate, Japan.

Flach, J. M., & Rasmussen, J. (2000). Cognitive engineering: Designing for situation awareness. In N. Sarter & R. Amalberti (Eds.), *Cognitive engineering in the aviation domain* (pp. 153–179). Mahwah, NJ: Lawrence Erlbaum Associates.

Hall, D., McNeese, N., & Llinas, J. (2010). H-space: Humans as observers. In D. Hall & J. Jordan (Eds.), *Human centered fusion* (pp. 59–84). Norwood, MA: Artech House.

Hutchins, E. (1995). *Cognition in the wild.* Cambridge, MA: MIT Press.

MacEachren, A. M., Cai, G., Fuhrmann, S., McNeese, M. D., & Sharma, R. (2004, May). *GeoCollaborative crisis management (GCCM): Building better systems through advanced technology and deep understanding of technology-enabled group work (project highlight paper).* Proceedings of the 5th Annual NSF Digital Government Conference, Seattle, Washington, 24–26. Publisher: Digital Government Society of North America.

McNeese, M. D. (1986). Humane intelligence: A human factors perspective for developing intelligent cockpits. *IEEE Aerospace and Electronic Systems, 1*(9), 6–12.

McNeese, M. D., McNeese, N. J., Endsley, T., Reep, J., & Forster, P. (2016). Simulating team cognition in complex systems: Practical considerations for researchers. *Proceedings of the 7th International Conference on Applied Human Factors and Ergonomics (AHFE 2016) and the Affiliated Conferences.* Orlando, FL. In K. S. Hale & K. M. Stanney (Eds.), *Advances in Neuroergonomics and Cognitive Engineering* (pp. 255–267). Switzerland: Springer International Publishing.

McNeese, M. D. & Vidulich, M. (Eds.). (2002). *Cognitive systems engineering in military aviation environments: Avoiding cogminutia fragmentosa.* Wright-Patterson Air Force Base, OH: Human Systems Information Analysis Center (HSIAC) Press.

Perkins, D. N. (1986). *Knowledge as design.* Hillsdale, NJ: Lawrence Erlbaum Associates.

Rasmussen, J. (1986). *Information processing and human-machine interaction: An approach to cognitive engineering.* New York, NY: North Holland Publishers.

Sanderson, P. M. (2003). Cognitive work analysis. In J. Carroll (Ed.), *HCI models, theories, and frameworks: Toward an interdisciplinary science* (pp. 225–264). New York, NY: Morgan-Kaufmann.

Vicente, K. J. (1999). *Cognitive Work Analysis.* Mahwah, NJ: Lawrence Erlbaum Associates.

Woods, D. D., Johannesen, L. J., Cook, R. I., & Sarter, N. B. (1994). *Behind human error: Cognitive systems, computers, and hindsight.* Wright Patterson AFB, OH: CSERIAC.

Zhang, J., & Patel, V. L. (2006). Distributed cognition, representation, and affordance. *Pragmatics & Cognition, 14*(2), 333–341.

Author Index

A

Abowd, G., D., 97
Abrahamsson, P., 334
Adams, K. A., 134
Adibhatla, V., 243
Ahmadun, F., 133
Ahn, H., 163
Ahuja, J., 143
Akhondi, M., 82
Aldwin, C. M., 165
Allen, G., 216
Allison, L. K., 242
Altay, N., 136–138
Altman, I., 7
Ancona, D. G., 183
Anda, B., 331
Anderson, J. R., 223, 224, 233, 236
Ang, C. S., 184
Angles, J., 200
Argote, L., 51
Ariff, M. I. M., 49–51
Arthur, W., 242
Artman, H., 251
Atkinson, P., 90
Austin, J. R., 50–51
Ayoub, P. J., 92, 114

B

Baber, C., 201
Badke-Schaub, P., 184
Badler, N. I., 200
Baek, Y., 279
Bains, P., 12
Balakrishnan, B., 12, 243
Balogun, O., 134
Banbury, S., 252
Bannon, L., 9, 107
Barab, S. A., 32–33, 39
Barker, R. G., 7
Barrows, H. S., 32
Barsade, S. G., 167, 170
Bartel, C. A., 184
Bartlett, F., 343
Beal, S. A., 188
Beale, R., 97
Beck, K., 70, 334, 337
Behring, A., 289
Bell, B., 163, 242, 269
Bell, B. S., 183

Bell, S. T., 242
Belz, M., 253
Benbasat, I., 201
Bennett, J. M., 222, 234
Berntson, G. G., 168
Berry, K. A., 234
Biven, L., 223
Blumenfeld, P. C., 38
Blumenthal, S. H., 343
Boff, K. R., 11, 343
Boonthanom, R., 171
Boos, M., 253
Bootzin, R. B., 164
Borràs, J. M., 124t
Botero, I. C., 253
Boulos, M. N. K., 68
Bowman, J. M., 143, 254
Boyd, E. M., 293
Boyd, J. R., 321
Boyle, E., 134
Brandon, D. P., 50, 57
Brannick, M. T., 166
Bransford, J. D., 7–8, 19, 32, 35, 38, 190
Brant, J., 70
Braswell, R., 279
Braunstein-Bercovitz, H., 169
Breslow, L., 273
Breton, R., 252
Brewer, I., 12
Brief, A. P., 167
Brown, A. L., 19, 32–33
Brown, C., 6, 32
Brown, C. E., 9, 92
Brown, J. S., 7, 33, 36
Bruce, H., 120
Bryant, N., 344
Buie, E., 330
Buley, L., 329
Bunderson, J. S., 142
Burke, C. S., 184
Burns, C. M., 265, 364
Burtscher, M. J., 249
Busemeyer, J. R., 201
Busuttil, L., 279
Byrne, M. D., 224

C

Cabeza, R., 72
Cacioppo, J., 168
Cai, G., 368

Camilleri, V., 279
Campbell, G. E., 248
Campione, J. C., 33
Cañas, A. J., 304
Cannon-Bowers, J. A., 162, 168, 184, 186, 205, 213, 242
Carlsen, S., 82
Carper, B. A., 293
Carraher, T., 139–140
Carrithers, J. R., 133
Carroll, F., 68
Carroll, J. M., 68
Carroll, K., 68
Carteaux, R., 39
Cassenti, D. N., 231
Caulfield, M., 74
Cellier, J. M., 269
Chan, I. Y. S., 165
Chao, J. T., 68
Chen, C., 36
Chen, C. H., 36
Chen, X., 51
Chi, E. H., 76
Chi, M. T. H., 32
Childress, M. D., 279
Chin, G., 335
Christoffersen, K., 274
Citera, M., 9, 92, 200, 346
Clair, J. A., 134
Clark, L. A., 169
Clark, M. A., 185
Clark, R. E., 118
Clore, G. L., 167, 169
Cobb, P., 33
Cockburn, A., 67, 70
Cocking, R. R., 19, 32
Cohen, D., 182
Cohen, M. A., 236
Cohn, M., 337–338
Collins, A., 7, 33, 36
Comfort, L. K., 142, 162
Confrey, J., 33
Connors, E., 12, 206
Connors, E. S., 140, 206, 214, 243, 245
Converse, S., 213
Converse, S. A., 139
Cook, R. I., 303, 364
Cooke, N. J., 49, 113–114, 139, 163
Coombs, W. T., 134, 136
Cooper, C. L., 162
Coppola, D. P., 137–138
Coulson, R. L., 32
Cox, L. J., 201
Craik, K., 343
Crandall, B., 118–120
Cummings, M. L., 263
Cunningham, W., 69, 73

D

Dahm, W., 342
Dancy, C. D., 12
Dancy, C. L., 226, 234
Darzi, S. A., 184
Dauer, L. T., 162
Davies, M., 331
Davis, E. A., 35
Davis, K. A., 36
Day, E. A., 242
De Graaff, N., 82
DeAngelis, T., 344
DeChurch, L. A., 142–144, 184, 189, 242, 248–249, 253
DeHart, R. E., 133
Denning, S., 184–185, 190
Derryberry, D., 169
Dervin, B., 111
Dew, K., 124t
Dewe, P. J., 162
Dhukaram, A. V., 201
Dickerson, J. A., 210
Dickinson, T. L., 139
Dickson, M. W., 242
Diederich, A., 201
Dikmen, M., 263, 265
Dimentman-Ashkenazi, I., 169
DiSessa, A., 33
Dismukes, R. K., 266
Dix, A., 97
Dodge, T., 39
Dolcos, F., 72
Dominguez, C., 89
Dourish, P., 108
Dowell, J., 301
Draper, S. W., 16
Driskell, J. E., 168
Drury, J. L., 215
Du, J., 36
Düchting, M., 350
Duffy, T. M., 32–33
Duguid, P., 7, 33
Dul, J., 343
Dumville, B. C., 184, 213, 248, 253
Duncan, S., 290
Duran, J. L., 163

E

Echo, R., 343
Edwards, B. D., 242
Eggleston, R. G., 10, 301, 305
Ellis, D., 111
Ellwart, T., 249
Elm, W. C., 301
Elster, J., 165

Ender, T. R., 330
Endsley, M., 250–252
Endsley, M. R., 142
Endsley, T., 139
Endsley, T. C., 183
Engelmann, T., 242
Entin, E. B., 172
Entin, E. E., 166
Er, N., 35, 37
Ergener, D., 10
Erickson, J., 38
Espinosa, J. A., 185
Evans, B. M., 109
Eyrolle, H., 269

F

Fales, A. W., 293
Fan, W., 215
Fan, X., 367
Faulkner, B., 134
Feliciano, J., 68
Feltovich, J., 32
Feltovich, P. J., 32
Ferzandi, L., 12, 183, 242
Ferzli, M., 69
Fidel, R., 109, 111, 120
Filho, E. M., 242
Fincham, J. M., 233
Finck, M., 331
Finlay, J., 97
Fiore, S. M., 139, 184–185
Fischer, A., 133
Fischer, J. H., 164
Fisher, R. P., 115
Fiske, S. T., 7, 139
Fitts, P. M., 343
Flach, J., 112
Flach, J. M., 10, 89
Flentge, F., 289
Fletcher, J. D., 213
Flores, F., 112
Floridi, L., 315
Fogarty, W. M., 162
Folkman, S., 165
Forster, P., 139, 183
Fowler, M., 70
Foy, L., 251
Fu, W., 224
Funge, J., 200
Funke, J., 133

G

Gaffney, G., 348
Garbis, C., 251
Garcia, D., 73

Garland, G., 169
Garrett, J. J., 329
Gartenberg, D., 273
Garud, R., 184
Gasper, K., 169
Ge, X., 35–40
Geaneas, P. Z., 133
Geertz, C., 41
Geiselman, R. E., 115
George, J. M., 269
Gerosa, L., 109
Gershgoren, L., 242
Ghani, I., 331
Giacobe, N., 266
Giacobe, N. A., 11–12, 55,
    99, 244
Giambatista, R. C., 248
Gibson, D. E., 170
Gibson, J. J., 84, 89
Gick, M. L., 35, 38
Gilbreth, F. B., 343
Gilbreth, L. M., 343
Gilson, L., 248
Gilson, L. L., 244
Glantz, E. J., 12, 300, 304–306
Glaser, R., 32
Gluck, K. A., 223
Goldstein, E., 201
Golovchinsky, G., 110
González-Ibáñez, R., 108, 111
Goodman, P. S., 183
Goodstein, L. P., 301
Goodwin, G. F., 184, 205
Gorman, J., 175
Gorman, J. C., 163
Granger, D. A., 234
Grant, T., 321
Gray, S., 201
Gray, S. A., 201
Gray, W. D., 200
Gredler, M., 205
Green, D., 251
Green III, W. G., 136–138
Greenbaum, J. M., 308
Greenberg, S., 57, 171, 173
Greene, B., 38
Greiff, S., 133
Grötzbach, L., 350
Gualtieri, M., 332
Guarino, B., 222
Guba, E. G., 41
Guitouni, A., 252
Gunzelmann, G., 222, 226
Gupta, N., 248
Gurtner, A., 185, 191
Gutwin, C., 57, 171, 173
Guzdial, M., 68

H

Hager, R. S., 141
Hajdukiewicz, J., 262
Hakonen, M., 182
Hall, D., 99, 140, 366
Hall, D. L., 11, 122, 140, 243–244, 328
Hall, R. J., 207
Hamilton, K., 183, 185, 186, 189, 191–192,
        242–243, 248–250, 266
Hammersley, M., 90
Hammond, K. R., 165
Hancock, P. A., 163
Hannafin, M., 32
Hansen, P., 109
Harrison, D., 185
Hart, P., 135
Hart, S., 345
Hart, S. G., 169
Hartson, R., 333
Hauland, G., 252
Haynes, S. R., 236
Healey, A. N., 184
Heavey, C., 51
Heffner, T. S.
Hellar, B., 243
Hellar, D. B., 26, 55–56, 173, 186, 189, 244
Hellhammer, D. H., 234
Hellmann, T. D., 335
Helm, E. E., 249
Helmreich, R. L., 135
Helton, W. S., 169
Henly-Shepard, S., 201
Henningsen, D. D., 253
Herndon, B., 50–51
Hesse, F. W., 168
Hester, R. L., 234
Hill, J., 57
Hinden, R. M., 343
Hirokawa, R. Y., 107
Hmelo-Silver, C. E., 32
Ho, A., 263
Hodgetts, H. M., 266
Hoffman, R., 173, 344
Hoffman, R. R., 173
Holland, D., 69
Holland, J., 85
Hollands, J. G., 343
Hollenbeck, J. R., 107
Hollingshead, A. B., 49–50, 57, 253
Hollnagel, E., 6, 16, 112, 301, 330, 340, 343
Holsti, O. R., 133
Holyoak, K. J., 35, 38
Hornof, A. J., 229
Horvitz, E., 110
Hove, S. E., 331
Hsu, J. S.-C., 51

Huang, K., 36
Hudlicka, E., 163
Hutchins, E., 6–7, 9, 16, 48, 85, 140, 303, 364
Hyldegård, J, 110
Hyrkkanen, U., 331

I

Isenhour, P. L., 68

J

Jackson, M., 60
Jacobs, L. F., 217
Jacobsen, A. A., 342
Jadidi, M., 73
Jansen, B. J., 109–110, 338
Järvelin, K., 109
Jefferson, Jr., T., 206
Jefferson, T., 243
Jeffries, R., 69
Jenkins, B. M., 141
Jenkins, D., 349
Jenkins, D. P., 251
Jentsch, F., 184
Jett, Q. R., 269
Jimenez-Rodriguez, M., 143
Johannesen, L. J., 200, 205, 303, 364
Johnson, G. J., 345
Johnson, M. D., 142
Johnston, J. H., 162
Jonassen, D. H., 32
Jones, D. G., 142, 243, 247, 252
Jones, D. M., 266
Jones, R. E., 343
Jones, R. E. T., 55, 140, 200, 203, 205–208,
        212–214
Jones, R. K., 135
Jurca, G., 335–336

K

Kaber, D., 344
Kammen, D. M., 82
Kaplan, S., 253
Kapucu, N., 162
Karlsson, K., 82
Karsten, H., 108
Karunakaran, A., 106, 108–110
Kase, S. E., 222
Kauffman, D., 36–37
Keane, T. M., 169
Keel, P. E., 57
Keeman, V., 204
Kennedy, D., 248
Kerick, S. E., 231
Kessler, R. R., 69

Kieras, D. E., 223, 228–229
Kim, B., 279
Kimble, C., 10
Kirlik, A., 182
Kirmeyer, S. L., 269
Kirschbaum, C., 234
Kirsh, D., 85
Kitchin, R. M., 201
Kittur, A., 76
Klausen, T., 140
Klein, G., 10, 185, 317
Klein, G. A., 165, 173, 303–304, 307
Klein, G. L., 215
Klein, L. C., 222, 234
Klimoski, R., 242
Klimoski, R. J., 249
Knight, P., 201
Knott, B. A., 53
Kolbe, M., 249
Komis, V., 278
Konradt, U., 249
Kooter, B., 321
Kop, R., 68
Kosko, B., 201–204, 210
Kosslyn, S. M., 233
Kozlowski, S., 242
Kozlowski, S. W. J, 183
Krishnappa, R., 110
Kuhlthau, C. C., 110–111
Kumar, V., 70
Kyng, M., 308

L

Lai, Y., 215
Laird, J. E., 223, 227
Lamb, B., 123, 124
Lamb, J. N., 123
Land, S., 32
Land, S. M., 36, 38
Laraia, W., 141
Larson, J. R., 254
Lauche, K., 184
Lave, J., 6
Lawrence, B. S., 183
Lawton, G., 75
Lazarus, R. S., 163–165, 169
LeDoux, J. E., 233
Lee, F. J., 224
Lee, J. D., 182
Lehman, J. D., 35
Lehrer, R., 33, 38
LeRouge, C., 332
Letsky, M., 107, 139, 248
Leung, M.-Y., 165
Leupp, D. G., 10
Lewis, J. R., 345

Lewis, K., 50–51
Liang, D. W., 50
Liao, J., 51
Lichacz, F. M. J., 252
Lim, A., 183
Lin, X., 35
Lincoln, Y. S., 41
Linn, M., 35
Liotti, M., 233
Lipshitz, R., 317
Lisetti, C., 163
Littlepage, A. M., 51
Littlepage, G. E., 51
Liu, B., 215
Liu, M., 38
Long, J., 301
Lothian, J., 292
Lubow, R. E., 169
Ludovice, P., 68
Luege, T., 137
Luminet, O., 164
Lund, H., 82

M

Ma, J., 332
Mace, W. M., 6, 22
MacEachren, A. M., 368
Mack, A., 154
Magill, S. A. N., 269
Mancuso, V., 24–25, 99, 139, 183, 185, 189–192, 244, 266
Mancuso, V. F., 53–55, 139, 243–244, 300
Manns, J., 201
Manser, T., 249
Maramba, I., 68
Marczyk, J. L., 215
Marks, M. A., 246–247
Mathiesen, B. V., 82
Mathieu, J., 205, 248
Mathieu, J. E., 182
Matthew, M. D., 188
Matthews, M. D., 251
Maurer, F., 335
Mavor, A. S., 222
Max, D., 164
Maybury, M., 53
Maynard, M. T., 182, 242, 248
MaZaeva, N., 134
Mazeland, H., 172
McBride, M. E., 134
McComb, S., 116–117, 248
McComb, S. A., 184
McCormick, E. J., 15–16
McCrickard, D. S., 68
McCurry, J. M., 273
McDaniel, R., 184

McDougal, W., 170
McDowell, K., 231
McGrath, J. E., 242, 244
McGuinness, B., 251
McHugh, A. P., 135
McMullen, S. A. H., 328
McNally, R. J., 164
McNeese, M., 122, 364–368
McNeese, M. D., 5, 6, 9, 122, 139, 303–306, 328, 331, 364–368
McNeese, N., 306
McNeese, M., 286
McNeese, N. J., 6, 10, 15, 99, 108, 111, 117
Meister, D., 343
Mell, J. N., 49
Mendoza, S. H. V., 278
Meshkati, N., 344
Mesmer-Magnus, J., 184, 189, 191
Mesmer-Magnus, J. R., 142–144, 242, 248–249, 253
Meyer, D. E., 228
Michinov, E., 51
Michinov, N., 51
Mikolajczak, M., 164
Milanovich, D. M., 248
Milgram, P., 303
Militello, L. G., 92
Miller, C., 69
Minotra, D., 263, 265–269, 272–274
Minsky, M., 165
Mitchell, P. J., 263
Mitchelson, J. K., 242
Mkrtchyan, L., 204
Moffett, J., 332
Mogg, K. K., 164
Mohammed, S., 13, 25, 108, 183–185, 189, 191–193, 195, 213, 242, 248–250, 252–254
Molich, R., 352
Monsell, S., 265
Montebello, M., 279
Montero, J., 204
Montero de Juan, J., 201
Moogk, D. R., 336
Moon, B., 114
Moore, L., 135
Moore Jr., L. R., 223
Moreland, R. L., 49–50
Morley, T., 68
Morris, C. S., 163
Morris, M. R., 109–110
Moskowitz, J. T., 165
Mullaney, T., 344
Mumaw, R. J., 302
Murase, T., 242
Murray, D., 330
Myaskovsky, L., 49–50

Myers, C. W., 163
Myers, G. J., 336
Mysirlaki, S., 278

N

Nadel, L., 201
Nadkarni, S., 185
Nägele, C., 186
Neale, D. C., 68
Nebe, K., 350
Nemeth, C. P., 48
Neocleous, C., 204
Neumann, A., 184
Newell, A., 227
Newman, S. E., 36
Nielsen, J., 345, 352
Nilsson, H., 247
Niskanen, V. A., 204
Norberg, A. L., 343
Norman, D., 343
Norman, D. A., 7, 16, 48, 165, 301
Novak, J. D., 304
Ntuen, C. A., 134, 137, 142

O

Obendorf, H., 331
Oliver, K., 32
Opdyke, W., 70
Orlitzky, M., 107
Ortony, A., 165
Oyserman, D., 139

P

Panksepp, J., 223, 233
Papaioannou, M., 204
Papargyris, A., 279
Paraskeva, F., 278
Parasuraman, R., 263
Park, H., 279
Parker, K. R., 68
Parr, A., 190
Parsons, W., 136
Patadia, S., 82
Patel, V. L., 364
Patterson, E. S., 242, 302, 307
Paul, S. A., 110–111
Pearson, C. M., 134
Pedersen, H., 269
Pejtersen, A.M., 120
Pendleton, B. A., 76
Pennebaker, J. W., 171
Peres, A. L., 333
Perkins, D. N., 8, 19–20, 92, 99–100
Perry, D. G., 343

Perryman, R., 248
Perusich, K., 10, 183, 200, 203, 206, 242, 331
Petersen, S., 82
Pew, R. W., 222
Pfaff, M., 12, 25, 87
Pfaff, M. S., 117, 139, 141, 170–174, 215, 223,
    243, 247, 278, 280, 292
Phillips, C., 200
Picard, R. W., 163
Pickens, J., 109
Pierce, L. G., 110
Pirke, K.-M., 234
Pirzadeh, A., 166–167, 171
Planas, L. G., 35
Poltrock, S., 109
Post, T. A., 32
Potter, S. S., 301
Poulymenakou, A., 279
Prades, J., 124
Prekop, P., 110
Prince, C., 166
Pyla, P. S., 333

Q

Qin, Y., 233
Quinn, D. M., 162

R

Rack, O., 249
Rafferty, L. A., 242
Randall, K. R., 10, 242, 249
Rao, G. S. V. R. K., 184
Rapoport, A., 134
Rapp, T., 248
Raser, J. C., 247
Rasmussen, J., 6–7, 16, 91, 94–95, 205, 274,
    301, 364
Rasmussen, L. J., 112, 120
Realff, M., 68
Reddy, M., 12, 25, 108–110
Reddy, M. C., 108, 110–111, 115, 117
Reed, M. A., 169
Reep, J., 99, 255, 365
Reep, J. A., 12, 25
Reeves, J. R., 69
Regli, S. H., 26, 346, 348
Reifers, A. L., 222
Ren, Y., 51
Rentsch, J., 10
Rentsch, J. R., 183, 200, 206–207, 242, 249, 331
Resick, C. J., 242, 249
Resnick, L., 140–141
Reuveni, Y., 249
Reynolds, A. M., 248
Riley, J. M., 212

Ritter, F. E., 222, 234, 236
Roberts, D., 70
Rock, I., 154
Rodriguez, J. T., 204
Rodriguez-Repiso, L., 204
Ronkainen, J., 334
Rosen, M. A., 139, 184
Rosenthal, U., 135
Rosson, M. B., 68
Roth, E., 344
Roth, E. M., 301–302, 307
Rouse, W. B., 186, 205
Rousseau, R., 252
Rovira, E., 263
Roy, E., 164
Ruan, D., 201
Ruiz-Casado, A., 124t
Russ, A. L., 343
Russell, D. M., 162, 165, 174
Russell, J. A., 162, 165, 174
Rutledge, K., 98

S

Said, A. M., 133
Salas, E., 107, 139, 163, 168, 184, 186, 205,
    242, 248
Salmeron, J. L., 204
Salmon, P., 251, 349
Salmon, P. M., 252
Salo, O., 334
Salomon, G., 6
Salvucci, D. D., 223
Sampson, D. G., 278
Sanders, E. B.-N., 351
Sanderson, P. M., 364
Santos, E., 242
Sarter, N. B., 200, 205, 303, 364
Sauro, J., 345
Schank, R. C., 185
Schauble, L., 33
Schauenburg, B., 253
Schenk, F., 201
Schinke, R. J., 242t
Schizas, C., 204
Schmidt, K., 9, 107
Schoelles, M., 222
Schoelles, M. J., 222
Schraagen, J. M., 173
Schraagen, J. M. C., 117
Schreiber, M., 242
Schwaber, K., 336
Schwarz, N., 139, 167
Scott, C. P. R., 248
Sellers, J., 58
Selvaraj, J. A., 6
Semin, G. R., 139

Semmer, N. K., 186
Serfaty, D., 166
Setchi, R., 204
Sevdalis, N., 184
Shah, C., 108–109, 111
Shaluf, I. M., 133
Sharma, R., 368
Shea, P., 68
Sheppard-Sawyer, C. L., 164
Shirkey, E. C., 163
Shou, W., 215
Shuffler, M., 143, 248
Sieck, W. R., 135
Simsek, Z., 51
Sinnott, J. D., 32
Smarandache, F., 204
Smart, P. R., 135, 246
Smith, E. R., 139
Smith-Jentsch, K. A., 248
Sneha, S., 332
Snyder, D. E., 8, 91, 305
Sohrab, S. G., 253
Sonnenwald, D. H, 110
Spence, P. R., 110
Spielberger, C. D., 169
Spies, K., 168
Spiro, R. J., 32
Stahl, G., 168
Stanton, N., 242, 251, 349
Stanton, N. A., 242, 251, 349
Stasser, G., 143, 146, 253
Staveland, L. E., 169
Stein, B. S., 8, 198
Steinkuehler, C., 290
Stephens, R. J., 242
Stewart, D., 253
Stocco, A., 233
Storey, M., 68
Stout, R., 205
Stout, R. J., 242
Strack, M., 253
Strater, L. D., 212, 213
Strauch, B., 135
Strauss, H. W., 162
Strohmaier, M., 73
Suh, B., 76
Sun, B., 367
Sun, H., 165
Sun, S., 367
Sutcliffe, A., 332
Sutcliffe, K. M., 142
Sycara, K., 246

T

Taatgen, N. A., 224
Taber, R., 202, 204

Talevski, A., 82
Tamblyn, R. M., 32
Tang, Y., 246
Tannenbaum, S. I., 139, 182
Tassinary, L. G., 168
Tausczik, Y. R., 171
Taylor, C., 124
Taylor, F. W., 343
Taylor, R. M., 188
Taylor, S. E., 7
Taylor, W., 204
Tellegen, A., 169
Tenenbaum, G., 242
Terrell, I. S., 243
Terzopoulos, D., 200
Tesler, R., 185, 191–193, 248, 250, 252, 255
Tesler, R. M., 183, 191–193, 248
Thomas, L. C., 269
Thomas, M., 39
Thomas, M. K., 38–40
Thomson, D. M., 48
Tinapple, D., 307
Titus, W., 143, 146
Todd, P., 201
Tolle, K., 332
Trafton, G. J., 226
Trafton, J. G., 273
Tremblay, S., 252
Tremoulet, P. D., 26, 346, 348
Tschan, F., 58, 186
Tu, X., 200
Tulving, E., 48
Turner, A., 134
Tushman, M. L., 183
Tuttle, R. M., 162
Tuzun, H., 39
Tyworth, M., 11–12, 53, 54, 62, 99, 244, 266

U

UDell, J., 73
Uitdewilligen S., 249
Undre, S., 184

V

Van Bergen, A., 269
Van den Bossche, P., 113
Van Dijck, J., 328
Van Vugt, M. K., 231, 236
Vartiainen, M., 182, 331
Vasantha-Kandasamy, W. B., 204
Vashdi, D. R., 249
Verstrynge, V., 164
Vicente, K., 7
Vicente, K. J., 92–93, 95, 120, 262, 274, 301–303, 317, 349, 364

Vidulich, M., 5, 10–11, 364
Vincent, C., 12, 184
Vitoriano, B., 201, 204
Von Clausewitz, C., 140
Voss, J. F., 32
Voulgari, I., 278

W

Wacker, J., 249
Wagner, C., 73
Walker, G., 242, 251, 349
Walker, G. H., 251
Walker, M. C., 141
Waller, M. J., 248–249, 253
Wang, W, 68
Ward, J. L., 73, 331
Warner, N., 139
Warsta, J., 334
Watson, D., 169
Weber, J. G., 82
Weber, S. G., 289
Wegner, D. M., 49, 51, 55, 59
Wei, M., 82
Weick, K. E., 111
Weiss, H. M., 167
Wellens, A. R., 10–11, 184, 245, 250–252
Wenger, E., 32–33
Westbrook, J. I., 269
Westermann, R., 168
Wheeler, S., 68
Whetzel, C. A., 234
Whitaker, R., 6, 9, 92, 200, 301, 346
Whitaker, R. D., 6, 9–10
Wickens, C. D., 264, 343
Wiebe, E., 69
Wildman, J., 143, 248
Wildman, J. L., 183

Williams, L., 67, 69–70
Willig, C., 146
Willis, G. B., 115
Wilson, D., 10
Wilson, T. D., 111
Winograd, T., 112
Wittenbaum, G. M., 143, 253–254
Wohl, J. G., 321
Wood, J. R., 349
Wood, L. E., 349
Woods, D. D., 6, 10, 16, 20, 99, 112, 200, 205,
    301–303, 307, 330, 343, 364
Wulf, T., 69

X

Xiao, Y., 303
Xie, K., 36

Y

Yang, K., 69
Yen, J., 367
Young, N. C., 162
Yu, J., 165

Z

Zaff, B. S., 9–10, 28, 91–92, 200, 301, 304, 346
Zagalsky, A., 68
Zajonc, R. B., 165
Zanzonico, P., 162
Zhang, J., 364
Zhao, Y., 68
Ziegert, T., 289
Zijlstra, F. R. H., 249
Zimmermann, D., 350
Zsambok, C. E., 10

Subject Index

Note: Page numbers followed by f and t refer to figures and tables, respectively

A

Active errors, 323
ACT-R, 224–227, 224f
   conflict-resolution cycle, 225–226, 225f
   memory systems, 224
   perceptual-motor modules, 224
   subsymbolic activation value, 226
Adaptable interface, 187
Adaptive Architectures for Command and
   Control (A2C2), 172
Adaptive team model, 166
Advanced Interface Design Lab (AIDL)
   simulator, 269–273
   display panels, 269
   interface design of, 269
   interface layout, 270f
   interruption, 272–273
   predictability, 269, 272
   task load, 272
   volatility, 269
Advanced knowledge and design acquisition
   methods (AKADAM), 8–9, 304–305
Agile development, living laboratory and,
   333–338, 335f
   principles, 334
   software development, 336–337
   sprint, 336–338
   success criteria, 337
   user feedback evaluation, 338
AIDL simulator. See Advanced Interface Design
   Lab (AIDL) simulator
Air Force Research Laboratory (AFRL), 4
Analog storytelling, 190, 192
Analyses of variance (ANOVA's), 322
Analysis, CTA, 118
Artificial-intelligence (AI) approach, 200
Attention processes, 6

B

Backseat driving approach, 69
Beta activity, 231
Big data, 328
Black box, 25
Blood oxygen level dependent (BOLD), 231
Boston Marathon bombing, 141, 368
Building cognitive systems, 133

C

Canvas LMS, Federated Wiki, 68, 77
Card-sorting, 115–116, 349–350
Catastrophic events, 133
CCRINGSS program, 82–85
Character, WoW, 279t
Chat panel, NeoCITIES interface, 245
CIS. See Collaborative information
   seeking (CIS)
CITIES game, 243
CITIES simulation, 10, 11
Class, WoW, 279t
Client server architecture, 194
Coagmento tool, 111
Cognition process, 140
   CIS, 107–108
   CSE, 112
   CTA, 118
Cognitive architectures, 221–236
   ACT-R, 224–227, 224f
   and driving story, 229–230
   EPIC, 228–229
   overview, 221–223
   psycho-physiological measures and,
     230–234
     EDA, 233–234
     EEG, 230–231
     fMRI, 231–233
   Soar, 227–228
   stressed serial subtraction model,
     234–235
Cognitive artifacts, transactive memory, 48
Cognitive engineering and decision
   making, 5, 6
Cognitive ergonomics, 6
Cognitive mapping, 201
Cognitive modeling, 199–214
   definition, 200–201
   FCMS, 201–214
     ECM teams study, 204–207
     SA–FCM modeling approach, 207–214
   overview, 200–201
Cognitive readiness, 213
Cognitive science, 5
Cognitive systems approach
   to natural gas. See Natural gas exploration/
     exploitation

Cognitive systems engineering (CSE), 5, 52, 112,
    133–134, 163, 301–303, 343
  functional action, 302–303
  information technology relevance, 302
  intrinsic constraints, 302
  MCT, 124t
  proportionality of, 22–23
  workaround activities, 302
Cognitive task analysis (CTA), 6, 117–119
  CIS, 119–121
  CWA, 120–121
  KE, 117, 118
Cognitive work analysis (CWA), 120–121
Collaboration and timing, 190
Collaboration, natural gas exploitation, 83–85
Collaborative and human–autonomous systems, 8
Collaborative decision-making process, 186
Collaborative design technologies (CDT), 9
Collaborative ecosystem, 8
Collaborative information seeking (CIS),
    108–109
  Coagmento, 111
  cognition, 107–108
  CSE, 112
  CTA, 119–121
  CWA, 120–121
  definitions, 108–109
  ISP model, 110
  KE, 114, 116–117
    conceptual techniques, 114–116
    CTA, 117, 118
    definition, 113
    observations and interviews, 113–115
    process tracing, 113–115
    team mental model, 116–117
  LL approach, 121–123
    ethnographic studies, 122
    KE methods, 122
    MCT, 123–124, 124t
    scaled-world simulations, 122
    support tools, 122
  sensemaking, 111
  social and technical, 108
  social stream, 110
  team cognition, 111
  team decision making, 107
  technical stream, 110–111
Colombian aircraft, 135
Combat exposure scale (CES), 169
Combined critical decision method (CDM), 304
Command and control (C2) centers, 141
Command, control, communications, and
    intelligence (C³I), 9
Commercial off-the-shelf (COTS)
    computers, 343
Common operational picture (COP), 134,
    137, 141

Communication
  and completion rates, 147, 148t, 151t
  Manchester University case-study context,
    147–149
  Penn State case study context, 151–152
Comprehension stage, 142
Computer simulation for team research (CSTR),
    246–248
Computer-supported collaborative learning
    (CSCL), 71
Computer-supported cooperative work (CSCW),
    10, 71, 301
Computing technology, 343
Concept mapping, 115, 191
Conceptual knowledge representation, 8–9
Conflict-resolution cycle, 225–226, 225f
Context(s)
  comparison, 154–155
    communications, 154
    performance, 155
    sharing hidden knowledge, 155
  crisis management, 135–136
  cultural, 134
  Manchester University case-study, 146–151
  Penn State case study, 151–154
Context switching, 90, 367
Contextualistic-social constructivist
    perspectives, 9
COPE (C³ Operator Performance Engineering), 9
Credit-based curriculum, 145
Crew Awareness Rating Scale (CARS), 251
Crisis
  cycle, 136–137
    team activities, description, 138, 138t
  defined, 133–134
  inherent chaotic nature, 143
Crisis environments, team performance,
    280–281
  after the raid, 284–286
  prior to the raid, 281–282
  during the raid, 282–284
Crisis management
  cognitive systems engineering, 133–134
  impacted by culture, 134–144
    challenges teams face, 140–141
    context, 135–136
    COP, 141
    crisis cycle, 136–137
    cultural composition, 144
    hidden knowledge, 143–144
    information sharing, 142–143
    team cognition, 139–140
    teams, 138–139
    team situation awareness, 141–142
Critical cognitive process, 184
Cross-pollinating effect, 10
Crowd sourcing, 366

CSCL. *See* Computer-supported collaborative learning (CSCL)
CSCW. *See* Computer-supported collaborative work (CSCW)
CSE. *See* Cognitive systems engineering (CSE)
CTA. *See* Cognitive task analysis (CTA)
C'Thun, raiding, 290–291
Cultural context, 134
CWA. *See* Cognitive work analysis (CWA)
Cyber analysts, 365
Cyber operations
  transactive memory, 53
    ATM-I capabilities, 58–59
    distributed teams, 60
    functional domains, 54–55, 54t
    hands-on approach, 53
    human–computer interaction approach, 60
    intrusion/threat analysis, 55
    knowledge structures, impact of, 59
    scaledworld simulation, 54
    teamNETS, 55–57, 56f, 60
    transactive memory assessment, 55
    virtual feedback system, 59–60

D

Daily standup, 337
DART. *See* dual-task attention research testbed (DART)
Data collection, CTA, 118
Data preparation, CTA, 119
Data structure, CTA, 119
Decision-aiding support methodology, SA–FCM, 209–211
  map, validate, 211
  node, extend, 211
  relationship between nodes, 209
  SA requirements to FCM node translation, 209
  weight identification, 209–210
  weight to fractional values conversion, 210
Decision-aids, 200
Decision-making, sub processes of, 201
Declarative memory system, ACT-R, 224, 226
Defense Advanced Research Projects Agency (DARPA), 342
Demographic information, 315
Department of Health, 123
Department of Homeland Security, 137
Design research approach, 33–34
Design seeds, 307
Digital *vs.* analog recordings, audio messages, 316–317
Directory-updating stage, transactive memory, 49
Discovering meaning, CTA, 119
Disruptiveness of interruptions, 262
Distributed cognition, 7, 9

Distributed team cognition, transactive memory, 48, 57–58, 60
Dragon kill points, WoW, 279t
DreamVision, 331–333, 336, 337
Dry-erase markers, 304, 305, 308, 310
Dual-task attention research testbed (DART), 263–268
  in cybersecurity, 266–267
  in living laboratory framework, 267–268
  NETS-DART, 266
  operator task, 263
  priorities change with time, 264f
Dungeon/instance, WoW, 280t
Dynamic prioritization, 264
Dysfunctional reasoning, 165

E

ECM. *See* Emergency crisis-management (ECM)
EDA. *See* Electrodermal activity (EDA)
EEG. *See* Electroencephalography (EEG)
Ecological settings and cognitive design practice, 7–8
Ecologic interface design (EID), 317
Educational researchers, 32–34
Elaborative prompts, 35
Electrodermal activity (EDA), 233–234
Electroencephalography (EEG), 230–231
Emergency crisis-management (ECM), 11, 186, 204–207
  characteristics, 204
  domain type, 205
  information requirement, 206
  NeoCITIES task, 206
  utilizing, 206
Emergency operations center (EOC), 318–325
  findings, 322–325
    accurate reporting rates, 323f
    errant reporting rates, 324f
    session aggregated response timing, 323f
  interface screen-shot, 319f, 320f
  overview, 318–320
  requirements and hypotheses, 321–322
  text-to-speech software, 319
Emotional contagion concept, 167
Emotional factors, stress, 163
Emotion, stress, and collaborative systems, 161–173
  midway breather, 167
  overview, 162–163
  research approaches, 167–173
    manipulations and measures, 168–169
    team-level emotion and stress research, applications, 172–173
    team-level measures, 170–172
  theoretical perspectives, 164–167
    and individual cognition, 164–166
    and team cognition, 166–167

Environmental psychology, 7
Environmental Services/Army, 147
EPIC architecture, 228–229
    cognitive processor, 229
    information handling, types, 228
    perceptual processors, 229
Ergonomics, 342–343
Ethnographic fieldwork, LLF
    collaborative systems/technologies, 91
    context switching, 90
    fracking process, 91
    functional analysis, 91
    observation learning, 90–91
    opportunities, 91–92
    supply chain, 91
    team of teams interaction, 91
Ethnography, 305, 306–304
    LL approach, 122
    MCT, 124t
Etymology, 348–349
Event-related potentials (ERPs), 230, 231
Event tracker, NeoCITIES interface, 246
Experimental-cognitive psychology, 5
Experimental methodology, 186–189
    general experimental procedure, 188–189
        dependent measures, 188–189
    NeoCITIES
        rationale, 186–187
        scenario development, 187
        team simulation, 186

F

Federated Wiki, 73, 74f. *See also* Wikipedia
    advantages, 77–78
    Canvas LMS, 77
    Google Docs, 76–77
    idea mining, 74, 75f
    Slack, 77
    vs. Wikipedia, 76, 76f
Fork and Pull model, 71–72
Formative evaluation, 355
Fracking process, 91
Freeze probe technique, 188
Fukushima Daiichi nuclear power disaster, 162
Functional action, 302
Functional knowledge representation, 9
functional magnetic resonance imaging (fMRI),
    231–233
Functional requirements, 350
Fuzzy cognitive maps (FCMS), 25
    activation value, 202
    ECM teams study, 204–207
    SA–FCM modeling approach, 207–214
        for data fusion, 214
        for decision-aiding support, 208–211

    for measuring cognitive readiness, 213
    for predictive assessments, 212–213
    sample, 202f
Fuzzy logic, 201

G

Gameplay, NeoCITIES, 244, 246
Gamma rhythms, 231
Germanwings Flight 9525, 137
GitHub/GitLab, 68, 70–72
Global positioning system (GPS), 301
Google Docs, Federated Wiki, 76–77
Graphical user interface (GUI), 319
Group support systems, 287–289
Guided individual reflexivity, 191
Guided team reflexivity, 191
Guild, WoW, 280t

H

Hands-on approach, 53
Hazardous materials, 147, 186
Health and Homeland Alert Network (HHAN),
    141
Healthcare.gov, 344
Hemodynamic response function (hrf), 231
Heuristic evaluators, 352
Hidden knowledge, 365
Hidden-profile task, 253
Hierarchical Task Analysis (HTA), 349
Highfidelity mock-up, 333
Human–agent interaction, 19, 23
Human-centered approach, 85–87
Human–computer interaction (HCI), 13, 60, 163,
    301, 314
Human factors, 300, 342–343
Human Factors and Ergonomics Society
    (HFES), 5
Human factors/cognitive system terms,
    15–17
Human-in-the-loop simulations, 230
Human-in-the-loop testing, 97
Human–social cognition, 7
Human systems engineering, 341–357
Human-systems integration (HSI), 13
Hyperventilating patient, 316

I

Idea mining, Federated Wiki, 74, 75f
IDEAS. *See* interaction design and
    engineering for advanced systems
    (IDEAS)
idsNETS, 266
Ill-structured problems, 32

Independent variable manipulations, 189–191
  reflexivity, 191
  storytelling, 190
    collaboration and timing, 190
    metaphorical *versus* analog, 190
India's intelligence network, 141
Individual memory, transactive memory, 49
Individual reflexivity, 191
Industrial/cognitive systems engineering, 13
Industrial/Organizational Psychology literature,
    193
Information allocation stage, transactive
    memory, 49
Information communication technologies (ICTs),
    143
Information Sciences and Technology (IST),
    MINDS Group, 84, 86
Information Search Process (ISP), 110
Information sharing, 142–143, 188
Infosphere, 315
Integrated living laboratory framework (LLF), 5,
    363–368. *See also* Living laboratory
    framework (LLF)
  objectives, 363–364
  perspective, 364
  practical considerations, 367
  problem-centered challenges, 364–366
  technological advancements, 366–367
Integrated technologies, 141
Integrative perspective, value, 193
Interaction design and engineering for advanced
    systems (IDEAS), 346–353
  definition, 347–348
  design and engineering, 350–351
  evaluation, 353
  implementation, 353
  interface review, 351–353
  needs analysis, 348–350
  origin, 346–347
  real-world application, 354–356
  requirements generation, 350
  six steps of, 348f
  with software development methods, 355–356
Interdisciplinary perspective, value, 193–194
Interface, NeoCITIES simulation, 245–246
  chat panel, 245
  dispatch panel, 245
  event tracker, 246
  team monitor, 246
  unit monitor, 246
Intermediate Object Oriented Design and
    Applications (IST 311), 72
Intermodal interfaces, 313–325
  application domains, 318
  definition, 314–315
  design challenges, 318

EOC, 318–325
  findings, 322–325
  overview, 318–320
  requirements and hypotheses, 321–322
  naturalistic decision making and situational
    awareness, 317
  straight digitalization, case against, 315–317
Intrinsic constraints, 302
ISchools (Information Schools), 13
ISP. *See* Information Search Process (ISP)
Ispatch panel, NeoCITIES interface, 245
Iterative design process, 35–36

J

Java design and development course (IST 311), 71

K

KE. *See* Knowledge elicitation (KE)
Knowledge elicitation (KE), 114, 116–117,
    307–308
  conceptual techniques, 114–116
  CTA, 117, 118
  definition, 113
  difficulty, 118
  generality, 118
  LL approach, 122
  LLF
    individual cognition, 92
    mutual learning and enhancement, 93–94
    operations, 94
  observations and interviews, 113–115
  process tracing, 113–115
  realism, 118
  team mental model, 116–117
Knowledge representation, CTA, 119
Knowledge/skill acquisition, 6
Knowledge, skills, abilities (KSAs), 213
Knowledge structures, 6

L

Large-scale datasets, 366
Learning management system (LMS), 68
Lesser offenses, 306
Linguistic Inquiry and Word Count (LIWC), 171
Living ecosystem, 23
Living laboratory (LL) approach, 33–35, 121–123
  applications, 52–53
  cognition, 33
  design research, 33–34
  educational researchers, 32–34
  ethnographic studies, 122
  ill-structured problems, 32
  KE methods, 122

Living laboratory (LL) approach (*Continued*)
    MCT, 123–124, 124t
    OLEs, 32
    scaled-world simulations, 122
    students-as-designers
        knowledge, 38
        project-based learning, 38–40
        roles, researchers, 40–41
        stakeholders, 41
        TRE, 41, 40
    support tools, 122
    transactive memory. *See* Transactive memory
    web-based cognitive support system, 35
        contribution, learners, 37
        iterative design process, 35–36
        prompts, 35
        roles, researchers, 37
        think-aloud technique, 37
Living laboratory framework (LLF), 4, 8, 68, 73,
        89, 132, 139
    agile development, 333–338
        principles, 334
        software development, 336–337
        sprint planning, 336
        sprint review, 337–338
        success criteria, 337
        user feedback evaluation, 338
    brief description, 329t
    cognitive systems engineering, 328–329
    dynamic complex systems, 88
    elements of, 20–23, 20f, 21f
    ethnographic fieldwork
        collaborative systems/technologies, 91
        context switching, 90
        fracking process, 91
        functional analysis, 91
        observation learning, 90–91
        opportunities, 91–92
        supply chain, 91
        team of teams interaction, 91
    integrated, 5, 363–368
        objectives, 363–364
        perspective, 364
        practical considerations, 367
        problem-centered challenges, 364–366
        technological advancements, 366–367
    interpretation, 4–5
    knowledge elicitation
        individual cognition, 92
        mutual learning and enhancement, 93–94
        operations, 94
    mutual learning, multisource knowledge on,
        88
    natural gas exploration/exploitation, 87–89
    optimal performance, 89
    principles of, 23

problem-based learning, 88
prototypes
    challenges/research, 99–100
    cyclic framework, 99
    envisioned designs, 99
    NeoCITIES architecture, 99
    opportunities, 100
    scaled-world environment, 99
relevance, 17–23
scaled-world simulation
    challenges/research, 97–98
    convergence, 96
    decisions and actions, 97
    human-in-the-loop testing, 97
    kinetic cycling, 96
    opportunities, 98
    supply chains, 96
theoretical foundations
    challenges/research, 95
    distributed information, 94
    information fusion, 94–95
    opportunities, 96
    problems experienced, 95
and UX, 328–333
    dual-faced challenge, 330
    knowledge elicitation to sense-making,
        330–331
    mitigating risk through personas, 331–332
    mock-ups and prototypes, 333
    problem identification, 329–330
    product vision building, 332
LL approach. *See* Living laboratory (LL)
        approach
LLF. *See* Living laboratory framework (LLF)
LMS. *See* Learning management system (LMS)
Long-term memory structures, Soar, 227
Look-ahead time (LAT), 269
Low-fidelity mock-up, 332

M

Macroawareness, 8
Macrocognition, 6, 8
Manchester University case-study context,
        146–151
    communication, 147–149
        and completion rates, 147, 148t
        sharing hidden knowledge, 149–150
Manual motor processor, EPIC, 229
Massively multiplayer online games (MMOGs),
        278–279
Massive Online Open Course (MOOC) settings,
        68
MCT. *See* Multidisciplinary cancer teams (MCT)
Mental models, 6
Metacognition, 6

Metaphorical storytelling, 190
Metaphorical *vs.* analog, 190
Method focus, CTA, 118
Midway breather, 167
MINDS. *See* Multidisciplinary initiatives
        in naturalistic decision systems
        (MINDS)
MINDS Group, College of Information Sciences
        and Technology (IST), 84, 86
Mindware inferencing, 8
Minimum viable product (MVP), 336
Minotra, D.
    high-task load scenario
        with critical events, 268f
        without critical events, 267f
    NETS-DART in, 266f
Mission awareness rating scale (MARS),
        188–189, 251
Mitigation stage, 136
MMOGs. *See* massively multiplayer online
        games (MMOGs)
Mock-up, 351
Model Based Systems Engineering approach, 347
MOOC settings. *See* Massive Online Open
        Course (MOOC) settings
Mozilla, Fork and Pull model, 71
Multidisciplinary cancer teams (MCT),
        123–124, 124t
Multidisciplinary initiatives in naturalistic
        decision systems (MINDS), 5, 11, 266
    The Pennsylvania State University, 11–13
Municipal Police Department (MPD), 305

N

National Governors Association Emergency
        Preparedness Project, 136
Natural gas exploration/exploitation, 83, 89
    affordance, 84
    CCRINGSS, 82–85
    collaborative ecology, 83–84
    collaborative technology, 84
    distributed cognition theory, 85
    domains, 86
    dynamic complex systems, 88
    ecological system, 84
    effectivity, 84
    ethnographic fieldwork
        collaborative systems/technologies, 91
        context switching, 90
        fracking process, 91
        functional analysis, 91
        observation learning, 90–91
        opportunities, 91–92
        supply chain, 91
        team of teams interaction, 91

human-centered approach, 85–87
individual operations and analysis, 87
information flows, 85
knowledge elicitation, 92–93
    individual cognition, 92
    mutual learning and enhancement, 93–94
    operations, 94
living laboratory framework (LLF), 87–89
open architecture computing, 85
optimal performance, 89
organization and enterprise architecture, 87
premise, 85–87
prototypes
    challenges/research, 99–100
    cyclic framework, 99
    envisioned designs, 99
    NeoCITIES architecture, 99
    opportunities, 100
    scaled-world environment, 99
scaled-world simulation
    challenges/research, 97–98
    convergence, 96
    decisions and actions, 97
    human-in-the-loop testing, 97
    kinetic cycling, 96
    opportunities, 98
    supply chains, 96
supply chain, 86–87
teamwork and collaboration, 87
theoretical foundations
    challenges/research, 95
    distributed information, 94
    information fusion, 94–95
    opportunities, 96
    problems experienced, 95
Naturalistic decision making (NDM), 317
Needs analysis, 348–350
    card-sorting, 349–350
    ethnography, 348–349
    semistructured interviews, 349
    sources of information, 348
    task analysis, 349
Negative affectivity (NA), 169
NeoCITIES, 10, 136, 144, 186–188, 206
    analog story, 190
    piloting, 188
    rationale, 186–187
    scenario development, 187
    simulation, 146
        communication codes, 148t
    team performance, 189
    team simulation, 186
NeoCITIES 3.0, 194
NeoCITIES architecture, 99
NeoCITIES Experimental Task Simulator (NETS),
        55–57, 244

NeoCITIES simulation, 243–248, 266
  as CSTR, 246–248
  evolution, 243–244
  gameplay, 244, 246
  interface, 245–246
    chat panel, 245
    dispatch panel, 245
    event tracker, 246
    team monitor, 246
    unit monitor, 246
  3.1 interface, 245f
  overview, 243
  simulation structure, 244–245
    modifiable components, 244
    scoring model, 244–245
    scripting of events, 244
  as test bed for team cognition research, 248
  traditional roles, 244
NetBeans integrated development environment
  (IDE), 70
NETS. *See* NeoCITIES Experimental Task
  Simulator (NETS); NeoCITIES
  Experimental Task Simulator (NETS)
NETS-DART, 266, 266f, 267, 274
Next Generation (NextGen) Air Transportation
  System tools, 212
Nontrivial task, 185

O

Objective-dependent variables, 187
Object oriented programming (OOP) languages,
  PBL, 69, 70
Observation learning, 90–91
Observation-sniper, 307
Observe-Orient-Decide-Act (OODA) loop
  framework, 321
Oculomotor processor, EPIC, 229
Offense seriousness, 306
Open learning environments (OLEs), 32

P

Paired comparison testing, 116
Pair programming approach, 69, 70
Pair-wise comparison ratings, 249
Paper prototyping, 333
Participatory design process, 305, 351
Paul, S.A., 111
PBL. *See* Problem-based learning (PBL)
Peer-interaction process, 36
Penn State, 151–154
  communication, 151–152, 151t
  sharing hidden knowledge, 152–154
Pennsylvania Act 235 (domain training), 306
Perception stage, 142
Perceptual-motor modules, ACT-R, 224

Perceptual processors, EPIC, 229
Personas, mitigating risk, 331–332
Pilot's Associate program, 366
Planning stage, crisis cycle, 136
Police cognition and participatory design,
  299–310
  literature review, 301–303
  living lab framework, 300, 309f
  methodology, 304–310
    ethnography, 305, 306–304
    knowledge elicitation, 307–308
    scaled worlds, 308
Positive affectivity (PA), 169
Positive and negative affect schedule (PANAS), 169
Predictability, AIDL simulator, 269, 272
Prekop, P., 110
Preparedness, crisis cycle, 136
Primary investigators (PIs), 145
Problem-based approach, transactive memory, 53
Problem-based learning, 7
Problem-based learning (PBL)
  collaborative tool, 67–68
  group assignments
    backseat driving, 69
    Federated Wiki. *See* Federated Wiki
    Fork and Pull model, 71–72
    GitHub, 68, 70–71
    idea mining, 74, 75f
    IST 311, 72
    LLF, 73
    OOP languages, 69, 70
    pair programming, 69, 70
    PHP language, 75, 75f
    wiki model, 73. *See also* Federated Wiki
  LLF, 88
Problem solving, 7, 32, 35–39, 85, 88, 300
Procedural memory system, ACT-R, 224, 226
Procedural prompts, 35
Process tracing techniques, 113–115
Production rule system, EPIC, 229
Project-based learning, 38
  elementary *vs.* high school students, 39–40
  QA Quests, 39
Projection stage, 142
Prompts, web-based cognitive support system, 35
Prototypes, LLF
  challenges/research, 99–100
  cyclic framework, 99
  envisioned designs, 99
  NeoCITIES architecture, 99
  opportunities, 100
  scaled-world environment, 99
Psycho-physiological measures and cognitive
  architectures, 230–234
  EDA, 233–234
  EEG, 230–231
  fMRI, 231–233

Q

QA. *See* Quest Atlantis (QA)
QAP correlations, 189
Qualitative communications analysis methods, 172
Quest Atlantis (QA), 39, 40
Question prompts, 36, 37

R

Raid and crisis management environments, 291–292
Raid leader, WoW, 280t
Raid, WoW, 280t
Rasmussen, J., 112
Real-time intermodal data, 321
Recognition-primed decision (RPD) model, 304
Reflection prompts, 35
Reflexivity, 191
    storytelling and, 185–186
Research and development (R&D), 342
Research experience (RX), 365–366
Response stage, crisis cycle, 136
Retrieval coordination stage, transactive memory, 49
Risk-management process, 133
Rules-based action, 139

S

Scaled worlds, 308
Scaled-world simulations, 54, 261–274
    AIDL simulator, 269–273
        display panels, 269
        interface design of, 269
        interface layout, 270f
        interruption, 272–273
        predictability, 269, 272
        task load, 272
        volatility, 269
    DART, 263–268
        caveats, 265
        in cyber security, 266–267
        dynamic prioritization, 264
        in living laboratory framework, 267–268
        NETS-DART, 266
        operator task, 263
        primary subsystem, 266–267
        priorities change with time, 264f
        secondary subsystem, 267–268
    LL approach, 122
    LLF
        challenges/research, 97–98
        convergence, 96
        decisions and actions, 97
        human-in-the-loop testing, 97

kinetic cycling, 96
    opportunities, 98
    supply chains, 96
    MCT, 124t
    teamNETS, 55–57, 56f
Schraagen, J.M.C., 117
Science/technology/engineering/math (STEM) majors, 322
Scoring model, 244–245
Scrum methodology, 336
Segments, ethnographic methods, 306
Semi-structured cognitive interview, CIS, 115
Semistructured interviews, 349
Sensemaking, CIS, 111
Serious offenses, 306
Shared repository approach, 71
Short stress state questionnaire (SSSQ), 169
Simulation
    interface, 146, 147f
    NeoCITIES, 144, 146
    scaled-world, 144
Situated cognition, 7, 139–140
Situational awareness, development and maintenance, 321–322
Situational awareness rating technique (SART), 188–189
Situation assessment records (SAR), 307
Situation awareness (SA), 250–252
    definition, 250
    levels of, 250–251
    limitation, 251
    SAGAT, 251
Situation awareness (SA), 6, 137, 141–142, 182, 184, 188, 193, 207
Situation awareness–FCM (SA–FCM) modeling approach, 204, 208f
    for data fusion, 214
    for decision-aiding support, 208–211
        methodology, 209–211
    goal level, 207
    for measuring cognitive readiness, 213
    for predictive assessments, 212–213
Situation awareness global assessment technique (SAGAT), 251, 321
Situation awareness global awareness technique (SAGAT), 188–189, 193
Slack, Federated Wiki, 77
Soar (cognitive architecture), 227–228
    goal, 227
    long-term memory structures, 227
    problem–space operator representation, 227–228
    short-term memory, 227
    structure of, 228f
Sociocognitive Simulations, 11
Socio-cognitive technologies, 368
Sociotechnical systems, 133

Specialization of expertise, transactive
memory, 51
Specialization (spec), WoW, 280t
Spielberger State–Trait Anxiety Inventory
(STAI), 169
SPM fMRI analysis output, 232f
Sprint, 336
planning, 336
retrospective, 337–338
review, 337–338
Squashing function, 202
Stabilization stage, 137
Statistical analyses, 194
Stimulus-hypothesis-option-response (SHOR)
model, 321
Storytelling, 190
and reflexivity, 185–186
Stress, 162
Students-as-designers
knowledge, 38
project-based learning, 38–40
roles, researchers, 40–41
stakeholders, 41
TRE, 41, 40
Subject matter experts (SMEs), 347
Subsymbolic activation value, 226
Summative evaluation, 356
Support tools
LL approach, 122
MCT, 124t
System testing, 353

T

Tactical Decision Making under Stress
(TADMUS) project, 162
Task, CTA, 118
Task, expertise, and people (TEP) unit, 50
Team cognition, 139–140, 166, 184–185, 241–254
commercially software, 242
and culture, 144–155
comparison of contexts, 154–155
Manchester University case-study
context, 146–151
NeoCITIES simulation, 146
Penn State case study context, 151–154
differential effects of multiple types, 191–192
effectiveness of interventions designed,
192–193
information sharing, 253–254
measures of, 242
NeoCITIES simulation, 243–248
as CSTR, 246–248
evolution, 243–244
gameplay, 244, 246
interface, 245–246
interface components, 245

NETS, 244
overview, 243
simulation structure, 244–245
as test bed for team cognition research,
248
traditional roles, 244
SA, 250–252
simulation-based studies, 242
temporal focus, 192
temporality, 185
TMM, 248–250
Team cognition, transactive memory, 48–49
Team mental models (TMMs), 182, 184, 189, 193,
213, 248–250
content types, 248
evaluation methods, 248–249
KE, 116–117
pair-wise comparison ratings, 249
types, 248
Team monitor, NeoCITIES interface, 246
TeamNETS simulation, 55–60, 56f
Team reflexivity, 191
Team situation awareness, 141–142
Technocentric approaches, 200
Technological disruptions, 366
Technology-rich ethnography (TRE), 40, 41
Temporal mental models, 185
TEP unit. *See* Task, expertise, and people
(TEP) unit
Testing environment, 135, 145
Text-to-speech software, 319
Theoretical foundations, LLF
challenges/research, 95
distributed information, 94
information fusion, 94–95
opportunities, 96
problems experienced, 95
Think-aloud technique, 37
Threshold function, 202
TMMs. *See* team mental models (TMMs)
Top-down approach, 170, 209
Transactive memory, 6, 49–52
cognitive artifacts, 48
collective encoding, 50
computing and technology, 51
cyber operations, 53
ATM-I capabilities, 58–59
distributed teams, 57–58, 60
functional domains, 54–55, 54t
hands-on approach, 53
human–computer interaction
approach, 60
intrusion/threat analysis, 55
knowledge structures, impact of, 59
scaledworld simulation, 54
teamNETS, 55–57, 56f, 60
virtual feedback system, 59–60

definition, 49
dimensions of effective, 50–51, 50t
distributed cognition, 48
higher- and lower-order knowledge, 49
research, 51–52
specialization of expertise, 51
stages, 49
team cognition, 48–49
teamNETS, 58–59
TEP unit, 50
TRE. *See* Technology-rich ethnography (TRE)
Trends, AIDL, 269
Trier Social Stressor Task (TSST), 234

U

Ultima Online, 278
United States Coast Guard (USCG), 141
Unit monitor, NeoCITIES interface, 246
Urban search and rescue (USAR), 137
Usability, 13
    requirements, 350
    testing, 353
User-centered requirement framework, 350
User experience (UX), 13, 328–333, 365–366
    dual-faced challenge, 330
    knowledge elicitation to sense-making,
        330–331
    mitigating risk through personas, 331–332
    mock-ups and prototypes, 333
    problem identification, 329–330
    product vision building, 332
User interface requirements, 350

User science and engineering (USE) Laboratory, 11
Utterances sequences, 172
UX engineer, 329

V

Variance, AIDL, 269
Vicente-style abstraction hierarchy diagram, 305
Virtual feedback system, 59–60
Visual processor, EPIC, 229
Vocal motor processor, EPIC, 229
Volatility, AIDL simulator, 269

W

Waterfall development methodologies, 334
Web-based cognitive support system, 35
    contribution, learners, 37
    iterative design process, 35–36
    prompts, 35
    roles, researchers, 37
    think-aloud technique, 37
Web Emergency Operations Center (WebEOC),
    141
Web 2.0 technologies, 68
Weltanschauungs, 6
Wikipedia, 73. *See also* Federated Wiki
    vs. Federated Wiki, 76, 76f
Wikis, 68. *See also* Federated Wiki
Wireframe, 351
Workflow requirements, 350
World of Warcraft (WoW), 278–279, 278–280
    and NeoCITIES, 286–287